# Handbook of Finite Element Analysis

# Handbook of Finite Element Analysis

Edited by **Connie Mcguire**

NY RESEARCH PRESS
New York

Published by NY Research Press,
23 West, 55th Street, Suite 816,
New York, NY 10019, USA
www.nyresearchpress.com

**Handbook of Finite Element Analysis**
Edited by Connie Mcguire



International Standard Book Number: 978-1-63238-250-4 (Hardback)



Printed in China.

# Contents

# Preface

The concept of finite element analysis has been described in this exhaustive book. In the past few years, the Finite Element Method (FEM) has evolved into a crucial technology in the modeling and simulation of several manufacturing systems. This text presents the latest study and growth discoveries on this extensive subject through unique and pioneering research studies providing a variety of examination tracks of FEM in electrical, civil and biomedical engineering. This book is a compilation of several well researched data which will help the experts and students in accomplishing these respective researches. We intend to provide some of the vital facts regarding FEM to our readers.

All of the data presented henceforth, was collaborated in the wake of recent advancements in the field. The aim of this book is to present the diversified developments from across the globe in a comprehensible manner. The opinions expressed in each chapter belong solely to the contributing authors. Their interpretations of the topics are the integral part of this book, which I have carefully compiled for a better understanding of the readers.

At the end, I would like to thank all those who dedicated their time and efforts for the successful completion of this book. I also wish to convey my gratitude towards my friends and family who supported me at every step.

**Editor**

Section 1

# Recent Advances of Finite Element Analysis in "Bio-Engineering"

Chapter 1

# Finite Element Analysis in Dental Medicine

Josipa Borcic and Alen Braut

Additional information is available at the end of the chapter

## 1. Introduction

Studying dental structures and surrounding tissues in the oral cavity presents the basis for understanding the occurrence of pathological process and enables the correct approach and treatment. Oral rehabilitation is inherently difficult, due to the functional and parafunctional forces within the mouth that result in extremely complex structural responses by the oral tissue [1]. The success of restorative materials depends on their properties to withstand and resist occlusal forces and successfully support the remaining oral structure [2]. Studies examining the biomechanical behavior of oral structures require sophisticated simulations of the fundaments of the stomatognathic system [3].

There were numerous ways and attempts of experimental research, but due to the complexity of dental structures, composed of various tissue materials mechanically and chemically interconnected, and due to complex tooth morphology and surrounding structures, these attempts failed to obtain precise and reliable results. Researches have used photoelastic methods, computer simulation methods and finite element analysis to conduct stress analyses of sound and restored teeth in order to predict their fracture resistance. Conventional methods such as photoelasticity and the strain-gauge methods are inadequate to predict reliable stress distribution in the tooth [4]. The use of traditional load-to-failure bench-top testing is unable to recreate the failure mechanisms seen clinically; hence the use of FEA is gaining popularity because of its ability to accurately asses the complex biomechanical behavior of irregular prosthetic structures and heterogeneous material in a non-destructive, repeatable manner [5].

## 2. Finite element analysis

Finite element analysis (FEA) is a numerical method of analyzing stresses and deformations in structures which originated from the need for solving complex structural problems in civil and aeronautical engineering. In order to achieve this goal, the structures are broken down into many small simple segments or elements, each with specific physical properties (figure 1). Than, an operator uses a computer program in order to obtain a model of stresses produced by various loads [6,7]. A major advantage of finite element analysis (FEA) is its ability to solve complex biomechanical problems for witch other study methods are inadequate. Stress, strain and some other qualities can be calculated in every point throughout the structure. FEA is also being used as part of the design process to simulate possible structure failure, as a mean to reduce the need for making prototypes, and reducing a need for performing actual experiments, that are usually expensive and time-consuming [8]. This method allows researches to overcome some ethical and methodological limitations and enables them to verify how the stresses are transferred throughout the materials [9].

In the area of dentistry, FEA has been used to simulate the bone remodeling process, to study internal stresses in teeth and different dental materials, and to optimize the shape of restorations. Because of the large inherent variations in biological material properties and anatomy, mechanical testing involving biomaterials usually require a large number of samples. With FEA the necessity of traditional specimens can be avoided, and by using a mathematical model it also eliminates the need for large number of experimental teeth. It has been used to represent simulated tooth mechanical behavior under occlusal loads in details [8].

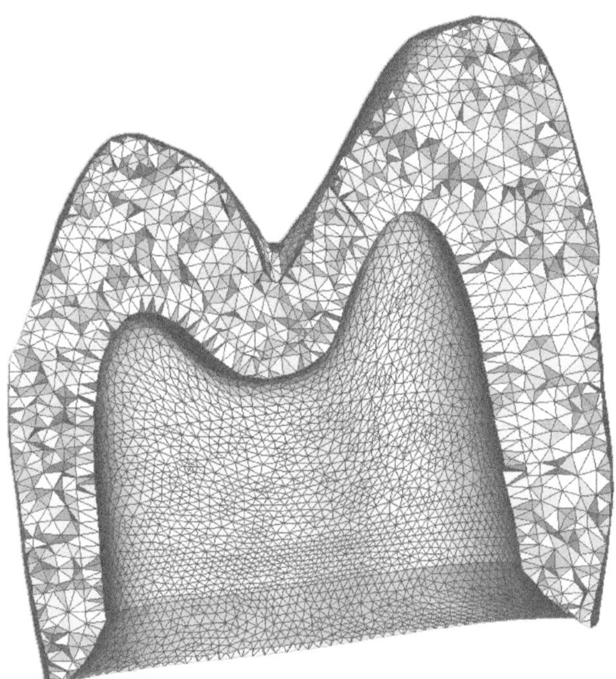

**Figure 1.** Elements of an FEA model.

## 2.1. Finite element model

The decision to use 2D or 3D models to investigate biomechanical behavior of complex structures, by FEA, depends on many inter-related factors, such as the complexity of the geometry, material, properties, mode of analysis, etc. Although 2D models are simpler, easier to build and less time consuming, they do not represent the complexity of the real problem. 2D model might be considered when studying the qualitative biomechanical behavior, but for the quantitative stress analysis the 2D models overestimate stress magnitudes and do not represent the realistic model. 3D model may provide more reliable data that more accurately represent non-linear and anisotropic materials. 3D models should be carefully created with appropriate mesh density [3]. Khera et al. were the pioneers in the utilization of 3D models. The models were obtained from sectional images of human mandible, but this is no longer required due to the use of a computerized tomography (CT) [10].

The 3D geometry of the tooth (figure 2) can be reconstructed in two ways. The old traditional method consists of embedding the tooth in red epoxy and sectioning it perpendicularly to the long axis by a precise saw (figure 3). Each section is than digitally photographed and the 3D geometry of the tooth is being constructed from these cross-sections using specialized computer program. The solid model is transferred into a finite element analysis program, where a 3D mesh is being created, and subsequently the stress distribution analysis performed (figure 4) [4, 25].

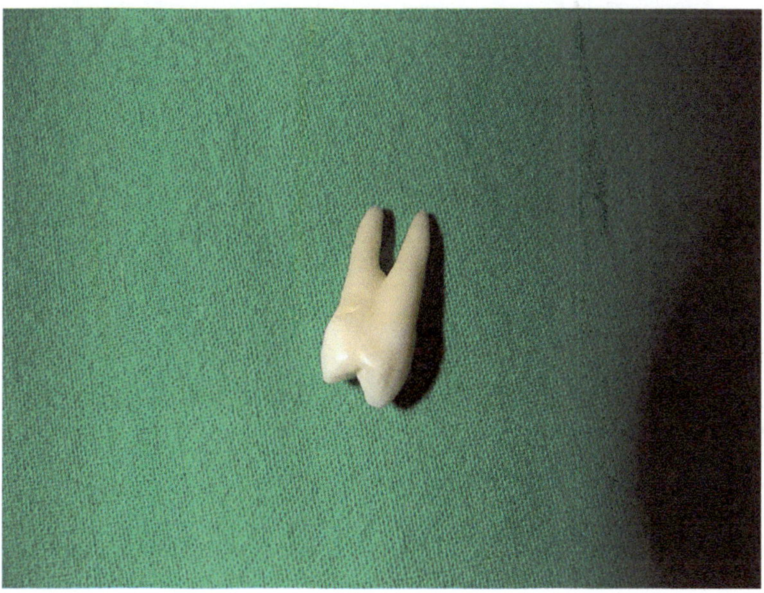

**Figure 2.** Natural tooth

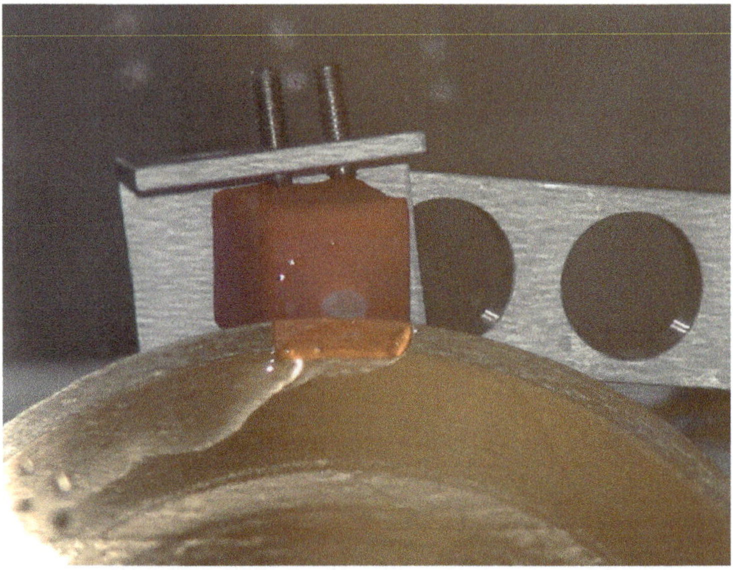

**Figure 3.** Embedded tooth in red epoxy.

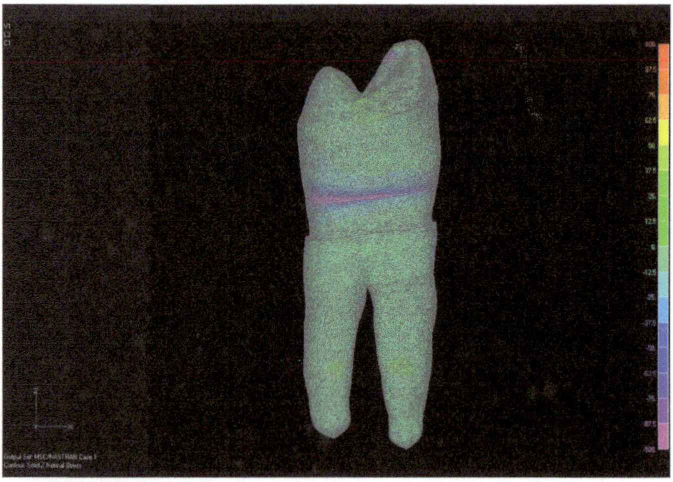

**Figure 4.** 3D model of the sound tooth

The second, latest method of reconstructing a 3D tooth model is performed with the aid of CT. It facilitates and speeds up the acquisition and produces more accurate model. With this method the surrounding soft structures can be also included, larger areas scanned and reconstructed, while the structures itself still remain in the patient mouth. The next big advantage of CT model rendering consists of the possibility to scan the same structure before and after the performed therapy procedures, and periodical follow-ups of the therapy success. Technologies such as micro-CT scanning open up the possibility for complex 3D modeling [11]. However, the process of going from image to mesh involves a number of processing steps, each with potential geometric errors [12].

## 2.2. Interpretation of the FEA results

The results obtained from a FEA on the restored system contain information about the stress distribution of each component of the restoration, instead of only a single value of failure load typical of in vitro results. A correct interpretation of FEA results should be based on the stresses and strength of each component of the system. To obtain accurate conclusions from these interpretations, three conditions must be fulfilled. First, FEA should adequately represent the real stress values; second, strength of the different materials must be known; third, an adequate failure criterion must be used [13].

It is not possible to implement the results from FEA directly into a clinical situation, but it has to design the model in such a way that is mimics the real situation as closely as possible. FEA analysis must be interpreted with a certain amount of caution. Most of the researches modeled dental structures as isotropic and not othotropic. The finite element model represents a static situation at the moment of load application and not an actual clinical situation. In reality, the loading of the structure is more dynamic and cyclic. The materials of the various tooth structures were assumed to be isotropic, homogenous and elastic, and that they remain such under applied loads. More precise measurements can be obtained if the material properties are set as anizotrophic and non-homogeneous, but such setup requires much more complex mathematical calculations. It is better to use a non-linear elastic-plastic material model than the linear models that are used in most FEA studies [14].

The values from finite element analysis are presented as maximum and minimum principal stresses. Most of the previously published studies have analyzed the results from Von Mises maximal stress [15-19]. This is probably associated to the fact that this is the normal criterion for the most engineering analyses, which usually deal with ductile materials such as steel and aluminum [13]. It is known that the Von Mises criterion is only valid for the ductile materials with equal compressive or tensile strength, but materials exhibiting brittle behavior such as ceramics, cements or resin composites presents reported values of compressive strength significantly greater than tensile strength [20]. Positive and negative values indicate that the corresponding regions are subjected to tensile or compressive stresses (figure 5) [21].

The response of the structure is different if asymmetrical loading is considered. When the tooth is compressively loaded, displacements do not appear to be significant because of the rather large compressive yield strength. The situation is different if the asymmetrical loading is considered, when the tensile stress occurs. The dental tissues are more resilient to compressive than tensile forces. Any occlusal contact that can create tensile stress, also creates the possibility to create a lesion in tooth structure. When lateral loads are applied, tensile stresses generated in the areas are of higher values than when vertical loads are applied onto the same areas. The increase in the load does not cause a change in the overall stress pattern, but increases the values. The loading, that the tooth is subjected to, may cause cracks in the tooth, but not necessarily its immediate failure. Most of the failures of dental materials used for tooth restorations are caused by tensile stress. Precise occlusal adjustments of teeth occlusal surfaces should be performed to prevent such events. The average chewing force varies between 11 and 150 N, whereas force peaks are 200N in the anterior, 350N in the posterior and 1000N with bruxism [22].

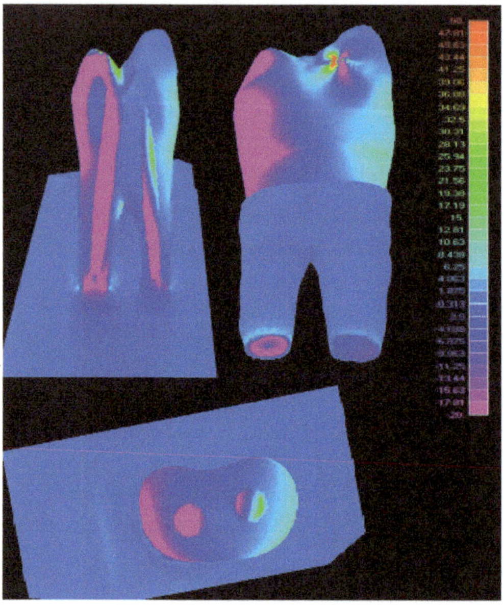

**Figure 5.** FEA model of a restored apicotomysed tooth

## 3. Materials and types of reconstructions in dental medicine

The use of different materials for restoration substantially modifies the stress distribution of an originally healthy tooth. The difference between the elastic modulus of tooth and restorative material may be a source of stress in the dental structures. If the stress exceeds the yield strength of the materials, fracture of the restorative materials or the tooth may occur. The occlusal force leaning against the tooth or dental implant axis causes the structure to bend, and the higher tensile stresses are produced. The oblique force loading on the dental structure is the major cause of dental damage and the further attention should be paid to the importance of the occlusal adjustment [4, 7, 25].

The way the chewing force application is much more important than the dentine and the enamel properties, or even the properties of the restorative materials. The consequences of the same chewing force for different teeth also need to be highlighted because structural changes can occur depending upon the magnitude of the force, which can affect the tooth morphology in extreme (premature contacts) or repetitive cases (fatigue) [11].

### 3.1. Natural tooth

The properties of tooth are not homogenous, but are anisptropic like dentin (due to its capillary morphological structure) or enamel (due to its prismatic structure) [23]. Various studies have shown that the failure was confined mostly to the occlusal walls and margins, and was usually seen on the buccal surfaces of lower molars and premolars (figure 6 and 7) [24,25]. Excursive mandibular movements place the buccal cusps in tension or in compression and open up the occlusal margins (figure 8). Enamel near the cemento-enamel junction is highly stressed because the reactive forces have to flow into and through this thin wedge of tissue for it to be transmitted into the root of the tooth and subsequently into the supporting alveolus bone [2]. This is the reason why the restorations inserted into the cervical region can be subjected to high compressive stresses even though these areas are not susceptible to direct contact during mastication [26,4].

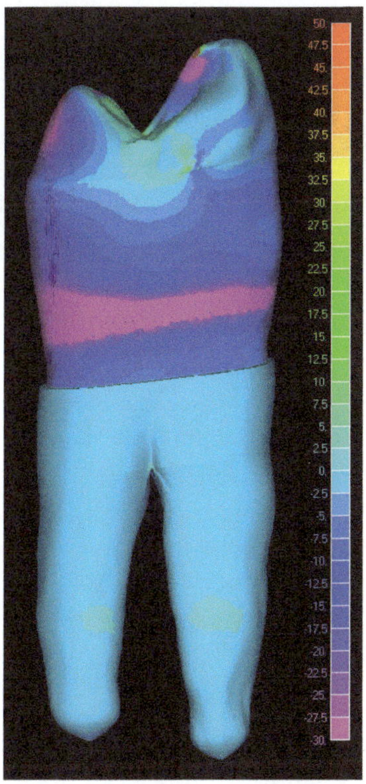


**Figure 6.** FEA analysis in sound tooth in normal occlusion looking from outside

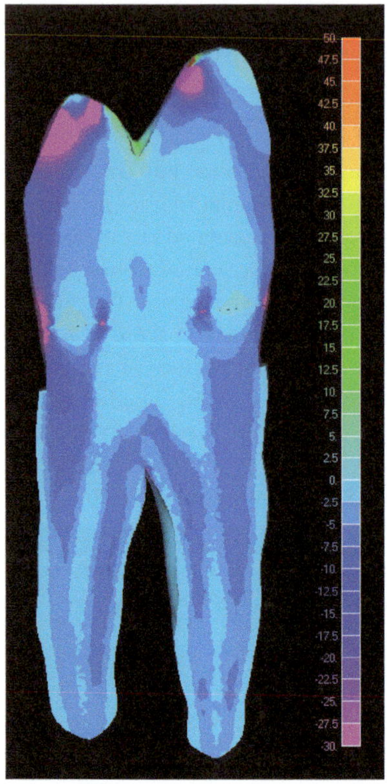


**Figure 7.** FEA analysis of the sound tooth (cross section)

### 3.2. Dental restorations

Many detrimental effects during restorative procedures are reported to be produced because of lack of understanding of biomechanical principles underlying treatment. Biomechanical studies are crucial in order to highlight the behavior of restored tooth to functional forces [27].

It was earlier thought that the only forces that dislodge the tooth restorations were the pulling forces of sticky foods, while little thought was given to the biomechanics of the tooth structure. Later, it was seen that forces applied on the occlusal surface of the tooth could induce stresses in a restoration remote from the point of application of the force [2]. Heymann suggested that two mechanisms operate and cause failure. One is the lateral excursive movements resulting in lateral cuspal movements which generate tensile stresses along tooth restoration interface, and the other one are heavy forces in centric occlusion which cause vertical deformation on the tooth leading to compressive and shear stresses [24]. The presence of an occlusal restoration weakens the tooth structure and increases the stresses. Especially the depth is more critical than the width [26,28,29]. This restoration may also influence the retention of class V restoration because of the tooth flexure theory. Breakdown of the margins of class V restorations may be the result of occlusal loading [2].

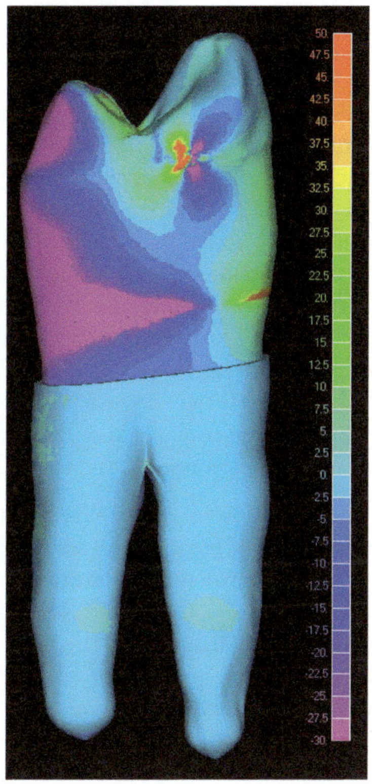


**Figure 8.** FEA analysis in the sound tooth in the case of malocclusion

A growing interest in aesthetic dental restorations has led to the development of innovative materials for aesthetic restorations of teeth. These new systems have focused on physical properties, such as modulus of elasticity, that are more closely matched to natural tissue, in order to decrease stress concentrations within the dental structure and reduce the incidence of failure. The development of adhesives has created a need to measure the adhesive bond strength of restorative materials to mineralized tissue. Several methods for studying have been developed but FEA capable of quantifying the effect of each tested parameter on bond strength [30].

Composite occlusal restorations have been shown to reduce cuspal flexure compared to an occlusal amalgam restoration [26]. Composite resin in combination with the acid-etch technique and adhesive systems have been used for the restoration of tooth caries and cervical lesions that we commonly call „dental composite restoration". Evaluation of marginal integrity at the composite resin-tooth interface is required for clinically successful restorations. Polymerization contraction occurs during light curing and may cause marginal disintegration [31]. The maximum stresses due to the shrinkage of the cement layer may cause debonding of the cement layer. This debonding on one side will cause relaxation of stresses at the other side of the restoration and will cause (micro) leakage with all its detrimental effects [14].

The fracture load of the final restoration is the result of the combined effects of bonding between the underlying tooth, the ceramic restoration, and the resin composite cement. Clinical stress distribution in ceramic dental restorations may be quite complex. Several factors are associated with crack initiation and propagation, including the shape, microstructural no homogeneities, the size and distribution of surface flows, residual stresses, ceramic-cement interfacial features, thickness of restorations, different elastic modulus and the magnitude and orientation of the applied load. On the structural factors, the connector areas are the most influential in failure [22]. Traditional load-to–failure testing has proved irrelevant in predicting the clinical performance of ceramics, largely because they cannot recreate the failure mechanisms seen in clinical specimens [5]. The FEA was used to determine the optimal stress distribution in the ceramics bridges that would reduce the risk of connector fracture. The points of greater stress were found within, or near the connector [22].

The FEA demonstrated that with the use of an idealized inlay preparation form and an optimized bridge design emphasizing a broadening of the gingival embrasure, the forces derived from mastication can be adequately distributed to levels which are within the fracture strength of current ceramics [5]. Tensile stresses tend to be more critical than compressive stresses for ceramic materials. The strength of ceramic restorations is significantly affected by the presence of flows or other microscopic defects [32]. Tensile stress concentration at cementation surface of the ceramic layer was suggested to be the predominant factor controlling ceramic failure [33]. Fea showed lower tensile stress levels at the cementation surface than in the area under and between the load points, which could explain the occlusal to cervical direction of fracture seen in the fractographic analysis. Although the polymer crown had a higher fracture resistance than ceramics, a larger amount of the occlusal load was transferred through the tooth, resulting in catastrophic fracture of the tooth. This fracture behavior can limit the use of polymer crowns when compared to ceramic systems [32]. Molar crowns made of stiffer materials are less prone to debonding and crowns made of higher elastic modulus material protect the tooth structures from damage [23].

Veneers used in restorative rehabilitations for anterior teeth are retained by the adhesive systems and resin cements. These restorations are mechanically not strong, because they are made up of a brittle material, but they have good retention due to the resin-dentine bonding. The remaining tooth tissue is the most important factor for the longevity of the veneers where the buccal, cervical region is the most critical region. Teeth totally recover their properties when veneers are placed as a partial enamel substitute. The use of ceramic was a key element since the elasticity modulus of porcelains is matched well with enamel [11].

The widely used method for treatment of structurally weakened teeth is the post and core system. This system can be classified into two basic core system, metal posts and cores that are custom cast as a single piece, and two element designs composing a prefabricated post to which other materials core is subsequently adapted [34]. The difference between the elastic modulus of dentine and the post material may be a source of stress for root structures. Debonding of posts because of contraction stress of the cement was found as the most common mode of failure [27]. The effect of post design is also very important for dentinal stress distribution since the placement of a post can create stresses that lead to root fracture (figure 9) [1]. Increased intracanal stresses below the level of crestal bone would explain the higher incidence of deep root fractures in teeth restored with post-retained crowns Horizontal loads generate more dentinal stress than vertical loads. Shorter posts are associated with more dentinal stress concentration around the post apex. Consequently, extending the apical post beyond the level of alveolar bone is essential to avoid stress concentration in the region of the post apex. However, very long posts are associated with higher intracanal stress values. A higher amount of radicular dentin around the post is important in order to reduce dentinal stress concentration within the root [35]. The use of post materials conflicts with the mechanical resistance of teeth because of mismatch in the stiffness with the residual dental structure [36]. Many studies have shown that fiberglass posts give better biomechanical performance. Titanium posts concentrate stress close to the post-cement interface, promoting weakness of restored tooth. Akkayan [37] observed that the fractures occurring with the use of fiberglass and quartz posts systems could be repaired, whereas this was not the case with zirconium and titanium posts. Thus, fiberglass post can be considered a very good choice because they offer good biomechanical performance, provide excellent aesthetics, and exhibit good adhesion to cementing agents [38].

Clinicians generally agree that NiTi rotary files have good properties to produce desirable tapered root canal forms, but also have a risk of fracture during instrumentation. These instruments have been developed to overcome the rigidity of stainless steel instruments [39]. Design of an instrument is the main factor in their mechanical behavior. Cyclic fatigue, which is a failure process associated with repetitive stressing, and torsion have been reported as dominant factors in file fracture [40].

With the application of adhesive technology to endodontics, the term monoblock has become familiar. Monoblock units can be created in a root canal system either by adhesive root sealers in combination with a bondable root filling material or adhesive post systems. The concept of creating mechanically homogenous units within the root dentine is excellent in theory, but accomplishing these ideal monoblock in the canal space is challenging because bonding to dentine is compromised by volumetric changes in resin-based materials, high cavity configuration factors, debris on canal walls, and differences in regional bond strengths [27].

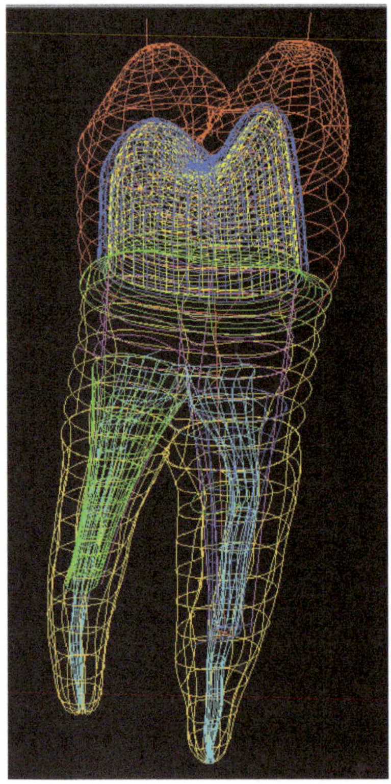

**Figure 9.** 3D modeling of post and core system.

### 3.3. Dental implants and anchorage systems for tooth movements and bone fracture

Dental implants are widely used to replace decayed teeth or to support prostheses. The failure is associated with bone loss around an implant neck. Bone loss can be activated by excessive implant loading, as by bacterial infection or trauma [41]. Mechanical stress can have positive and negative consequences for bone tissue and for maintaining osseointegration of an oral implant [9]. The prognosis for stress concentration at the bone-implant interface is of the utmost importance in dental implant research [41]. FEA has been widely used in the field of oral implantology to estimate peri-implant stress and strain [42]. The relation between implant design and load distribution at implant-bone interface is important in the search for optimal implant configuration with minimum stress peaks. Another significant factor is bone quality, in mechanical terms; this is determined by bone strength. Increase in implant length and diameter leads to reduction of stress magnitudes within the cortical bone [41]. FEA study shows that non-submerged implants showed higher stress values in the peri-implant bone than submerged ones and the use of soft liner materials considerably reduces the stress levels in the peri-impant bone interfaces. Different heights and the use of soft liners were relevant in the stress distribution to the bone adjacent to the implants. Better distribution of the stresses will provide a more predictable osseointegration [9].

Prosthesis retention remains a much debated topic in the implant literature. Clinical studies comparing cement- and screw- retained implant restorations reveal no differences in outcomes. There is evidence from laboratory and FEA studies that implants with an internal-type connection exhibit better stress distribution with off-axis loading [43]. The combined use of implants and teeth has been questioned because of the differences of mobility between the abutments. Several authors have concluded that the tooth-implant bond does not have a negative influence on the marginal bone and soft tissues, but special care must be taken in planning in order to compensate for the differences in biomechanical responses between the implant and the tooth [44].

The biomechanical background of orthodontic tooth movement has been explored by many authors, and orthodontic movement principally depends on stress and strain in periodontal ligament (PDL). PDL is a thin connective tissue between the root and bone and play a key role in tooth mobility [12]. Accurate FEA model creation of a tooth and PDL is possible due to the use of micro-CT. Anchorage control in orthodontic treatment is an important factor in treatments outcome. Miniscrews and miniplates are being widely used because of their small size and superiority over endosseous implants due to the fact that they can be immediately loaded. Miniplates have the same features with the plates used in maxillofacial surgery [45]. Good treatment results have been reported by using miniscrews for orthodontic anchorage in various malocclusions, but major problem is their high failure rate. Unlike dental implants, mechanical interdigitation at the cortical bone rather than osseointegration is required for the stability of miniscrews. The placement angle, the type of miniscrews, and the direction of forces significantly affect the distribution area and the amount of stress [46]. Inadequate design and non-homogenous force distribution can cause stress directly effecting on the screws and may impair screws stability. Mobile plates can irritate the surrounding tissue and may cause inflammation. The FEA study revealed that the new miniplates are highly efficient in reducing stress on the fixation screws [45].

Fractures of the mandibular angle are the most problematic in the facial region because of the high frequency of complications and difficult surgical access to the site [47]. Infection and nonunion are commonly reported after rigid internal fixation of these fractures [48]. The stress analyses obtained from FEA modeling can provide information regarding interactions between hardware and bone during normal patient functioning. A single tension band on the superior borders provided more angle fracture stability than a single bicortical plate placed inferiorly. This results support the use of the single tension band configuration as a less invasive fixation approach to fractures [47].

## 4. General guidelines

The results of the finite element analysis must be interpreted with a certain amount of caution. Most of the researches modeled dental structures as isotropic and not othotropic.

The finite element model represented a static situation at the moment of load application and not an actual clinical situation. In reality, the loading of the structure is more dynamic and cyclic. More precise measurements could be obtained if the material properties are set as anisotropic and non-homogeneous, but such setup requires much more complex mathematical calculations.

To obtain better understanding of the tooth lesions, which is important for the clinical treatment and restoration of damage, analyses of stress distribution in the oral cavity under various loading condition are highly desirable. FEA is a valuable tool for investigation of stress distribution within various types of reconstructions and prosthodontic appliances in dental medicine.

The dental profession is influenced by various sources of information, which may be considered as "evidence-based" (controlled clinical studies) and "expert opinion". A realistic approach is to identify the strengths and weaknesses of the available clinical data and combine it with clinical experience [43]. Most researchers in FEA assumed that all materials used were homogenous, isotropic and linearly elastic. However, this assumption does not reflect the exact situation. The periodontal ligament has nonlinear mechanical properties and the bone is inhomogeneous [9,35]. The 3D analysis permits high efficiency when the biomechanical behavior of the structure needs to be evaluated under different loading conditions. In the last four decades many studies have shown how the Finite Element Analysis applied to dental mechanics has become a popular numerical method to investigate the critical aspects related to stress distribution. The use of more detailed 3D models could be helpful in understanding critical problems related to the restorative material choice and optimal application procedures. Improved computer and modeling techniques render the FEA a very reliable and accurate approach in biomechanical applications [9].

The results from FEA confirm the concept that the interfaces of materials with different module of elasticity represent the weak point of restorative systems. Restorations with material having a similar elastic modulus to tooth can save and strengthen the remaining tooth structure [27]. Combining fatigue experiments with FEA may eliminate, or at least minimize, experimental limitations by correlating fatigue failure to stress instead of specific testing configuration.

## 5. Conclusions

There are numerous ways and attempts of experimental research, but due to complexity of dental structures, composed of various tissue materials mechanically and chemically interconnected, and due to complex tooth morphology and surrounding structures, most of these attempts fail to present precise and reliable results.

The 3D analysis permits high efficiency when the biomechanical behavior of the structure should be evaluated under different loading conditions. In the biomedical fields, the FEA is an important tool since it can avoid the necessity of traditional specimens, and by using a mathematical model it eliminates the need of large number of teeth. The use of more detailed 3D models is helpful in understand critical problems related to the restorative material choice and optimal application procedures. Improved computer and modeling techniques render the FEA a very reliable and accurate approach in biomechanical applications.

When the tooth is compressively loaded, displacements do not appear to be significant because of its rather large compressive yield strength. The situation is different when asymmetrical loading is considered and tensile stress occurs. The dental tissues are more resilient to compressive than tensile forces. Any occlusal contact that can create tensile stress, also creates the possibility to create a lesion in tooth structure. Most of the failures of dental materials used for tooth restorations are caused by tensile stress. Precise occlusal adjustments of teeth occlusal surfaces should be performed to prevent such events. The difference between the elastic modulus of tooth and restorative material may be a source of stress in the dental structures. If the stress exceeds the yield strength of the materials, fracture of the restorative materials or the tooth may occur.

The FEA helps to improve preparation designs, indicates the right material or combination of materials to be used in various load and stress conditions in order to reduce material and/or tooth failure in clinical practice.

## Author details

Josipa Borcic*
*Department of Oral and Maxillofacial Surgery, Medical Faculty, School of Dental Medicine, University of Rijeka, Rijeka, Croatia*

Alen Braut
*Department of Restorative Dentistry and Endodontics, Medical Faculty, School of Dental Medicine, University of Rijeka, Rijeka, Croatia*

## 6. References

[1] Silva NR, Castro CG, Santos-Filho PCF, Silva GR, Campos RE, Soares OV, Soares CJ. Influence of different post design and composition on stress distribution in maxillary central incisor: Finite element analysis. Indian J Dent Res 2009;20:153-158.

[2] Vasudeva G, Bogra P, Nikhil V, Singh V. Effect of occlusal restoration on stresses around class V restoration interface: A finite-element study. Indian J Dent Res 2011;22:295-302.

[3] Poiate IAVP, Vasconcellos AB, Mori M, Poiate E Jr. 2D and 3D finite element analysis of central incisor generated by computerized tomography. Computer method and programs in biomedicine 2011;104:292-299.

[4] Borcic J, Antonic R, Muhvic Urek M, Petricevic N, Nola-Fuchs P, Catic A , Smojver I. 3-D Stress Analysis in Premolar, Coll. Antropol. 31 (2007) 4: 315–319.

* Corresponding Author

[5] Thompson MC, Field CJ, Swain MV. The all-ceramic, inlay supported fixed partial denture. Part 2. Fixed partial denture design: a finite element analysis. Australian Dental Journal 2011;56:301-311.

[6] Ding X, Zhu XH, Liao SH, Zhang XH, Chen H. Implant-bone interfaces stress distribution in immediately loaded implants of different diameters: a three-dimensional finite element analysis. J Prosthodont 2009;18:393-402.

[7] Zienkiewicz OC, Taylor RL. The Finite Element Method. 5th ed. Oxford, England: Butterworth-Heinemann 2000;1:1-20.

[8] Haiyan L, Jianying L, Zhenmin Z, Fok ASL. Fracture simulation of restored teeth using a continuum damage mechanics failure model. Dental Materials 2011;27:e125-e133.

[9] Santos MBF, Silva Neto JP, Consani RLX, Mesquita MF. Three-dimensional finite element analysis of stress distribution in peri-implant bone with relined dentures and different heights of healing caps. Journal of Oral Rehabilitation 2011;38:691-6.

[10] Khera SC, Goel VK, Chen RCS, Gurusami SA. A three-dimensional element model. Operative Dentistry 1988;13:128-137.

[11] Matson MR, Lewgoy HR, Barros Filho DA, Amore R, Anido-Anido A, Alonso RCB, Carrilho MRO, Ansuate-Netto C. Finite element analysis of stress distribution in intact and porcelain veneer restored teeth. Computer Methods in Biomechanics and Biomedical Engineering 2011;iFirst article:1-6.

[12] Hohmann A, Kober C, Young P, Dorow C, Geiger M, Boryor A, Sander FM, Sander C, Sander FG. Influence of different modeling strategies for the periodontal ligament on finite element simulation results. American Journal of Orthodontics and Dentofacial Orthopedics 2011;139:775-783.

[13] Perez-Gonzalez A, Iserte-Vilar JL, Gonzalez-Lluch C. Biomedical Engineering http://www.biomedical-engineering-online.com/content/10/1/44 (accessed 10 March 2012).

[14] Jongsma LA, Jager Ir. N, Kleverlaan CJ, Feilzer AJ. Reduced contraction stress formation obtained by a two-step cementation procedure for fiber posts. Dental Materials 2011;27:670-676.

[15] Pegoretti A, Fambri L, Zappini G, Biachetti M. Finite element analysis of glass fibre reinforced composite endodontic post. Biomaterials 2002;23:2667-2682.

[16] Asmussen E, Peutzfeldt A, Sahafi A. Finite element analysis of stresses in endodontically treated, dowel-restored teeth. Journal of Prosthetics Dentistry 2005;94:321-329.

[17] Genocese K, lamberti L, pappalettere C. Finite element analysis of a new customized composite post system for endodontically treated teeth. Journal of Biomechanics 2005;38:2375-2389.

[18] Sorrentino R, Aversa R, Ferro V, Auriemma T, Zarone F, Ferrari M et al. Three-dimensional finite element analysis of strain and stress distributions in endodontically treated maxillary central incisors restored with different post, core and crown materials. Dental Materials 2007;23:983-993.

[19] Gonzalez-Lluch C, Rodriguez-Cervantes PJ, Sancho-Bru JL, perez-Gonzalez A, barjau-Escribano A, Vergara-Monedero M et al. Influence of material and diameter of pre-

fabricated posts on maxillary central incisors restored with crown. Journal of Oral Rehabilitation 2009;36:737-747.

[20] Craig R, Powers JM. Restorative Dental Materials. St Louis Mosby 2002;11

[21] Lang L, Wang RF. Validation of finite element analysis in dental ceramics research. Journal of Prosthetics Dentistry 2001;86:650-654.

[22] Rezaei SMM, Heidarifar H, Arezodar FF, Azary A, Mokhtarykhoee S. Influence of Connector Width on the Stress Distribution of Posterior Bridges under Loading. Journal of Dentistry 2011;8:67-74.

[23] (N-23) Dejak B, Mlotkowski A, Langot C. Three-dimensional finite element analysis of molars with thin-walled prosthetic crowns made of various materials. Dental Materials 2012;28:433-441.

[24] Heymann HO, Sturdevant JR, Bayne S. Examining tooth flexure effects on cervical restorations.: a two year clinical study. Journal of American Dental Assosiations 1991;122:41-47.

[25] Borcic J, Anic I, Smojver I, Catic A, Miletic I, Pezelj Ribaric S. 3D finite element model and cervical lesion formation in normal occlusion and in malocclusion. Journal of Oral Rehabilitation 2005;32:504-510.

[26] Lee MR, Cho BH, Son HH, Um CM, Lee IB. Influence of cavity dimension and restoration methods on the cusps deflection of premolars in composite restoration. Dental Materials 2007;23:288-295.

[27] Belli S, Eraslan O, Eskitascioglu G, Karbhari V. Monoblocks in root canals: a finite elemental stress analysis study. International Endodontic Journal 2011;44:817-826.

[28] Hood JA. Biomechanical of the intact, prepared and restored tooth. Some implication an adaptive finite-element approach fo the analysis of dental restorations. International Dental Journal 1991;41:25-32.

[29] Goel VK, Khera SC, Gurusami S, Chen RC. Effect of cavity depth on stresses in restored tooth. Journal of Prosthetic Dentistry 1992;67:174-183.

[30] Ferreira RC, Caldas J, Paula GA, Albuquerque RC, Almeida CM, Vasconcellos WA, Caldas RB. Influence of surface area and geometry of specimens on bond strength in a microtensile test: an analysis by the three-dimensional finite element method. Journal of Prosthodontics 2011;20:456-463.

[31] Choi NS, Gu JU, Arakawa K. Acoustic emission characterization of the marginal disintegration of dental composite restoration. Composites 2011;42:604-611.

[32] Campos RE, Soares CJ, Quagliatto PS, Soares PV, Batista de Oliviera O, Santos-Filho PCF, Salazar-Marocho SM. In vitro study of fracture load and fracture pattern of ceramic crowns: a finite element and fractography analysis. Journal of Prosthodontics 2011;20:447-455.

[33] Dong XD, Darvell B. Stress distribution and failure mode of dental ceramic structures under Hertzian indentation. Dental Materials 2003;19:542-51.

[34] Adanir N, Belli S. Stress analysis of a maxillary central incisor restored with different posts. European Journal of Dentistry 2007;2:67-71.

[35] Al-Omiri MK, Rayyan MR, Abu-Hammad O. Stress analysis of endodontically treated teeth restored with post-retained crowns: A finite element analysis study. The Journal of the American Dental Association 2011;142:289-300.

[36] Ausiello P, Franciosa P, Martorelli M, Watts D. Mechanical behaviour of post-restored upper canine teeth: A 3D FE analysis. Dental Materials 2011;27:1285-1294.

[37] Akkayan B, Gulmez T. Resistance to fracture of endodontically treated teeth restored with different post systems. Journal of Prosthetic Dentistry 2002;23:2667-2682.

[38] Cooney JP, Caputo AA, Trabert KC. Retention and stress distribution of tapered-end endodontic posts. Journal of Prosthetics Dentistry 1986;55:540-546.

[39] LeeMH, Versluis A, Kim BM, Lee CJ, Hur B, Kom HC. Journal of Endodontics 2011;37:1152-1157.

[40] Sattapan B, Nervo GJ, palamara JE, Messer HH. Defects in rotary nickel-titanium files after clinical use. Jounal of Endodontics 2000;26:161-165.

[41] Demenko V, Linetskiy I, Nesvit K, Shevchenko A. Ulitmate masticatory force as a criterion in implant selection. Journal of Dental Research 2011;90:1211-1215.

[42] Geng JP, Tan KB, Liu GR. Application of finite element analysis in implant dentistry: a review of the literature. Journal of Prosthetic Dentistry 2001;85:585-598.

[43] Lewis MB, Klineberg I. Prosthodontic considerations designed to optimize outcomes for single-tooth implants. A review of the literature. Australian Dental Journal 2011;56:181-192.

[44] Lanza MDS, Seraidarian PI, Jansen WC, Lanza MD. Stress analysis of a fixed implant-supported denture by the finite element method (FEM) when varying the number of teeth used as abutments. J Appl Oral Sci 2011;19:655-661.

[45] Nalbantgil D, Tozlu M, Ozdemir F, Oztoprak MO, Arun T. FEM analysis of a new miniplate: stress distribution on the plate, screws and the bone. European Journal of Dentistry 2012;6:9-15.

[46] Suzuki A, Masuda T, Takahashi I, Deguchi T, Suzuki O, Takana-Yamamoto T. Changes in stress distribution of orthodontic miniscrews and surrounding bone evaluated by 3-dimensional finite element analysis. Americam Journal of Orthodontics and Dentofacial Orthopedics 2011;140:e273-e280.

[47] Kimsal J, Baack B, Candelaria L, Khraishi T, Lovald S. Biomechanical analysis of mandibular angle fractures. Journal of Oral and Maxillofacial Surgery 2011;69:3010-3014.

[48] Mathog RH, Toma V, Clayman L et al. Nonunion of the mandible: An analysis of contributing factors. Journal of Oral and Maxillofacial Surgery 2000;59:746.

# Chapter 2

# Finite Element Analysis of the Stress on the Implant-Bone Interface of Dental Implants with Different Structures

Liangjian Chen

Additional information is available at the end of the chapter

## 1. Introduction

Titanium and titanium alloys have become the preferred materials for dental implants owing to their good biocompatibility, excellent corrosion resistance and suitable mechanical properties. However, the existing titanium implants still have several drawbacks. Firstly, the bonding strength at the interface between the implant and the bone is not high enough and the biological fixation has not been achieved. Secondly, there exist mismatches between the elastic modulus of the implant and of the bone. A stress shielding or concentration can be easily induced on the interface and results in a potential risk to the long-term stability of the implant. The success or failure of an implant is determined by the manner how the stresses at the bone-implant interface are transferred to the surrounding bones [1,2]. The mandible has structural characteristic of an outer layer of dense cortical bone and an inner layer of porous cancellous bone. The elastic modulus and mechanical properties of cortical bones are different from those of cancellous bones. Nevertheless, current dental implants are mainly fabricated using dense titanium and titanium alloys, which have no features representing the difference between the inner and outer layers of the mandible or that between their elastic modulus. And therefore, the incompatibility of the mechanical properties between the implant and the bone was encountered. The use of porous metal implants for medical applications has two main advantages. One is the similar elastic modulus to the bone, which helps to prevent the stress shielding effect at the bone interfaces. The other is that it can provide a structural condition for the bone ingrowth to achieve biological fixation [3,4]. However, the low mechanical strength limits their further applications in the implanting industry. In this study, according to the structural characteristics of the mandible and the clinical requirements for the implant mechanical properties, a novel bio-mimetic design of implant is proposed for the titanium implants, which composes of a cortical bone zone with a dense structure and a cancellous bone zone with a porous outer layer and a dense core, as well as another three implants with different structures.

The finite element method is one of the most frequently used methods in stress analysis in both industry and science[5]. Three-dimensional (3-D) finite element analysis (FEA) has been widely used for the quantitative evaluation of stresses on the implant and its surrounding bone[6,7]. Therefore, FEA was selected for use in this study to examine the effect of the structure and elastic modulus of dental implant on the stress distribution at implant-bone interface. The 3-D models of the designed implants were constructed and the finite element analyses were carried out using Ansys Workbench 10.0. The stress distributions on implant-bone interface were investigated under static loading condition in order to provide design guidelines for the development of new implants. At the same time,the stress distributions on implant-bone interface were investigated in both dynamic and static loading conditions, and the fatigue behaviors of the bio-mimetic implant were analyzed based on fatigue theories and the formulas, in order to provide theoretical basis for the development of new implants.

## 2. Material and methods

### 2.1. Structur of the biomimetic implant

The biomimetic implant comprised of two layers, including the porous layer of open connected pores, which can provide the structure for bone ingrowth and has mechanical properties similar to the surrounding bones. The dense core ensures that the mechanical properties of implant meet the requirements of clinical applications (Fig.1).

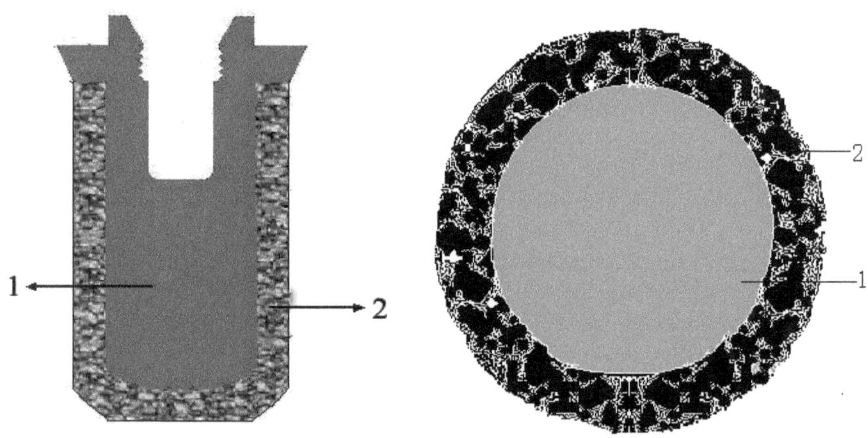


(A) (B)

**Figure 1.** Structure of the bio-mimetic porous titanium implant. A:section plane, B : cross section, 1: dense core, 2: porous layer

## 2.2. CAD and finite element modeling of the elements

A 3-D model of a mandibular section of bone with a missing second premolar and its superstructures were used in this study. A mandibular bone model was selected according to the classification system of Lekholm and Zarb. Trabecular bone was modeled as a solid structure in cortical bone. A bone block with dimensions of 20×14×35mm, representing the section of the mandible in the second premolar region, was modeled. It consisted of a spongy center surrounded by cortical bone of 2 mm.

Four implants models with dimensions of d4.1 mm×12mm were selected in this study. Those implants and abutment were assumed to consist of the same material. Implant No.1 was dense with a high elastic modulus. Implant No.2 was a bio-mimetic with a high modulus in the cortical bone zone and low modulus-outer and high modulus-interior in the cancellous bone zone. Implant No.3 had a high modulus in the cortical bone zone and a low modulus in the cancellous bone zone. Implant No.4 had a whole lower elastic modulus. The elastic modulus of the dense titanium (high modulus) was set as 103.4GPa. The elastic modulus of implant No.1 (low modulus) was set as 40% of the dense titanium. To investigate the effect of elastic modulus on the interface stress, modulus in the low modulus zone varied in the range of 80%,40% ,10% and 1.3% of the modulus of the dense titanium, i.e.1370MPa. Mechanical properties of the implants were shown in Table 1.

The 3-D model of the implants was constructed by the CAD software Pro/E. The finite element analyses were carried out using Ansys Workbench 10.0. Tetrahedron elements in implant and bone corresponding to SOLID45 type elements in ANSYS element library with each node had three degrees of freedom. The finite element model is shown Fig.2 and Fig.3. The physical interactions at implant–bone interfaces during loading were taken into account through bonded surface-to-surface contact features of ANSYS. Numbers of nodes and elements of implant and bones were shown in Table 2.

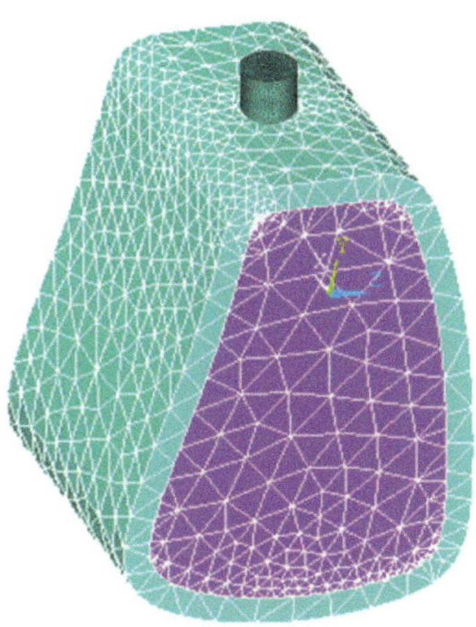

**Figure 2.** Finite element models of bone and implant

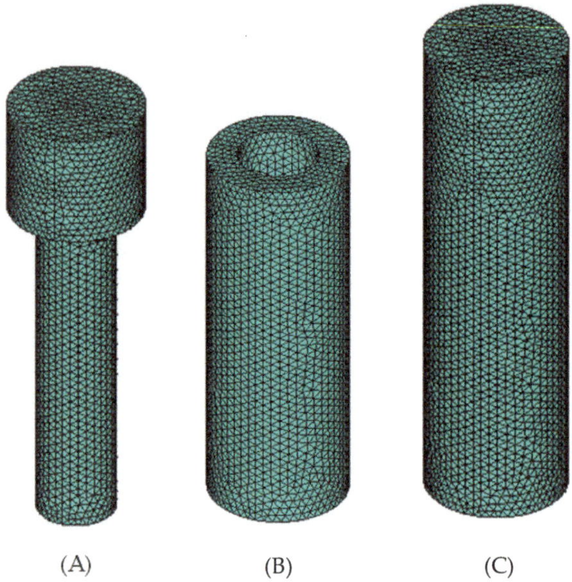


**Figure 3.** Finite element models of implant. A: dense body, B: porous layer, C: implant

| Material | Elastic Modulus/GPa | Poisson ratio ,υ |
|---|---|---|
| Lower modulus titanium | 41.36 | 0.35 |
| Dense titanium | 103.4 | 0.35 |
| Cortical bone | 13.70 | 0.30 |
| Cancellous bone | 1.37 | 0.30 |

**Table 1.** Mechanical properties of materials used in the study

| Implants | Number of nodes | | Number of elements | |
|---|---|---|---|---|
| | Dense | Porous | Dense | Porous |
| No.1 and No.4 | 15835 | - | 84828 | - |
| No.2 | 8880 | 9186 | 45486 | 42926 |
| No.3 | 3968 | 12313 | 20230 | 65703 |
| Cortical bone | 13329 - | | 65297 - | |
| Cancellous bone | - 5395 | | - 16324 | |

**Table 2.** Numbers of nodes and elements of implant and bones

### 2.3. Loads and boundary conditions

All materials were assumed to be homogenous, isotropic and linearly elastic. The bone-implant interfaces were assumed to be 100% osseointegrated. The sides and bottom of cortical and cancellous bones were set to be completely constrained, and the boundary conditions were extended to the corresponding node. Multi-constraining was imposed on implant from bottom to top, in order to limit the freedom of the roots.

Static loading was loaded to evaluate the implant-bone model. The implants were assumed to be under an axial force of 50-300N and a lingual force of 25 N in the angle of approximately 45° to the occlusal plane .

Static and dynamic analyses of the implant need to consider and ensure the safety in the design. In the literature, implants are often worked according to the results of static analysis. Under the same masticatory forces, dynamic effects may add 10–20% more loads to implant than static effects. This must be taken into account to safeguard the fracture or fatigue failure of the implant. Therefore, using dynamic loading during the evaluation of a new implant is more reasonable. In the simulation of the normal chewing motion, forces close to the masticatory forces of normal adults were loaded to implant-bone model. Time dependent masticatory load was applied. Time history of the dynamic load components for 5 s is demonstrated in Fig. 4. These estimations were based on the assumption that an individual has three episodes of chewing per day, each 15 min in duration at a chewing rate of 60 cycles per minute (1 Hz). This is equivalent to 2700 chewing cycles per day or roughly 106 cycles per year.

The von Mises stresses were used as the key indicators to measure stress levels and evaluate the stress distribution at implant-bone interface, as well as the maximum stress values on cortical bone. The main indicators are: 1) stress distribution in axial at the implant-bone interface, and 2) the maximum von Misese stresses.

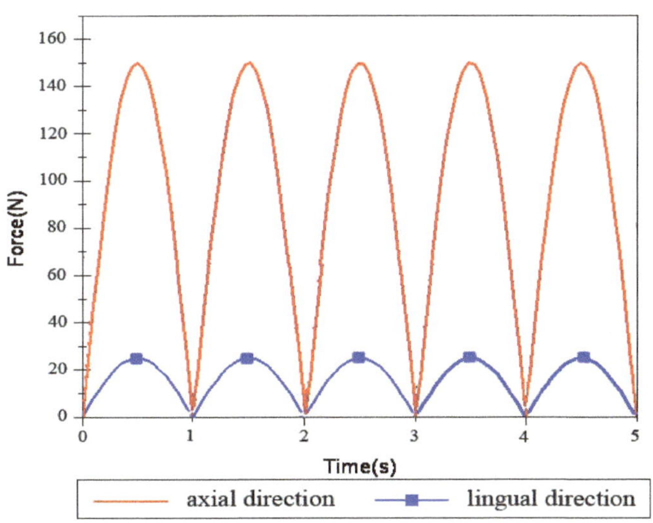


**Figure 4.** Dynamic loading in 5 seconds

## 2.4. Fatigue analysis

A good dental implant design should satisfy the maximum or an infinite fatigue life. This can only be ensured by physical testing or a fatigue analysis. In this study, the fatigue life of the dental implant was predicted using the finite element stress analysis with computer code of ANSYS/Workbench (ANSYS, 2003). Fatigue properties shown in Fig. 5 were used in fatigue calculations. Fig. 5 was known as S–N curves, showing fatigue properties of pure titanium in terms of alternating stress versus number of cycles. Fatigue life of prosthesis was calculated based on Goodman, Soderberg, Gerber and mean-stress fatigue theories which were illustrated in Table 3.

In Table 3, N indicates the safety factor for fatigue life in loading cycle, while Se is for endurance limit and Su is for ultimate tensile strength of the material. Mean stress σm and alternating stress σa are defined respectively as below, respectively.

$$\sigma_m = \frac{\sigma_{\max} + \sigma_{\min}}{2} \tag{1}$$

$$\sigma_a = \frac{\sigma_{\max} - \sigma_{\min}}{2} \tag{2}$$

Von Misses stresses obtained from finite element analyses are utilized in fatigue life calculations. All fatigue analyses were performed according to the infinite life criteria (i.e. N = $10^9$ cycles).

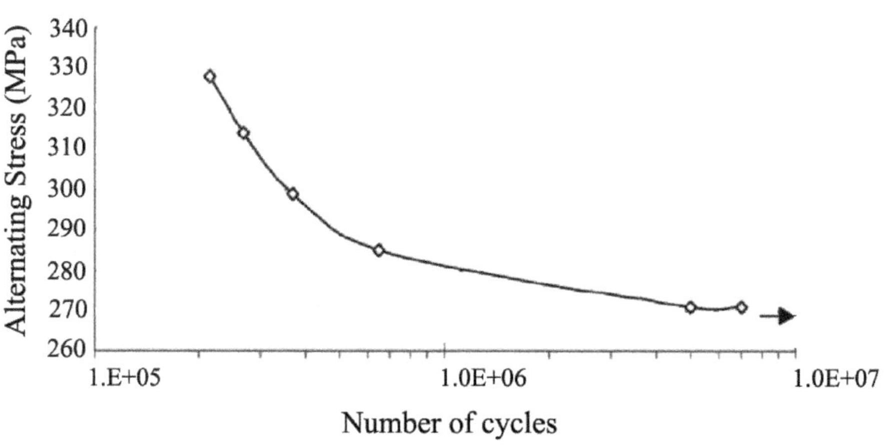


**Figure 5.** Fatigue curves (S-N curve) of pure titanium

| Fatigue theories | Fatigue formulas |
|---|---|
| Goodman | $\left(\frac{\sigma_a}{S_e}\right)+\left(\frac{\sigma_m}{S_u}\right)=\frac{1}{N}$ |
| Soderberg | $\left(\frac{\sigma_a}{S_e}\right)+\left(\frac{\sigma_m}{S_y}\right)=\frac{1}{N}$ |
| Gerber | $\left(\frac{N\sigma_a}{S_e}\right)+\left(\frac{N\sigma_m}{S_u}\right)^2=1$ |

**Table 3.** Fatigue theories and formulas used in fatigue life predictions

# 3. Results

## 3.1. Stress distribution on implant-bone interface under static loading condition

### *3.1.1. The maximum stresses at implant-bone interface*

Table 4 shows the maximum von Mises stresses of different structure implants. It can be seen that the interface stresses of implant No.3 are much higher than those of other implants. There is no obvious difference in the maximum stress between implant No.1 and No.4. Implant No.2 has the lowest maximum stress at both cancellous bone and root zone comparing with other implants. After the transferring of stress to the surrounding bones, the maximum stress in cortical bone is larger than that of cancellous bone in the surrounding bone tissue. Implant No.1 has the largest stress in cortical bone and No.3 has the largest stress in the root of cancellous bone.

| Implants | Stress/MPa | | | |
|---|---|---|---|---|
| | Cortical bone interface | Cancellous bone interface | Cortical bone | Cancellous bone |
| No.1 | 23.434 | 12.553 | 11.668 | 1.456 |
| No.2 | 23.451 | 8.261 | 9.685 | 1.525 |
| No.3 | 33.532 | 15.77 | 8.419 | 4.845 |
| No.4 | 23.453 | 14.482 | 9.012 | 1.799 |

**Table 4.** Maximum von Mises stresses of implants with different structures

### *3.1.2. Stress distribution at implant-bone interface of implants under static loading.*

Figure 6 represents the stress distribution at the implant-bone interface in an axial direction. It can be seen that the maximum stresses of the implant No.1, 2 and 4 show no difference in the cortical bone zone and the maximum stress zone is located at the marginal zone of cortical bone. The maximum stress zone of implant No.3 is located at the interface between cortical and cancellous bones. The area of the high stress zone and the value of interface stress of the implant No.2 are the smallest in both the cancellous bone and its root apex.

Figure 7 represents the stress distribution in the cancellous bone zone of the implants. In all cases, there are high stress zones in the junction of the porous layer and the dense body. Among them, implant No.2 has the lowest interface stress. In the cancellous bone zone, the interface stress decreases from top to bottom, and increases at the root apex. And once again, No.2 has the lowest stress at the root apex, while No.3 has an obvious higher value than the others. The maximum stress exists at the bone interface of the implant No.1, which was 42.96% higher than that of implant No.2.

It was demonstrated that the structure of the implants has a predominate influence on the interface stress. Implant No.3 has a high trend to cause the stress concentration, while implant No.2 can efficiently reduce the interface stress, facilitating the transportation of the interface stress to the surrounding bones, avoiding the stress shielding and concentration, which is beneficial for the long time stability of the implants.

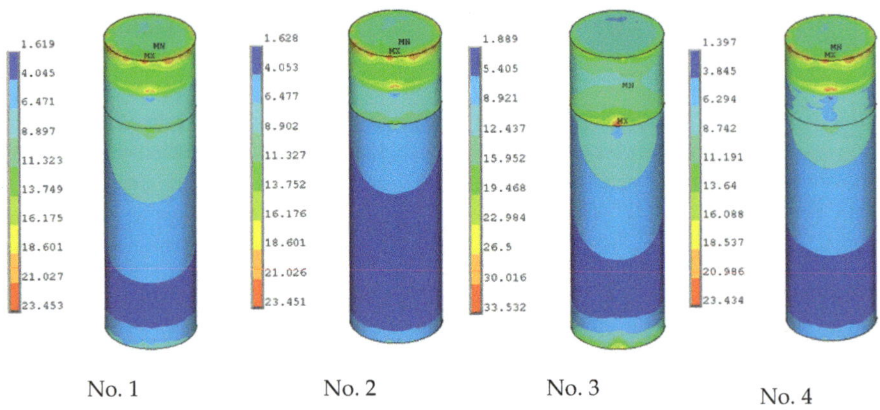


**Figure 6.** Stress distribution in axial direction at implant-bone interface of different structure implants

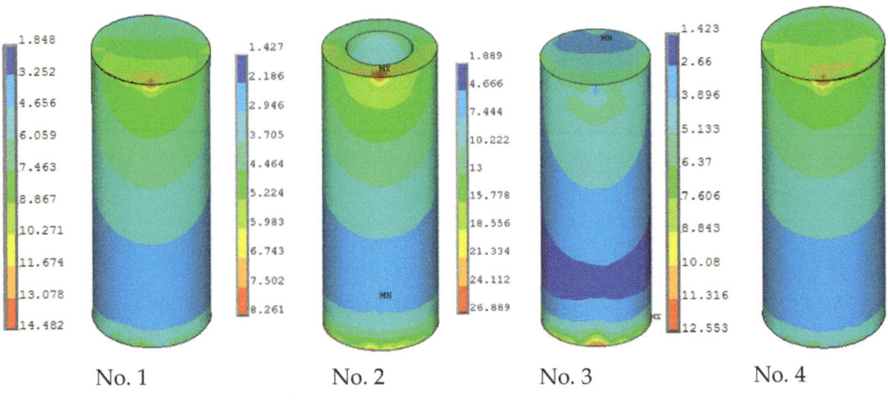


**Figure 7.** Stress distribution in spongy bone zone of different structure implants

### 3.2. Effect of elastic modulus on the interface stress distribution of implant No.2

It was demonstrated that implant No.2 has the lowest interface stress. Thus, it is chosen to study the effect of elastic modulus of low modulus zone on the interface stress distribution at the interfaces. The elastic modulus in the low modulus zone varies in the range of 80%, 40%, 10% 1.3% of the modulus of the dense titanium, i.e.1370MPa. Table 8 shows that the interface stress in cancellous bone decreases with the decrease of the modulus of the low modulus layer, while there is no significant change in the cortical bone zone. For the interface stress of surrounding bones, it can be seen that the stress increases and that at the root apex of cancellous bone decreases with the decrease of the modulus of the low modulus layer.

| Implants | Stress/MPa | | | |
|---|---|---|---|---|
| | Cortical bone interface | Cancellous bone interface | Cortical bone brink | Cancellous bone root apex |
| 80% | 23.452 | 12.725 | 9.172 | 1.739 |
| 40% | 23.451 | 8.261 | 9.685 | 1.525 |
| 10% | 23.451 | 3.733 | 11.224 | 1.094 |
| 1370MPa | 23.443 | 2.216 | 12.304 | 1.351 |

**Table 5.** Maximum von Mises stresses of implants

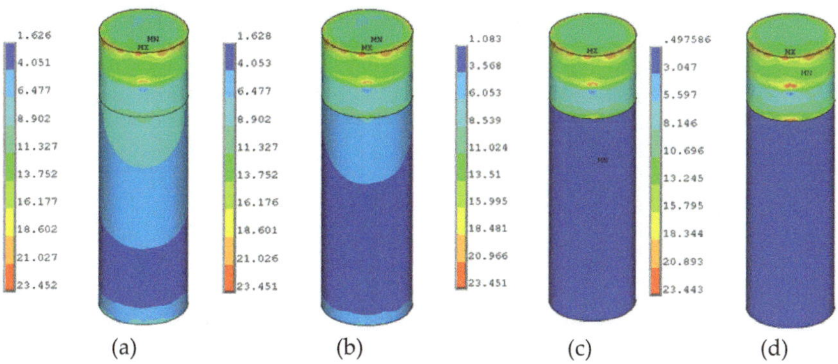


**Figure 8.** Stress distribution of implant No.2 in axial direction at implant-bone interface: (a)80%; (b)40%; (c)10%; (d)1.3%

Figure 9 represents the stress distribution at the implant-bone interface in the axial direction. It can be seen that, under the same loading, a decrease of the modulus at low modulus layer has no significant influence on the interface stress of cortical bone. Figure 5 shows the stress distribution at the interface between implant No.2 and cancellous bone. The interface stress varies significantly with the change of the modulus of the low modulus layer. As the modulus of the low modulus layer decreases, the area of the high stress zone reduces, and the volume of the interface decreases dramatically. When the modulus of the low modulus layer reduces to 10% of the dense value, a uniform distribution of the interfacial stress without any high stress zone is obtained. For the specimens with the modulus of 1370MPa, the interface stress is 2.216MPa, 82.6% smaller than that of 80% ones. With the decrease of the modulus, the interface stress between the dense core and the porous layer increases. Figure 10 represents the stress distribution in dense body of implants No.2. It can be seen that the high stress zone is located at the interface between the cortical bone and cancellous bone. For the specimens with the modulus of 10% of the dense ones, the maximum interfacial stress at the porous-dense core interface is 18.556MPa. And it reduced to 13.752MPa for those of 80% specimens.

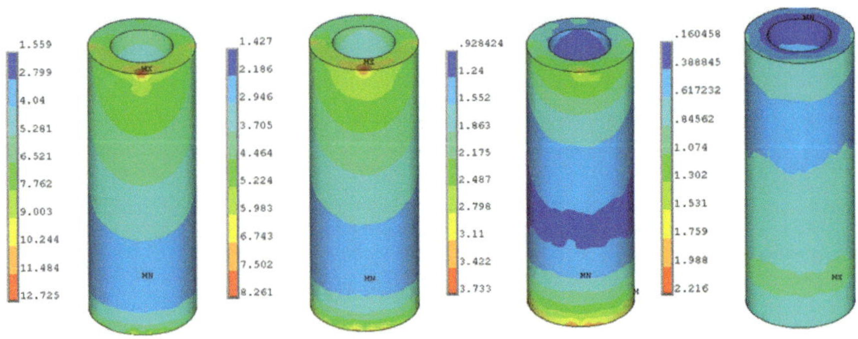


(a) (b) (c) (d)

**Figure 9.** Stress distribution at interface between implant No.2 and cancellous bone: (a)80%; (b)40%; (c)10%; (d)1.3%

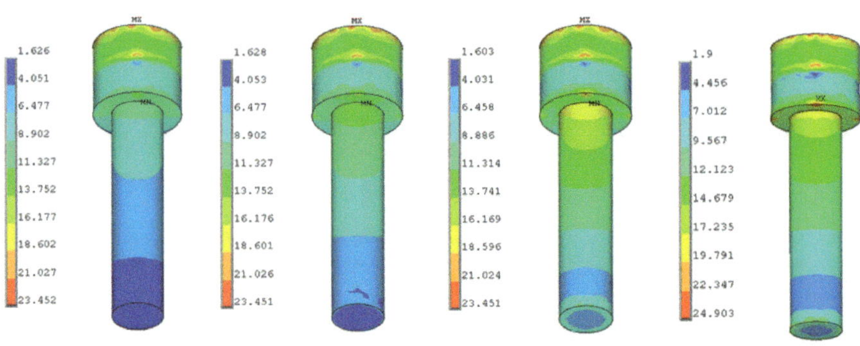


(a) (b) (c) (d)

**Figure 10.** Stress distribution in dense body of different modulus implants No.2: (a)80%; (b)40%; (c)10%; (d)1.3%

## 3.3. Effect of thickness of low modulus zone on the interface stress distribution of implant No.2

In order to further optimize the structure of the implant, the effect of thickness of low modulus zone on the interface stress distribution of implant No.2 was carried out, by varying the thickness of the low modulus zone from 0.5, 0.75, 1 to 1.25mm and maintaining the same implant diameter of 4.1mm and a constant modulus of low modulus zone, i.e.1370MPa. Figure 11 represents the stress distribution at the implant-bone interface in the axial direction. It can be seen that, in all cases, the cortical bone are in high stress zone while the cancellous bone are in low stress zone. The change of the thickness of low modulus zone affects the stress distribution of cancellous bone a lot while it has little influence on cortical bone, as shown in Fig.12. With the increase of the thickness, the interface stress decreases, especially in the root apex. Moreover, the distribution of the interface stress becomes more uniform. When it comes to an optimal thickness suitable for the clinical application, the strength and ingrowth of the bone tissues should be considered, which need further verification of MADIT experiments.

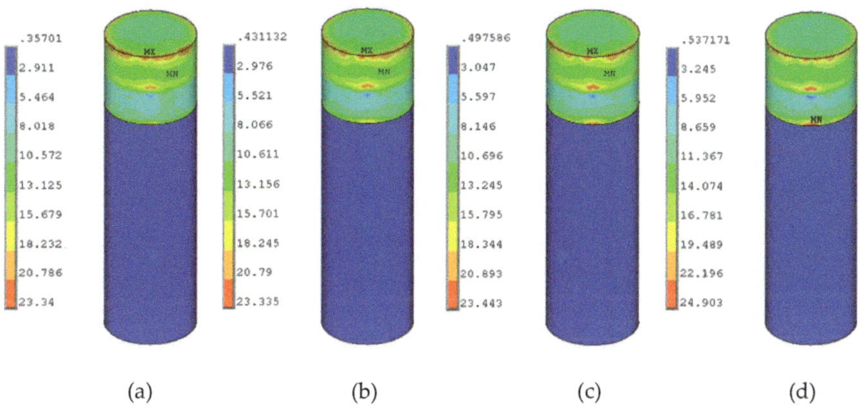


**Figure 11.** Stress distribution in axial direction ata implant-bone interface of implants No.2 with different thickness of low modulus layer: (a)0.5mm;(b)0.75mm;(c)1mm;(d)1.25mm

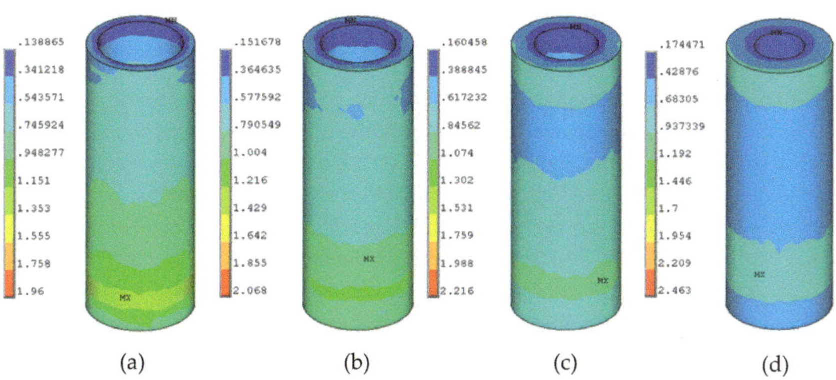


**Figure 12.** Stress distribution at cancellous bone interface of implant No.2 with different thickness of low modulus layer:(a)0.5mm;(b)0.75mm;(c)1mm;(d)1.25mm

### 3.4. Stresses distribution on implant-bone interface under static and dynamic loading conditions

In order to compare the Von Mises stresses of the sense implant with that of the bio-mimetic implant under dynamic loading and static loading stations, the model structures of NO.1 and NO.2 are designed in the same way. The elasitc modulus of NO.1 and NO.2 dense body both are 103.4Gpa,that of NO.2 porous layer is 41.36Gpa, the poisson ratio of all three is 0.35.

#### *3.4.1. Maximum stresses*

As shown in table 6, the maximum stresses under dynamic loading conditions were 17.15% higher than that under static loading conditions. The maximum stresses in cortical bone of two implants were similiar. However, the maximum stress of the dense implant was 75.79% high than that of the bio-mimetic implant in spongy bone, and 22.46% higher in the root region. The maximum stresses at implant-bone interface were much smaller than the yield strength of pure titanium (462MPa).

| Loading region | Maximum Von Mises stresses (MPa) | | | |
|---|---|---|---|---|
| | No.1 static | No.1 dynamic | No.2 static | No.2 dynamic |
| Cortical bone region | 15.265 | 17.884 | 15.264 | 17.882 |
| Cancellous bone region | 9.962 | 11.671 | 5.661 | 6.632 |
| Root-end region | 4.973 | 5.826 | 4.069 | 4.767 |

**Table 6.** Maximum Von Mises stresses of the dense implant and the bio-mimetic implant under static and dynamic loading conditions

#### *3.4.2. Stress distribution within the cortical bone surrounding the implant neck and in implant-bone interface of implants No.1 and No.2 in static and dynamic loading*

Figure. 13 represent the stress distribution within the cortical bone surrounding the implant neck. The maximum stress occurred at the edge of the cervical cortical bone of implant No.2 were greater than those of implant No.1. For No.2, the maximum stresses were 7.192MPa in static loading condition and 8.428MPa in dynamic loading condition. For No.4, the maximum stresses were the maximum stresses were 6.67MPa in static loading condition and 7.814MPa in dynamic loading condition. The results indicated the implant No.2 had 7.85% and 7.67% higher stresses than the implant No.4 in dynamic and static loading conditions, respectively. The maximum stresses of the implant No.2 in static and dynamic loading conditions were only10.42% and 21.21% of the yield strength of cortical bone, 69MPa, respectively .

Figure.14 represent the stress distribution in the implant-bone interface in an axial direction. In both loading conditions, the maximum stresses at implant interfaces in the implant No.1

and 2 showed no difference in the cortical bone area, while the high stress zone of the implant No.1 was greater than that of the implant No.2 in the spongy bone area and around the root apex. The yield strength of pure titanium was 462MPa. In static and dynamic loading conditions, the maximum stresses of the implant No.2 at the interface were 15.264MPa and 17.882MPa respectively, and they were only3.3% and 3.87% of the yield strength of pure titanium.

Figure.15 represent the stress distribution in interface of spongy-bone implant. The stresses at the implant interface in dynamic loading condition were all higher than those in static loading condition. Both implant bodies had high stress zones in the junction of the cortical bone and spongy bone, and the stresses at the implant interface showed a declining trend from top to bottom but increased at the root apex. The interface stresses of the implant No.1 was higher than that of the implant No. 2, and the maximum stress at the bone interface of the implant No.1 was 75.97% higher than that of the implant No.2.

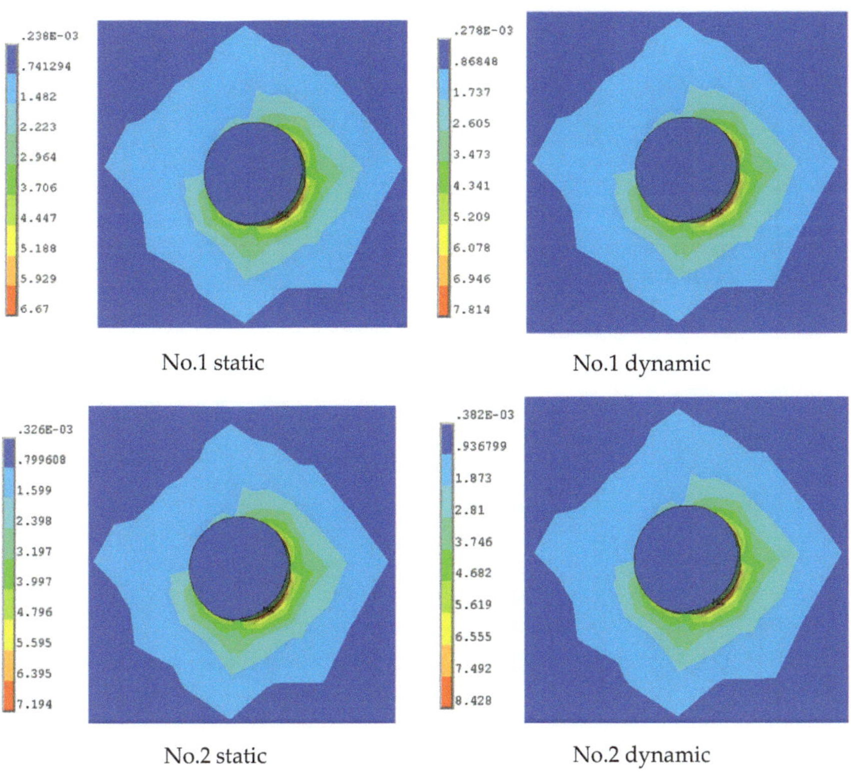


**Figure 13.** Stress distribution in the cortical bone of the implant No.1 and No.2 under static and dynamic loading conditions.

Figure. 16 represent the stress distribution in the dense body of the implants. There was a high stress zone of the dense body in the junction of the porous layer and the dense body of the implant No.2. The maximum stress of 12.306MPa in dynamic loading condition was higher than that of 10.504MPa in static loading condition. In the spongy bone area, the high stress zone of the dense body of the implant No.2 was greater than that of the implant No.1. The yield strength of pure titanium was 462MPa; the maximum stress at the interfaces of implant dense body did not reach the yield strength of pure titanium.

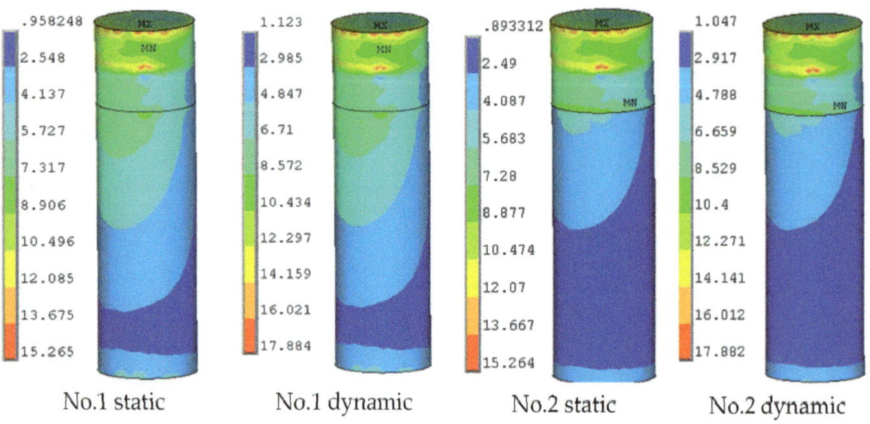


**Figure 14.** Stress distribution in the bone-interface of the implant No.1 and No.2 under static and dynamic loading conditions

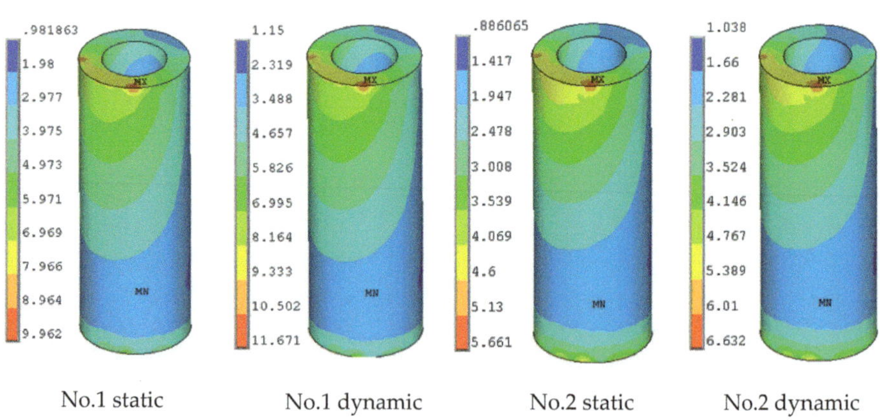


**Figure 15.** Stress distribution in the spongy bone-interface of the implant No.1 and No.2 under static and dynamic loading conditions

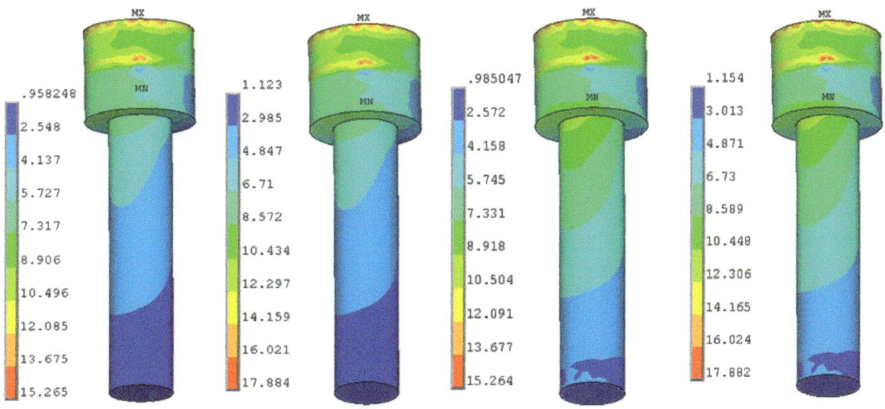


No.1 static No.1 dynamic No.2 static No.2 dynamic

**Figure 16.** Stress distribution in the dense body of the implant No.1 and No.2 under static and dynamic loading conditions

### *3.4.3. Fatigue analysis of bio-mimetic implant*

In the fatigue calculations, referring to the fatigue curves of pure titanium (S-N curves) shown in Figure 5, the fatigue life of implant was calculated based on Goodman, Soderberg, Gerber and Mean-Stess fatigue theories and formulas which were illustrated in Table 3. The endurance limit of pure titanium (Se) is 259.9MPa, and the yield strength (Sy) is 462MPa. The safety factors of different dense bodies of bio-mimetic implants with dynamic preload were calculated using the Soderberg formula in Table 3, as shown in Figure. 17. Figure. 17 represent the safety factors of different dense bodies of bio-mimetic implants when the dynamic preload was applied. Under an axial force of 50~300N and a lingual force of 45°25N in dynamic loading condition, the safety factors of dense body were all above 10. The results show that the bio-mimetic implant is safe against fatigue load.

Figure 18 showed the maximum stress at the interface of porous layer and the bone under different preloading conditions. With the increase of the loading, the interface stress of porous layer linearly increased. In dynamic loading condition with normal chewing force (axial 150N and lingual 45°25N), the maximum stress at the porous layer interface (σmax) was 6.632MPa, and the minimum (σmim) was 1.038MPa. When an axial force of 300N and a lingual force of 45°25N were applied, the maximum stress at the porous layer interface (σmax) was 11.38MPa, and the minimum (σmim) was 1.97MPa. According to the simulation results, it was predicted that the strength of the porous layer of the bio-mimetic implant and its bonding strength with the dense body interface should both be greater than the maximum interface stress (11.38MPa), which ensured the implant safety. The analyses of the implant interface stress provide a basis of mechanical properties for the preparation of porous layer of bio-mimetic implant.

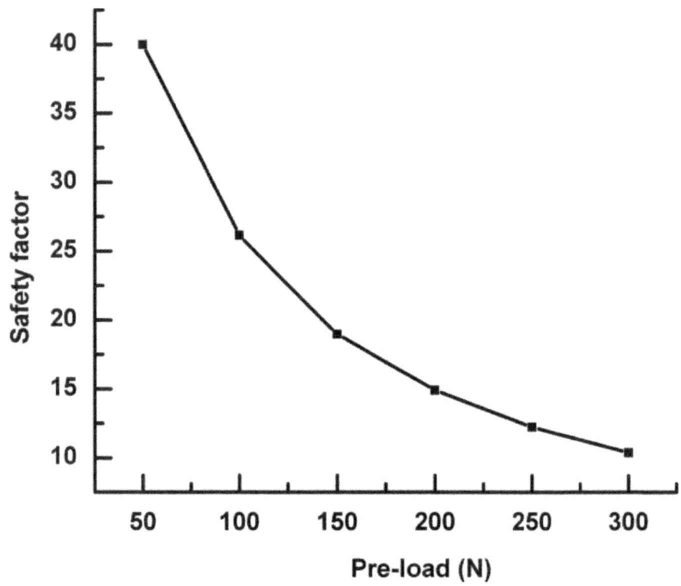


**Figure 17.** Safety factor for dense body of bio-mimetic implant under different dynamic loading

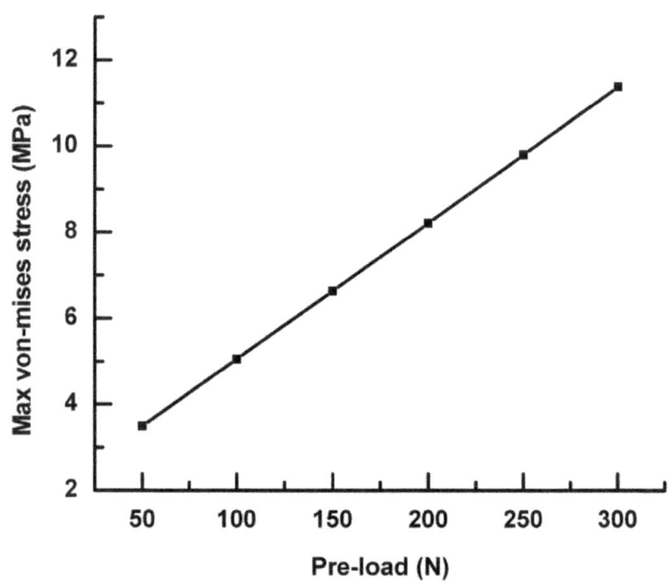


**Figure 18.** Maximum Von-mises stress for bone-interface porous layer of implant under different dynamic loading conditions.

## 4. Discussion

The functions of implant are mainly dependent on the direct bonding with the surrounding bones. The long-term success of an implant is determined by the reliability and stability of the implant bone interface, and the success or failure of an implant is determined by the manner that the stresses at the bone-implant interface transfer to the surrounding bones[1, 2]. The main factors contributing to the stability of implants include the structure of the implants, the distribution of the interface stress and the combination mode of the interface. In order to ensure the long-term stability of an implant, the implant should be designed according to two main principles. First, the load should be minimized to avoid exceeding its physiological tolerance as overloading can cause bone resorption or fatigue failure of the implant. On the other hand, underloading may lead to disuse atrophy and subsequent bone loss[3, 4]. Second, the contact zone with the bone should be increased to reduce the bone interface stress. The structural characteristic of the mandible shows an outer layer of dense cortical bone and an inner layer of loose cancellous bone. Both the elastic modulus and mechanical strength of cortical bone (10~18GPa) are higher than those of cancellous bone (1.3~4GPa). Current dense implants do not have the structure similar to that of the mandible, as well as modulus. As a result, the mechanical compatibility between the implant and the bone remains unresolved, and the modified active coating on the surface gets easily damaged in the implantation process. An implant with a low elastic modulus is believed to be beneficial to transferring the stress to the surrounding bones, resulting in a long-term stability[8,9]. The porous implant materials can tremendously improve the implant biocompatibility [10-12] by improving the adhesion and outgrowth of those osteoblasts, promoting the deposition of extracellular matrix, increasing the adsorption of nutrients and oxygen, and promoting the new bones' growth into pores to achieve biological fixation. The porosity can be changed to adjust the density, strength and elastic modulus of the material to achieve similar mechanical properties to the replaced hard tissues. Meanwhile, the porous structure can provide scaffold for the bioactive coating to promote osseointegration. In this study, according to the structural characteristics of the mandible and the advantages of the porous implant material, an idea of a bio-mimetic implant is proposed. It is a titanium implant composed of a cortical bone zone with a dense structure and a cancellous bone zone with a porous outer layer and a dense body. The cortical bone has a high modulus, and the porous outer layer of the cancellus bone zone has a low modulus. The dense body ensures the strength to meet the requirements of clinical applications. To optimize the structure of the bio-mimetic implants, the finite element analysis was carried out. The effects of implant structure, modulus and thickness of the low modulus layer on the distribution of the interfacial stress were studied.

The interfacial stress of the implants is mainly located at the interface between the implants and the surrounding bones,affecting the interface biological reactions such as bone resorption and remodeling. Cortical bone loss and early implant failure after loading are usually accompanied by the excess stress at the implant bone interface while a low stress may lead to disuse atrophy and subsequent bone loss [13,14]. It is indicated that, under the same situation, the smaller the bone surface area in contact with the implant body is, the greater the overall stress becomes [15]. Cortical bone, which has a higher modulus, higher strength and more resistance to deformation than cancellous bone [16], can bear more loading in masticatory movements [17-20]. In this study, it was supposed that the implant-bone osseointegration was 100%. Under the same loading condition, the stress distributions at the interface of four different structure implants were compared and analyzed, showing the change of the implant structure and modulus in the cancellous bone had significant effects on the stress distribution. In all cases, there are high stress zone at the interface between cortical and cancellous bone. In cancellous bone, the interface stress decreases from top to bottom, and increases at the root apex.

In the cortical bone zone, all implants present high stress values and the maximum stresses are in the same level. In the cancellous bone zone, the maximum stress of the dense implant interface was 75.58% higher than that of the bio-mimetic implant, and 22.21% higher than that in the root apex zone. The maximum stresses in cancellous bone and root region of implant No.2 are lower than those of other three implants. The maximum stress of implant No.4 is 42.96% higher than that of No.2. Implant No.3 has the highest stresses in root region. The stress distribution at bone-implant interface varied with elastic modulus of low elastic modulus layer. The maximum stresses of implant No.2 decreases with the decreasing of elastic modulus in cancellous bone region, while there is no significant difference in cortical bone region. When the modulus of the low modulus layer is reduced to 10% of the dense ones, a uniform distribution of interfacial stress without any high stress zone was obtained. With the increase of the thickness of the low modulus layer, the interface stress decreases, especially in the root apex. Moreover, the distribution of the interface stress becomes much uniform.

From the biomechanical point of view, a structure like implant No.2, a modulus matches the cancellous bone and a suitable thickness can effectively reduce the stress in the implant-bone interface and be beneficial to the transfer of interfacial stress to surrounding bones, which is favorable to the long-term stability of the implant. The structural characteristics of this implant are in line with those of the mandible, so that the elastic modulus of the porous zone can be reduced to make the elastic modulus of the implant match with that of the cancellous bone and thus help the interface stress tranferring. The structural characeristics of mandible of implants No.1and No.4 are ingored, which results in the un-uniform interface stress distribution and stress concentration in cancellous bone. Although implant No.3 has a mandible –like structure, the cancellous bone is a whole low modulus structure, which leads to stress concentration at both interface and root apex.

Implant No.2 has a low modulus-outer and high modulus-interior in the cancellous bone zone. The low modulus-outer can be realized by adjusting the porosity and pore size to match the mechanical properties, especially the elastic modulus, with the surrounding bones. Figure. 19 illustrates the stress distribution of the porous and dense implants under vertical loading. In the model, R refers to the radius of the implant, H refers to the height, and F refers to the vertical loading. Assuming that the compressive stress and shear stress are uniform, and the compressive stress and shear stress on porous and dense implants are σ1, τ1 and σ2, τ2, respectively. The porous implants provide more contact area with the bone than the dense implants. Assuming that A1 is the added contact area, the equilibrium equations of forces for porous and dense implants can be expressed as:

$$F = (\pi R^2 + A_1)\sigma_1 + 2\pi RH\tau_1 \tag{1}$$

$$F = \pi R^2\sigma_2 + 2\pi RH\tau_2 \tag{2}$$

Because the compressive strength at the interface is much larger than its shear strength, the values of σ1 similar to σ2, and the added zone A1 are larger, we can obtain τ1«τ2. It means that the shear force of porous implants is much smaller that that of dense ones, which is beneficial for the stability of the low strength cancellous bone.

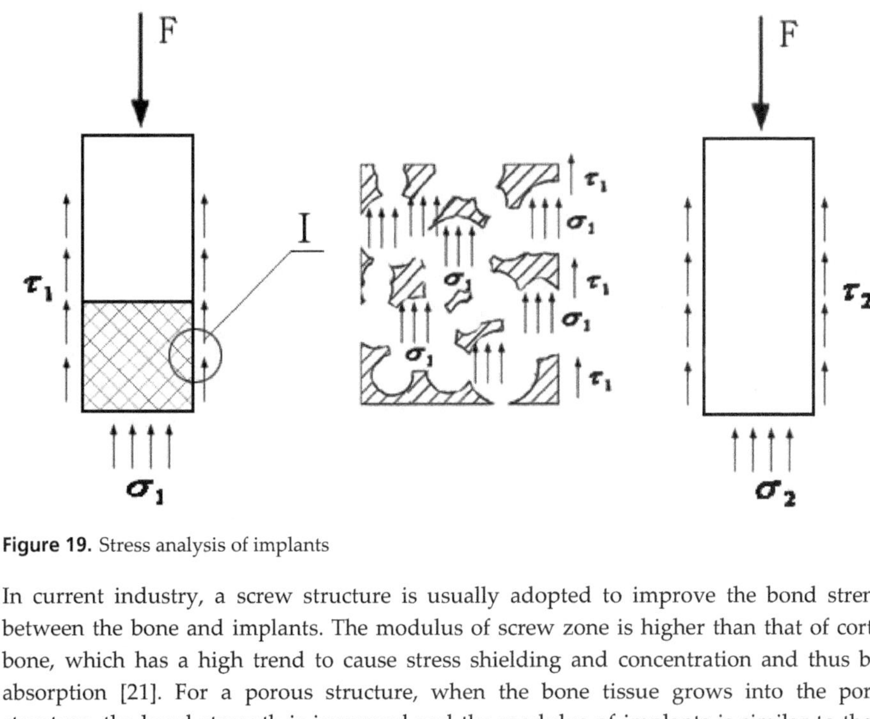


**Figure 19.** Stress analysis of implants

In current industry, a screw structure is usually adopted to improve the bond strength between the bone and implants. The modulus of screw zone is higher than that of cortical bone, which has a high trend to cause stress shielding and concentration and thus bone absorption [21]. For a porous structure, when the bone tissue grows into the porous structure, the bond strength is improved and the modulus of implants is similar to that of the surrounding bones. No bone absorption occurs under loading because part of the stress can be borne by bone tissues in the pore. In summary, biomimetic style implant No.2, with a high modulus in the cortical bone and low modulus-outer and high modulus-interior in the cancellous bone is superior in the stress teansferring. The porous structure can effectively reduce the shear force at the bone-implant interface, providing a suitable environment for bone tissue ingrowth, which is benefit for the longtime stability of the implants.

## 5. Conclusions

1. The distribution of interface stress is strongly depended on the structure of the implants. The bio-mimetic implant No.2 is favorable to transferring the interface stress from the cancellous bone and root apex bone to surrounding bones, avoiding stress shielding and concentration.
2. It is demonstrated that the interface stress varies significantly with the change of the modulus of the low modulus layer. The area of the high stress zone is reduced, and the value of the interface decreases dramatically. When the modulus of the low modulus layer is reduced to 10% of the dense value, a uniform interface stress distribution without any high stress zone was obtained.
3. The change of the thickness of low modulus zone affects the stress distribution of cancellous bone, while it has no significant influence on cortical bone. With the increase of the thickness, the interface stress decreases, especially in the root apex. Moreover, the distribution of the interface stress becomes much uniform.

## Author details

Chen Liangjian
*The Third Xiangya Hospital of Central South University, ChangSha, China*
*State Key Laboratory of Powder Metallurgy Central South University, ChangSha, China*

## 6. References

[1] Van Osterwyck H, Duyck J, Vander S, Vander PG, Decoomans M, Lieven S, Puers R, Naert L. The influence of bone mechanical properties and implant fixation upon bone loading around oral implants [J]. Clin Oral Implants Res, 1998, 9(6):407-412.
[2] Geng J, Tan K B C, Liu G. Application of finite element analysis in implant dentistry: a review of the literature [J]. J Prosthet Dent, 2001, 85(6):585–598.
[3] Vaillancourt H, Pillar RM, McCammond D. Factors affecting cortical bone loss with dental implants partially covered with a porous coating: a finite element analysis[J]. Int J Oral Maxillofac Implants, 1996, 11(11):351–359.
[4] Pilliar RM, Deporter DA, Watson PA, Valiquette N. Dental implant design effect on bone remodeling[J]. J Biomed Mater Res, 1991, 25(4):467–483.
[5] Bathe KJ. Finite element procedures[M]. Upper Saddle River (NJ): Prentice-Hall; 1996. p. 148–377.

[6] Sato Y, Wadamoto M, Tsuga K, Teixeira ER. The effectiveness of element down sizing on a three-dimensional finite element model of bone trabeculae in implant biomechanics[J]. J Oral Rehabil ,1999;26:288–91.

[7] Sahin S, Cehreli MC, Yalcm E. The influence of functional forces on the biomechanics of implant-supported prostheses—a review[J]. J Dent,2002; 20:271–82.

[8] Kayabas O , Yu¨zbasıog˘lu E , Erzincanli F. Static, dynamic and fatigue behaviors of dental implant using finite element method [J]. Advances in Engineering Software, 2006, 37(10):649–658.

[9] Meijer GJ, Cune M S, Vandooren M. A comparative study of flexible (polyactive versus rigid hydroxyapatite) permucosal dental implants: clinical aspects[J]. J Oral Rehabil, 1997, 24(2): 85-88

[10] St-Pierre J P, Gauthier M, Lefebvre LP, Tabrizian M. Three-dimensional growth of differentiating MC3T3-E1 pre-osteoblasts on porous titanium scaffolds [J]. Biomaterials, 2005, 26(35):7319–7328.

[11] Otsukia B, Takemotoa M, Fujibayashia S, Neo M, Kukubo T, Nakamura T. Pore throat size and connectivity determine bone and tissue ingrowth into porous implants: Three-dimensional micro-CT based structural analyses of porous bioactive titanium implants[J]. Biomaterials, 2006, 27(35):5892–5900.

[12] Takemoto M, Fujibayashi S, Neo M, Suzaki J, Kukubo T, Nakamura T. Mechanical properties and osteoconductivity of porous bioactive titanium[J]. Biomaterials, 2005, 26(30): 6014–6023.

[13] Misch CE. Contemporary implant dentistry [M]. 2nd ed. St. Louis: Mosby; 1998. p. 109–34, 207–17, 329–43, 595–608.

[14] Schroeder A. Oral implantology: basic, ITI hollow cylinder system [M]. New York: Thieme Medical Publishers; 1996. p. 60–65.

[15] Holmes DC, Loftus JT. Influence of bone quality on stress distribution for endosseous implants [J]. J Oral Implantol 1997;23(3):104–111.

[16] Misch CE. Density of bone: effect on treatment plans, surgical approach, healing, and progressive bone loading [J]. Int J Oral Implantol ,1990;6(2):23–31.

[17] Cochran DL. The scientific basis for and clinical experiences with Straumann implants including the ITI dental implant system: a consensus report [J]. Clin Oral Implants Res ,2000;11(11):33–58.

[18] Lekholm U, Zarb GA. Tissue-integrated prostheses [M]. In: Branemark PI, Zarb GA, Albrektsson T, editors. Tissue-integrated prostheses. Chicago: Quintessence; 1985. p. 199–209.

[19] CHEN Liang-jian, LI Yi-min. Influence of structure and elastic modulus of titaniun implant on implant-bone interfacial stress distribution[J]. Journal of Central South University(Science and Technology), 2009, 40(2):400-405.

[20] CHEN Liang-jian,GUO xiao-ping, LI Yi-min, LI Ting. Finite element analysis for interfacial stress and fatigue behaviors of bio-mimetic titanium implant under static and dynamic loading conditions. J cent south univ (Med sci),2010, 35(7):662-672.

[21] Gefen A. computational simulations of stress shielding and bone resorption around existing and computer-designed orthopaedic screws[J]. Medical and Biological Engineering and Computing, 2002,40(3): 311-322.

Chapter 3

# Finite Element Modelling of a Multi-Bone Joint: The Human Wrist

Magnús Kjartan Gíslason and David H. Nash

Additional information is available at the end of the chapter

## 1. Introduction

Computational models of biomechanical systems have been available for over 40 years. In the first issue of Journal of Biomechanics from 1968 there exists a paper by Marangoni and Glaser looking at the viscoelastic behaviour biological tissue and presented numerical results using a discrete model which can be thought of as a predecessor of the modern finite element models. In 1971 Rybicki et al published a paper on the mechanical stresses of the femur using the finite element method. Since then, published papers on finite element modelling increased yearly and now, 40 years later, the finite element method plays an important part on the analysis of geometrically complex structures. The hip has been researched extensively over these 40 year and numerous papers have been published from various different research groups on the mechanical response of the femur and total hip arthroplasty under various types of loading. What makes the hip an excellent candidate for finite element analysis is the fact that the geometry of the joint is well defined and can be easily extracted from CT or MRI scans but also the fact that the joint contact forces and musculoskeletal modelling of the hip joint has been extensively researched and measured (Bergmann et al 1993) giving a well defined loading condition during gait and other activities. The knee has also been researched using the finite element method where the joint geometry is well defined, but the loading conditions and the kinematics are more complex. Taylor et al (2003) have investigated the performance of total knee replacement using the finite element method.

Modelling of those joints is more complicated than of the hip and knee, due to complex bone geometry, soft tissue modelling as well as difficulty determining the physiologically relevant loading conditions acting on the joint.

The wrist and the ankle pose a challenge in biomechanical modelling due to the complex interactions between the many bones comprising the joint. Each bone will contribute uniquely to the high range of motion of the joint. The challenge in modelling of the multibone models is to capture the mechanism contributing to the stabilization of the joint. A stable joint is able to provide three-dimensional equilibrium under external loading which can also be interpreted as the ability of a joint to maintain a normal relationship between the articulating bones and soft tissue constraints under physiologic loads throughout the whole range of motion (Garcia-Elias et al. 1995). This implies that the joints need to be capable of distributing loads without generating abnormally high stresses on the articulating surface as well as being able to move within the joint's range of motion. Geometry of the bones also plays an important role in joint stability and the concavity or convexity of the articulating bones helps the bones to distribute stresses across the joint.

Work on finite element modelling of the wrist started in the 1990s with the works of Miyake et al and Anderson and Daniel who modelled the stresses on the radiocarpal joint using a plain strain contact model. That model contained the radius, scaphoid and the lunate as well as the extrinsic ligaments and the scapholunate ligaments. The TFCC was modelled using a series of spring elements. Albeit a two-dimensional model, it marked a beginning of further research interest in the numerical modelling of the wrist. Miyake et al (1994) published around the same time, a finite element model simulating the stress distribution of a malunited Colle's fracture. That same group later published a paper on the stress distribution in the carpus following a lunate ceramic replacement for Kienböck's disease (Oda et al 2000).

Other wrist models were published shortly afterwards and can be summarised in the following table.

| Author | Year | Type | Modelled |
|---|---|---|---|
| Miyake et al | 1994 | Finite element | Radius, scaphoid lunate |
| Anderson & Daniel | 1995, 2005 | Finite element | Radius, scaphoid, lunate, ulna |
| Schuind et al | 1995 | Rigid body | Whole carpus |
| Ulrich et al | 1999 | Finite element | Radius, scaphoid, lunate |
| Oda et al | 2000 | Finite element | Whole carpus excluding metacarpals |
| Carrigan et al | 2003 | Finite element | Whole carpus excluding metacarpals |
| Nedoma et al | 2003 | Mathematical model | Whole carpus |
| Gislason et al | 2009, 2010 | Finite element | Whole carpus |
| Guo et al | 2009 | Finite element | Whole carpus |
| Bajuri et al | 2012 | Finite element | Whole carpus |

**Table 1.** Previously published finite element models of the wrist

Carrigan et al published the first three dimensional wrist model where all the carpal bones were incorporated but not the metacarpals. Loading was applied onto the distal aspect of the capitate and was 15 N compressive force which is not representative of physiological in vivo loading on the wrist. Additionally the scaphoid needed to be constrained using unphysiological constraints in order to achieve convergence. In 2009 full three dimensional models of the wrist were published by Gislason et al and Guo et al incorporating the distal ends of the radius and ulna, all the carpal bones as well as the metacarpals. The Gislason model aimed to simulate load transfer behaviour of the wrist during gripping in three different subjects with the wrist in three different positions. The loading was determined on a subject specific basis where the forces and moments acting on the fingers were measured and by using a biomechanical model, the external forces were converted into joint contact forces acting on the metacarpals. The Guo model aimed to simulate the carpal bone behaviour after the transverse carpal ligament had been excised. The loading applied onto the Guo model was a combined 100 N compressive force acting on the the 2nd and 3rd metacarpal and some unphysiological constraints were applied to the model. Bajuri et al (2012) created a full three dimensional model simulating the effects of rheumatoid arthritis on the stress behaviour of the carpal bones.

Finite element models of the ankle also exist through the research of Chen et al (2003) and Cheung (2004) and although the chapter mainly discusses the creation of a finite element model of the wrist, there are many similarities in the methodology of creating a high quality finite element model of a multi bone joint, whether it be the wrist or the ankle.

The fundamental problems that researchers face in the creation of a finite element model of the wrist are the loading applied and the soft tissue constraints on the carpus. The wrist are a mechanically unstable joint so external constraints, in the form of ligaments, must be applied in order for the carpal bones to return to equilibrium whether they be modelled as spring elements or as separate geometrical entities.

With increased computational power and more enhanced software, it is possible to simulate more detailed structures to a higher degree of detail than before. With the current rate of software and hardware development, the user will soon become the limiting factor on the quality of the finite element models produced.

## 2. Image segmenting

A fully representative geometrical model is integral for the quality of the finite element model. With enhanced scanners and software it is possible to achieve high degree of resolution for the geometrical model. There exist many different image processing software packages that are capable of carrying out image processing and segmenting the scans in order to create three dimensional surface such as Mimics (Materialise), Simplware, Amira, 3D doctor, 3D slicer to name a few.

Segmentation of the wrist bones requires close attention to details as the geometrical features of each carpal bone can be highly irregular and can vary between individuals.

Using an automated segmentation from the abovementioned software packages sometimes can not be enough to capture the full three-dimensional geometry of the bones so manual segmentation is at times necessary. The importance of a high quality segmentation can not be underestimated in multibone modelling as the congruence of the articulating surfaces will play an important role in the contact formulation. Any rough edges on the articulating surfaces will cause penetration of nodepoints causing numerical instability and convergence problems once the finite element model will be run. It is therefore critical to the success of the computational model that the segmentation be carried out in an accurate manner. Another reason why the segmentation is the most critical aspect of the modelling, is the fact that once the geometry has been constructed and meshed, it is very difficult for the user to make any changes to it without starting from the beginning again.

The plane in which the segmentation should be carried out in, would be the plane with the highest resolution, which is primarily the axial plane. Using the sagittal and the coronal plane (or the other two planes with lower resolution) can also be beneficial in order to fine tune the segmentation in order to get a full three dimensional representation of the segmentation. Figure 1 shows segmentation of the carpal bones in axial and coronal planes.

Using the masks can also be a helpful tool in determining the distribution between cortical and cancellous bone. By eroding the mask of a given number of pixels, it is possible to create a hoop in each slice representing the two stiffness layers. Previously published papers have suggested that the thickness of the cortical shell in carpal bones is on average 2.6 mm (Louis et al 1995). Figure 2 shows the distribution between cortical and cancellous bone on the scans and in the finite element model.

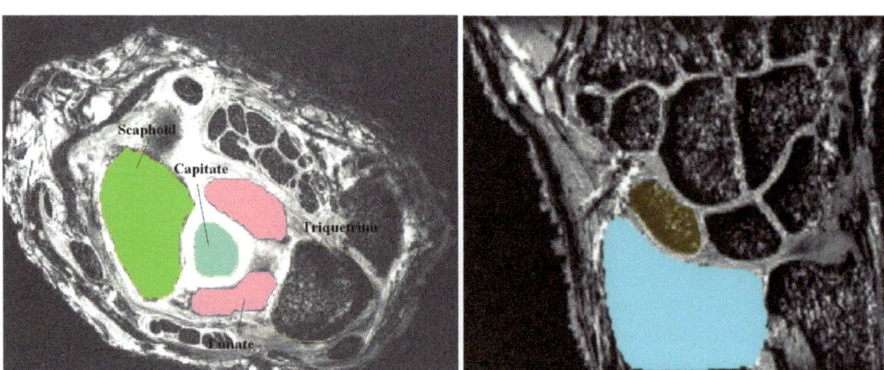


**Figure 1.** Segmentation of carpal bones in two planes

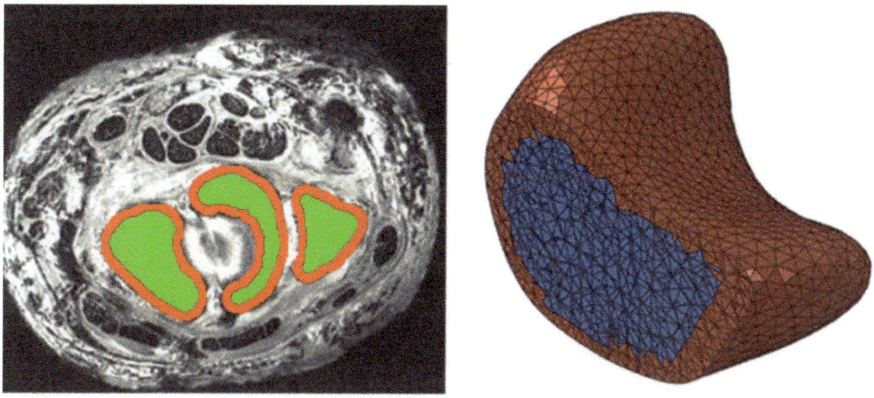

**Figure 2.** Distribution between cortical and cancellous bone in scans and on finite element model

Most software packages now offer the option of smoothing the three dimensional object. It is inevitable that unsmooth edges will occur from the image segmentation and will be more visible if some degree of manual segmentation is required. Figure 3 shows an example of how the radius bone will look like, before and after smoothing.

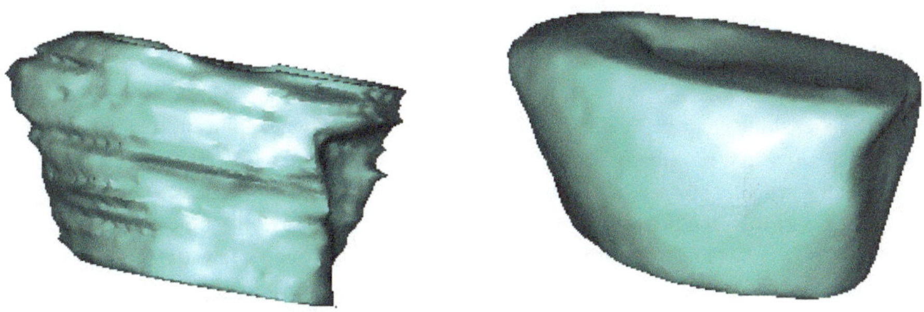

**Figure 3.** The radius bone before and after three dimensional smoothing

The smoothing is easily done within the software packages, but the user must be aware of the possible implications of the smoothing as it is possible to be too aggressive in the smoothing and therefore Lose volume whilst trying to obtain a good looking picture of the bone. Each iteration of the smoothing causes some changes in the volume of the three dimensional object although some software packages allow to compensate for the volume changes. A possible solution to these volume changes would be to recalculate the mask based upon and carry out manual adjustments of the mask and recalculate the three dimensional object and creating an iterative cycle until the smoothing will have negligible effects on the volume of the bone.

## 3. Meshing

The quality of the mesh of the finite element model will determine the quality of the solution. The process of meshing the three-dimensional objects using an automated meshing tool, which many of the image processing softwares packages discussed in previous chapter have incorporated, has significantly decreased the time and effort to create high quality meshes. The software packages then give the option of importing the meshes into finite element programs such as Ansys, Abaqus and others.

The versatility of the tetrahedral elements have made them popular candidates for the automatic meshing tools in the software packages. The tetrahedral elements are capable of capturing a high degree of geometric non-linearity and are the most popular elements used in biomechanical modelling research today. The problem with the tetrahedral elements is the stiffness of the 4 node tetrahedral element which can give too high stress values compared to the 10 node tetrahedral element. If using a 4 node tetrahedral element, the user must be confident that a sufficient number of elements is being used to capture the nonlinear geometry. For the presented models an average of roughly 430 thousand elements were used, resulting in an element density of about 10 elements/mm$^3$.

Hexahedral elements can also be used in biomechanical finite element models. In 2005 Ramos and Simões compared the performance of first and second order hexahedral elements and tetrahedral elements on a femur model and reported that there was little difference in the accuracy of the two types of tetrahedral elements. The tetrahedral elements were closer to a theoretical result, also calculated than the hexahedral elements. The hexahedral elements though showed a higher degree of stability and were less influenced by the number of elements.

As with other finite element models, the mesh quality will play a significant role in the overall solution quality. In a multibody analysis needing contact formulation, obtaining high element quality at the articular surfaces is important, as cartilage elements are soft and tend to deform to a greater extent than the bone elements. Therefore an ill shaped cartilage element, undergoing large deformations, is likely to be excessively distorted and cause divergence of the solution.

With increased computing power, the automatic meshing tool have become extremely powerful and have made it possible that the user will not need to spend much time on producing a high quality mesh, making it possible to model larger numbers of models and incorporating subject specific models.

## 4. Creation of the finite element model

During the creation of the finite element model, the best practice is to import each carpal bone individually allowing the user to keep control over whole assembly. Most of the image processing software packages will take into account the coordinates of individual pixels from the MRI or CT scans. Therefore the position of each carpal bone will be preserved after being imported into the finite element software.

## 4.1. Cartilage modelling

Modelling the cartilage is one of the greatest challenges faced by researchers working on joint modelling. Cartilage is not visible from CT scans, but can be identified using MRI scans. In clinical 3 Tesla scans it can be difficult to determine exactly where the cartilage boundary layer is located in three-dimensional space, making it difficult to create the cartilage layer via masking of the scans. In doing so, the researcher will need to interpolate the shape of the cartilage layer often resulting in an irregular shape causing meshing problems. Another aspect regarding incorporating the cartilage layer into the bone model is the scattering of stiff cortical bone elements and soft cartilage elements. That could cause numerical instabilities in the solution phase. A more practical approach is to extrude the external surfaces of the bones at the articulation and creating a solid volume layer representing the cartilage. Using this method will give a distinct boundary between the bone and the cartilage layer. Another possibility would be to extrude the elements directly creating a layer of wedge elements.

## 4.2. Material modelling

### *4.2.1. Bone*

Most finite element models of joints have used elastic material properties for both the cortical shell and for the cancellous bone. Bone is a viscoelastic material and its properties will depend on the strain rate. All published multibone joint finite element models have focussed on a quasi static analysis of the joint and therefore applying the loads slowly. The material properties used for bone material can be seen in Table 2 and are obtained from Rho et al (1997).

| Bone type | Young's modulus [MPa] | Poisson's ratio | Density [g/cm$^3$] | Ultimate tensile strength [MPa] |
|---|---|---|---|---|
| Cortical | $17*10^3$-$19*10^3$ | 0.25 | 2000 | 150 |
| Cancellous | 100-200 | 0.30 | 1500 | 20 |

**Table 2.** Bone material properties

The simplified material values presented in Table 2 will give an idea about the parameters that can be applied to a macroscopical finite element model of a bone. A more refined material model incorporating bone mineral density, the orthotropical behaviour and viscoelastic properties would add a substantial amount of complexity to the model.

### *4.2.2. Cartilage*

Many finite element studies have simulated the mechanical properties of the articular cartilage as elastic material which can be subjected to large errors. Articular cartilage is a complex material that has the properties of a fluid and a solid and has been researched extensively in the literature. Much of that research hasn't been applied into the finite element modelling of multibody joints, although many finite element models exist of cartilage only focussing on the material behaviour. Attempts have been made (Gislason 10 and Bajuri 12) to incorporate the non-linearities of the articular cartilage behaviour into the finite element models, by using Mooney-Rivlin hyper-elastic material properties using the material data obtained from Li et al (2007).

### *4.2.3. Ligaments*

Evaluating the material properties of ligaments pose a great challenge to researchers in multibone joint modelling as they operate only in tension and show viscoelastic material properties. In tension the ligaments show a non-linear characteristic at the initial stages of the load application (usually referred to as the toe region) but once a given reference strain or extension has been exceeded, the ligaments respond in a linear manner to loading. The reason for the non linearities in the toe region is due to the fiber orientation within the ligaments. The collagen fibers are placed in a "wavy" type of fashion and the initial load applied to the ligament goes to straighten the fibers and then they can be stretched in a linear fashion. Another reason is that the fiber lengths within the ligament differ and the initial loading goes to pull the fibers to the same length (Amis 1985). After the linear region then the ligaments follow another period of nonlinear behaviour where the stiffness decreases due to fibre failure until it reaches complete failure

The extrinsic ligaments are generally stiffer but weaker than the intrinsic ligaments which are elastic and strong. In 1991, Logan and Nowak carried out a study where two extrinsic ligaments (the radiocapitate (RC) and the radiolunate (RL)) and two intrinsic ligaments (the scapholunate (SL) and the lunotriquetrum (LT)) were tested to demonstrate the biomechanical difference between the two types of ligaments. Table 3 shows the findings from the study from Logan and Nowak.

| **Rate** | **SL [N]** | **LT [N]** | **RL [N]** | **RC [N]** |
|---|---|---|---|---|
| 1 mm/min | 197.1 ± 35.5 | 241.1 ± 41.8 | 50.8 ± 14.8 | 84.3 ± 16.0 |
| 100 mm/min | 232.6 ± 10.9 | 353.7 ± 69.2 | 107.2 ± 14.5 | 151.6 ± 23.0 |

**Table 3.** Results from Logan and Nowak on ligament material properties

From the table it can be seen that the loading rate primarily affects the extrinsic ligaments, making them stiffer and stronger under a rapid loading. This mechanism helps preventing ligament injury during fall, as the extrinsic ligaments anchor the mobile carpal bones to the radius and the ulna.

Tensile experiments on ligaments are difficult to carry out in practice. Wrist ligaments in particular are too short to be tested on their own, so the attaching bones are dissected along with the ligament and are held rigid in the tensile machine. It can be difficult to compare ligament tensile studies because they can be performed under different conditions which can have profound effects on the experimental results on which modellers of the joint rely. Other material studies have been carried out and published in the literature on wrist ligament properties (Berger 1997, Bettinger 1999).

### 4.2.4. Contact setup

Once the bones have all been incorporated into the finite element software and assembled together bone by bone and cartilage constructed, the contact formulation between the bones needs to be formulated. A surface-to-surface contact is most common method defining the contact between the bones, but node-to-surface configuration can also be implemented. Most finite element models will allow the user different contact models, such as the Lagrange method, the penalty method etc. The availability of these different contact models can be limited to the type of solution algorithm used. Additionally the user can determine the stiffness of the contact, but usually as "hard contact" is applied which is defined by

$$\begin{aligned} p &= 0, \quad h < 0 \\ h &= 0, \quad p > 0 \end{aligned}$$

Where p is the contact pressure, and h is the over closure between the two surfaces. Using kinematic contact method is generally preferred over the penalty contact as it introduces an additional stiffness to the system. Frictionless contact properties or friction using a low friction coefficient should also implemented on the articulating surfaces. By using frictionless contact, it is ensured that no shear stresses occur at the articulations.

It has been reported in the literature (Kauer 1986) that there is little or no movement between certain articulations, such as the articulations between the bones in the distal row of the wrist and the metacarpals (in the carpometacarpal joint). For those joints, it is possible to use a tie constraint so that no relative motion occurs between the two bones. That will help to simplify the model. The model can be seen in Figure 4

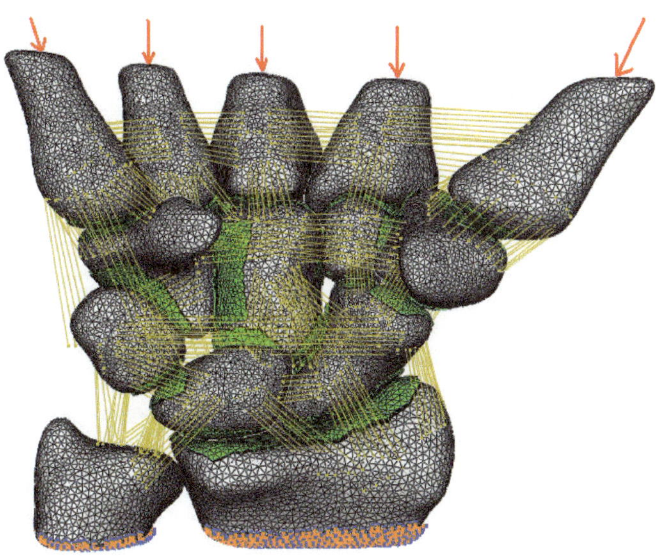

**Figure 4.** Finite element model

## 5. Soft tissue modelling

Due to the high mobility of joints such as the wrist and the ankle, they need to be constrained through a large and complicated set of ligaments to ensure structural integrity of the joint. Without any structural contribution from the ligaments, any finite model of the wrist or the ankle would diverge. As previously discussed then the material properties of each ligament will vary depending on its function and location.

The geometry of the wrist ligaments is complex and difficult to incorporate into a finite element model. Some ligaments wrap around the carpal bones without attaching to them, thus providing additional dorsal/volar constraints on the carpus. This can be seen for the dorsal radiotriquetral ligament which originates at the distal end of the radius and attaches to the proximal pole of the triquetrum, overlapping the lunate and adding to the transverse stability of the carpus.

Previous models have incorporated the ligaments as one dimensional spring elements (Carrigan, Gislason, Bajuri), which is the simplest approach of creating the geometry. Although this method will give a relatively good representation regarding the overall constraints of the carpus, the problem will persist that the spring elements will only constrain the carpal bones in the direction of the springs. Using non-linear springs, the user must make sure that the springs do not take any tensile forces. The literature gives a range of ultimate strength and strain values (Berger 1999, Nowak 1991) for various ligaments which can be used to recreate a non-linear stress-strain or force-displacement curve in the form of

$$F = \begin{cases} 0, & x < 0 \\ \dfrac{\alpha x^2}{2\varepsilon_{ref}}, & 0 \leq x < \varepsilon_{ref} \\ ax + b\,, & x \geq \varepsilon_{ref} \end{cases}$$

Where F is the ligament force, x is the strain and $\alpha$, a and b are constants. The force values can be converted into stress, by using measurements of the cross sectional areas of the ligaments as presented by by Feipel et al (1998).

Another possibility is to model the ligament as three dimensional surfaces using two dimensional elements, by identifying the insertion node points and creating the external lines of the ligament using splines, finally an area is defined from the lines and meshed using shell elements. Modelling the material behaviour can be modelled by implementing stress-strain curves for each ligament using hyperelastic material properties. The challenge in soft tissue modelling, beside the geometrical representation of the ligaments, is not over or under constraining the model. A figure of the model where ligaments are represented as three dimensional surfaces can be seen in Figure 5.

In a pilot study carried out on ligament modelling, it was seen that by using the elastic springs, there was a significant translation of the carpal bones, which decreased drastically by assuming linear elastic material properties of the ligaments. That over-constrained the system to a great extent and allowed extremely little bone movement under loading.

Modelling the ligaments as hyperelastic resulted in larger motion of the ligaments than allowed by using the elastic properties, but less using the non-linear springs. The springs are most probably under- constraining the whole system, but using three dimensional ligaments with elastic material properties are probably over-constraining the system. More research needs to be carried out on the soft tissue properties of multibone joints and the constraining effects various modelling techniques will have on the overall system.

**Figure 5.** Ligaments modelled as three dimensional surfaces.

## 6. Modelling of surgical procedures

With a computational model of the wrist in place, analysis of surgical procedures such as arthrodesis and arthroplasty can be carried out. Arthrodesis is a procedure that fuses together joints to reduce mobility. In the wrist and the ankle there are many individual joints and should just a single joint be fused, the procedure is called partial arthrodesis and a total arthrodesis if the whole joint is fused. This is a recognised surgical procedure to reduce pain and increase stability in the arthritic wrist. Simulating such procedures can be done using a finite element model, where instead of applying contact formulation a tie constraint is applied at the articulating joints. That will treat the two articulating bones as a single unit, not allowing any relative movement between them. After such a procedure it can be seen that the overall load transfer will be altered as additional constraints have been introduced to the system. This can be seen in particular on radiolunate fusion where high joint contact forces were seen on the capitolunate joint. Figure 6 shows the changes in load transfer in the midcarpal joint following radiolunate (RL), radioscaphoid (RS) and radioscapholunate (RSL) fusion compared to the untreated wrist

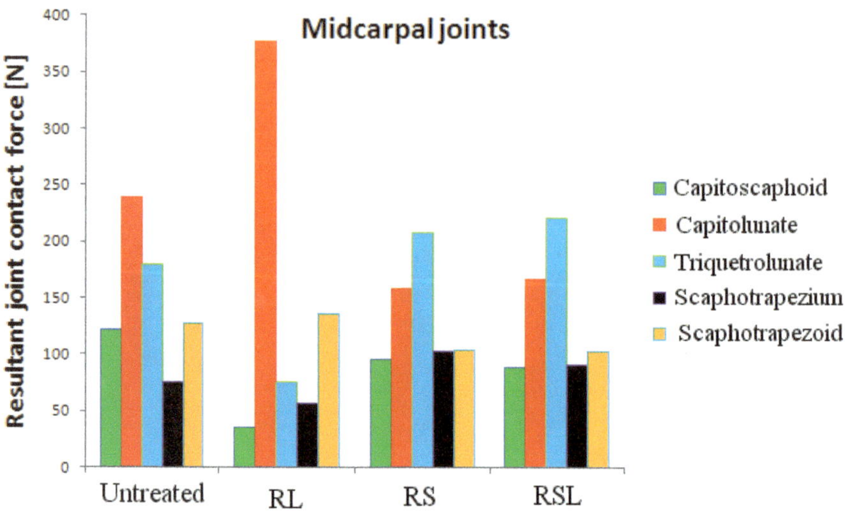


**Figure 6.** Changes in joint contact forces following a partial wrist arthrodesis.

Using the finite element method can be a useful tool to predict a possible surgical outcome, as can be seen with the radiolunate fusion, an extremely high force can be seen acting on the capitolunate joint. This can be explained by the fact that during gripping (and most other tasks) the thumb will be angled in such a way that the joint contact forces acting on the first carpometacarpal joint will tend to push the carpus ulnarly. This can be seen in Figure 4 how the thumb forces tend to ulnarly translate. With the lunate anchored to the radius and the capitate free to translate, it can be seen that under such ulnarly directed forces the capitate will be excessively constrained by the lunate thus causing such high joint contact forces. It can be seen that by fusing both the radius and the lunate, the model predicts more evenly distributed load through the midcarpal joints, however at the expense of a smaller range of motion.

Finite element models on total hip and knee arthroplasty have been prominent in the literature and extensive research has been carried out on the stress distribution in the femur following a total hip arthroplasty and has contributed to the clinical success of the joint replacements. Little has been written about total wrist arthroplasty and the effects it has on the distribution of load within the wrist. Grosland et al have reported on wrist implants in terms of design and carried out ex-vivo analysis, but a model is missing that captures a full three dimensional features of the implanted wrist. A preliminary model was created of the implanted wrist under physiological loading. It showed how the majority of the load was transmitted through the implant and onto the radius. The finite element model can be seen in Figure 7.

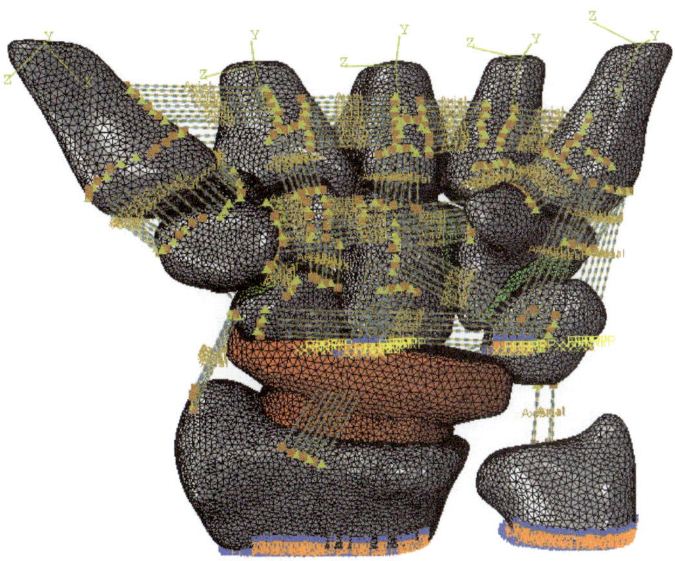

**Figure 7.** Finite element model of a total wrist arthroplasty

The stresses on the carpal bones and the implant can be seen in Figure 8

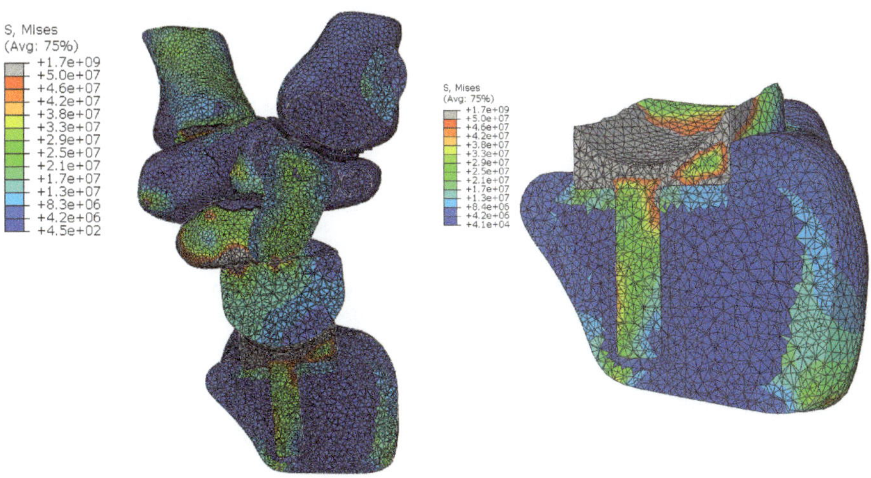


**Figure 8.** Load transmission through the implanted wrist

From a finite element perspective, modelling a total wrist arthroplasty is a simpler task than modelling the healthy wrist as a few of the carpal bones will be removed during the procedure which will decrease the number of contact surfaces. However problems regarding the fixation of the implant into the radius and the distal row will arise as well as contact between the proximal and distal part. In the pilot study, it was assumed that the implant was fully fixed in the radius as well as the distal component fully tied to the carpal bones in the distal row. There are many different types of wrist implants commercially available and the personal preference of the surgeon will in many cases determine which implant will be used. A finite element model will allow to virtually implant a prosthesis into the carpus and calculate the stresses under static loading. The main problem with carrying out such experiments is that the size and the location of the implant could be erroneous which will have a large impact on the overall solution.

The finite element method can be used as a tool to evaluate the different implant designs available on the market. Given the high failure rate of the implants, there is a demand to investigate closer the effects that a total wrist arthroplasty has on the overall load transfer through the wrist and what can be done to design for longevity and functionality of the implant.

## 7. Loading conditions

Applying in vivo loading conditions on the finite element model, is an extremely challenging aspect of the modelling, especially since there has been very little written about the biomechanical modelling of the wrist. Most studies have applied arbitrary loading conditions, 15 N compressive force acting on the distal end of the capitate (Carrigan et al), a combined compressive load of 100 N applied to the $2^{nd}$ and $3^{rd}$ metacarpal (Guo et al) and a combined 1000 N load acting on the scaphoid and lunate (Ulrich et al). The load cases are better defined when dealing with joints in the lower limb and the fundamental question, researchers must ask themselves is "what activity is characteristic for loading on the upper limb?". The answer to that is not clear cut and can range from compressive forces acting on the proximal part of the palm with subject trying to push an object to forces action on the fingers via gripping. There are many grip patterns defined in the literature (chuck grip, power grip, pinch grip etc.) which all contribute in a unique manner to the loading distribution through the fingers.

For the analysis presented in this chapter a grip pattern, seen in Figure 9 was used.

The gripping forces were obtained through a biomechanical study where the gripping strength of 50 subjects were measured using five 6-degrees of freedom force transducer (Nano 25-E and Nano 17, ATI Industrial Automation Inc, USA). Simultaneous collection of position data using an 8 camera motion capture system (Vicon, Oxford Metrics Ltd) was carried out to capture both the kinetic and the kinematic data. The external forces were converted in to joint contact forces acting on the metacarpals using a biomechanical model as described by Fowler and Nicol (2000). More detailed analysis on execution of the biomechanical trials can be found in Gislason et al (2009). The wrist models created were subject specific and the joint contact forces applied can be seen in Table 4

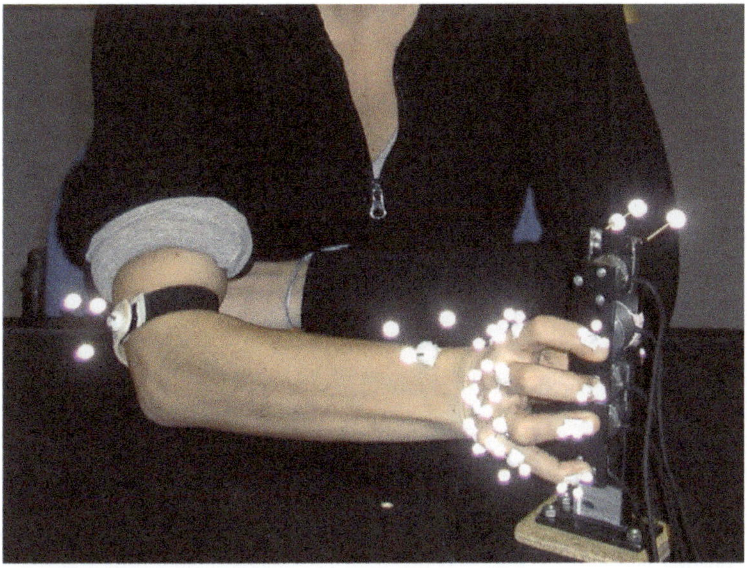

**Figure 9.** Grip pattern used for the analysis

| | Subject 1 | | | Subject 2 | | | Subject 3 | | |
|---|---|---|---|---|---|---|---|---|---|
| | Fx [N] | Fy [N] | Fz [N] | Fx [N] | Fy [N] | Fz [N] | Fx [N] | Fy [N] | Fz [N] |
| Digit 1 | 144.1 | -545.1 | -44.6 | 80.8 | -536.1 | -8.4 | 139.7 | -452.2 | -12.0 |
| Digit 2 | 253.2 | -270.7 | 141.8 | 84.1 | -294.2 | 10.5 | 110.7 | -156.8 | 87.4 |
| Digit 3 | 348.5 | -274.4 | 172.8 | 135.1 | -126.2 | 72.8 | 125.6 | -237.7 | 98.9 |
| Digit 4 | 117.3 | -236.1 | 29.2 | 67.0 | -94.0 | 54.7 | 113.7 | -198.0 | 78.5 |
| Digit 5 | 111.1 | -200.0 | -3.8 | 42.5 | -103.0 | 10.6 | 53.5 | -160.5 | 19.3 |

**Table 4.** Internal loading on the digits

Where

- Positive x-direction denotes ulnar direction
- Positive y direction denotes distal direction
- Positive z-direction denotes dorsal direction

As can be seen from Table 4, the contact forces were primarily directed, ulnarly, proximally and dorsally. The joint contact forces were applied to the model as nodal forces where a subset of nodes was chosen and the total force acting on each metacarpal was divided between the nodes.

The proximal ends of the radius and ulna were kept fixed and compressive forces applied to the distal end of the metacarpals.

Many studies have applied arbitrary boundary conditions onto the wrist, which will not give information about the possible in-vivo behaviour of the carpal bones under loading. By applying physiologically relevant loading conditions, it is possible to determine in more detail the mechanical features within the wrist that control the loading. Due to the extensive research carried out on the biomechanics of the hip and knee, modellers are able to apply physiologically relevant loading conditions onto their models and predict in-vivo loading.

## 8. Solution algorithms

For a multibody computational models, it is virtually impossible to solve an implicit model where convergence needs to be obtained for each contact surface for each loadstep. High residual forces at the boundaries of the contact surfaces are primarily seen that cause the solution to diverge. Damping can be introduced between the bones, which can be released gradually as the load step progresses and will be fully released when all of the loading has been applied. Experiments showed that the load step progressed well at the initial stages of the load step, but once the effects of the damping became less, cutbacks were seen in the solution process which increased as the solution reached towards the end of the load step. The solution never will reach the end of the loadstep. This is a classical behaviour of the proper contact not being established between the bones. It has been previously demonstrated in the literature how nonlinearities can cause divergence using the implicit code (Harewood 2007).

Most multibody analyses use the explicit algorithm to solve the model. The explicit algorithm assumes dynamic behaviour of the model and no convergence checks are carried out on the contact surfaces, which makes the explicit algorithm extremely robust in solving such a multi body system. The solution for time step t +$\Delta$t is based on the status of the model at the previous time step, t. In contrast for the implicit code the solution is based on the same time step. The time step in the explicit analysis is determined from the characteristic element length and material properties and is given by

$$\Delta t \leq \frac{2}{\omega_{max}}$$

where $\omega_{max}$ is the maximum eigenvalue in the system. Generally the time steps, $\Delta$t, are very small, resulting in long run times. The criteria for assuming a quasi static solution, is that the kinetic energy of the system does not exceed 5% of the strain energy.

## 9. Results

### 9.1. Finite element results

The results from the finite element model have shown that anatomical features play an integral role in the stress distribution through the wrist and therefore it is difficult to generalise about the results of a single standard model. However due to the complexity and time commitment creating the finite element models, it is not possible to generate a large cohort of models.

In the finite element models, the largest stress was seen at the in the cortical shell and were on average a magnitude higher than the stresses in the cancellous bone. On average the stresses in the cortical shell were around 18.6 MPa, and in the cancellous bone they were around 1.1 MPa. The stress distribution for one of the model can be seen in Figure 10

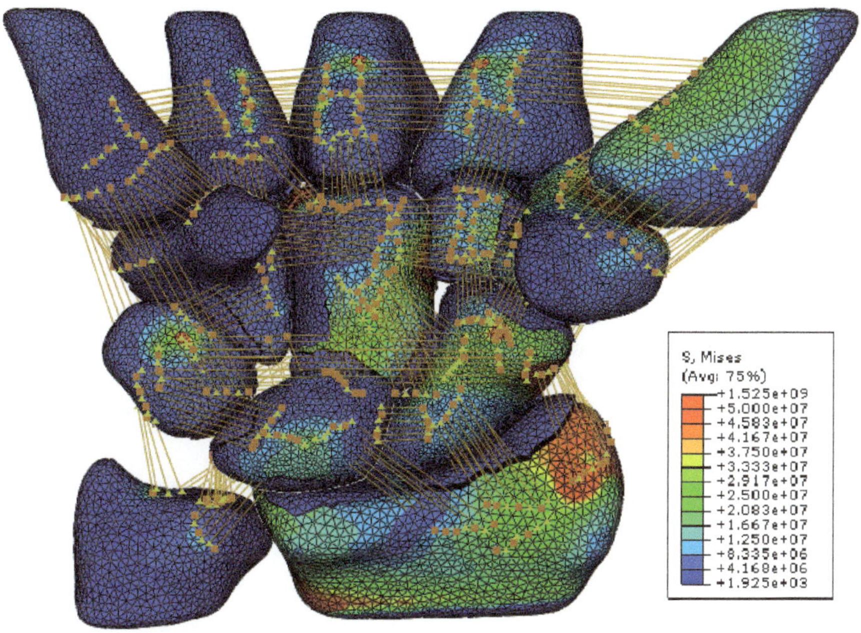


**Figure 10.** von Mises stresses in a single model

Ligaments opposing ulnar translation were more active than others in the model, in particular the dorsal radiotriquetral ligament which showed high degree of force going through it. That result is in agreement with the theoretical findings of Garcia-Elias (1995) who stated that in order to maintain stability, the dorsal radiotriquetral ligament would play an integral part in stabilisation of the carpus during gripping.

The force through the radius and ulna was distributed so that majority of the load was taken by the radius, ranging from 79-93% which is in agreement with the findings of Palmer and Werner (1984) who measured the load distribution between the two forearm bones using a load cell and reported that 80% of the loading was transmitted through the radius.

## 9.2. Validation

Validation is an important procedure to verify that the assumptions used for the computational model are correct. In 2005, guidelines were written by Viceconti regarding the methodology of producing a clinically relevant finite element model. There two important assessment tools for finite element models were introduced, *verification* and *validation*. The term verification is used to check numerical accuracy, that is how well the underlying equations are solved. To verify the model, the user can check that forces at all reactions sum up to give the input forces. Another example of verification can be seen when energy values are compared to check whether the solution is portraying quasi-static behaviour. The term validation is used to assess how well the underlying equations describe the physical phenomena. Validation must be carried out in the lab to test a specimen under the same conditions used in the computational model. Computational models are capable of creating complex load cases, so through validation some simplification generally must be done, which then can then be re-created through the computational model.

Validation of the computational model was carried out through two separate experiments. One measured the strain on the radius and ulnar with the carpus loaded through pull of the tendons (MacLeod 2007). The second measured the joint contact pressure of the radioscaphoid joint using a pressure sensitive film. Setup of the two experiments can be seen in Figure 11.

It was measured using the strain gauges on the radius and ulna that the load through the radius is around 70% and the remaining 30% through the ulna. These values are slightly lower than what the finite element model was predicting, but both recognise the radius as the main load bearing structure of the forearm.

The measurements of the contact pressure on the radioscaphoid joint showed that the joint contact pressure ranged between 4-5 MPa under a 600 N compressive load which is in agreement with the findings of the finite element model which predicted 6.5 MPa contact pressure on the joint under the same loading conditions.

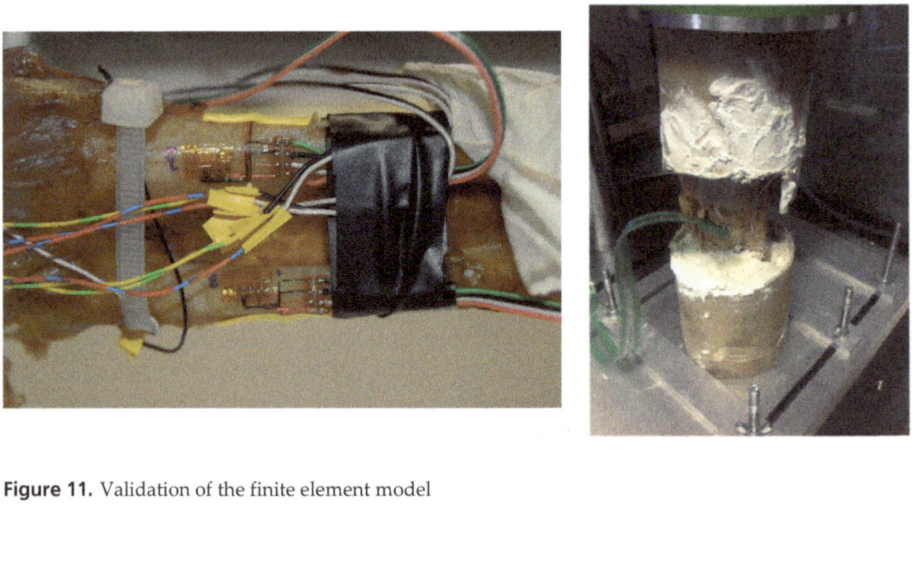

**Figure 11.** Validation of the finite element model

## 10. Conclusions

Creating a finite element model of the wrist and other multibody joints is a complex task where many different aspects of the modelling need to be addressed. The most important aspect contributing to a high quality finite element model is the construction of high integrity geometrical model and the soft tissue modelling. High integrity geometrical model of the articulating surfaces will aid the contact analysis, as a high degree of incongruence of the articulating surfaces can lead to element distortion, especially on soft cartilage elements. The external soft tissue constraints are important in order to maintain mechanical equilibrium as well as allowing the bones to translate and rotate under loading. These two factors will play an integral role in the success of the finite element model.

Finite element models of such complex joints such as the wrist and the ankle are likely to become more prominent in the future as computational power and modelling software quality increases. That will make modellers able to create models incorporation a higher degree of detail than previously has been published.

It is inevitable that errors are introduced in such complex models. The errors can either be within the control of the modeller or without. This chapter has discussed the procedures that the modeller can carry out to minimise the sources of errors in the model. However the modeller will have little control over errors that can be generated through using previously published material properties and geometrical representation of the ligaments and soft tissue.

Using the finite element method predicting the load transfer through the healthy and the pathological wrist can give clinicians important information regarding the choice of treatment which can lead to higher procedure success rates and improve the quality of life for many patients.

## Author details

Magnús Kjartan Gíslason and David H. Nash
*Department of Mechanical and Aerospace Engineering, University of Strathclyde, Glasgow, UK*

## 11. References

[1] Maragoni RD and Glaser AA: Viscoelastic properties of soft tissue model characterization, J Biomech, 1968, 1(1): 33-36.
[2] Rybicki EF, Simonen FA, Weis Jr EB: On the mathematical analysis of stress in human femur, J Biomech, 1972, 5(2):203-215.
[3] Bergmann G, Graichen F, Rohlmann A: Hip joint loading during walking and running, measured in two patients. J Biomech, 1993, 26(8):969–90.
[4] Taylor M, Barrett, DS: Explicit finite element simulation of eccentric loading in total knee replacement. Clinical Orthopaedics and Related Research, 2003, 414:162-171.

[5] Garcia-Elias M, Ribe M, Rodriguez J, Cost J, and Casas J: Influence of joint laxity on scaphoid kinematics. Journal of Hand Surgery, 1995, 20B(3):379–382.
[6] Anderson DD and Daniel TE: A Contact-Coupled Finite Element Analysis of the Radiocarpal Joint, Seminars in Arthroplasty, 1995, 6(1), 30-36.
[7] Miyake T, H. Hashizumea H , Inouea H, Shia Q, Nagayama N: Malunited Colles' fracture Analysis of stress distribution, Journal of Hand Surgery (European volume), 1994, 19(6): 737-742.
[8] Anderson DD, Deshpande BR, Daniel TE and Baratz ME: A Three-Dimensional Finite Element Model of the Radiocarpal Joint: Distal Radius Fracture Step-off and Stress Transfer, The Iowa Orthopaedic Journal, 2005, 25, 108-117.
[9] Schuind F, Cooney WP, Linscheid RL, An KN and Chao EYS: Force and Pressure Transmission Through the Normal Wrist: A Theoretical Two-Dimensional Study in the Posteroanterior Plane, Journal of Biomechanics, 1995, 28(5), 587-601.
[10] Ulrich D, van Rietbergen B, Laib A and Ruegsegger P: Load transfer analysis of the distal radius from in-vivo high resolution ct-imaging. Journal of Biomechanics, 1999, 32:821–828.
[11] Oda M, Hashizume H, Miyake T, Inoue H and Nagayama N. A stress distribution analysis of a ceramic lunate replacement for kienbok's disease. Journal of Hand Surgery (British and European Volume), 2000, 25B(5):429–498.
[12] Carrigan SD, Whiteside RA, Pichora DR and Small CF: Developement of a Three Dimensional Finite Element Model for Carpal Load Transmission in a Static Neutral Posture, Annals of Biomedical Engineering, 2003, 31, 718–725.
[13] Nedoma J, Klézl Z, Fousek J, Kestřánek Z, Stehlík J: Numerical Simulation of Some Biomechanical Problems, Mathematics and Computers in Simulation, 2003, 61, 283-295.
[14] Gislason M, Nash DH, Nicol A, Kanellopoulos A, Bransby-Zachary M, Hems TEJ, Condon B and Stansfield B.: A Three Dimensional Finite element Model of Maximal Grip Loading in the Human Wrist, Proc. IMechE Part H, Engineering in Medicine, 2009, Vol 223 (H7), 849-862.
[15] Gislason M, Stansfield B and Nash D: Finite element creation and stability considerations of complex biological articulations: The human wrist joint, Medical Engineering and Physics, 2010, 32:523-531.
[16] Guo X, Fan Y and Li ZM: Effects of Dividing the Transverse Carpal Ligament on the Mechanical Behaviour of the Carpal Bones under Axial Compressive Load: A Finite Element Study, Medical Engineering & Physics, 2009, 31, 188-194.
[17] Bajuria MN,Mohammed Rafiq Abdul Kadira, Murali Malliga Ramanb, Kamarul T: Mechanical and functional assessment of the wrist affected by rheumatoid arthritis: A finite element analysis, Medical Engineering and Physics, 2012, in press.
[18] Chen WP, Ju CW and Tang FT: Effect of Total Contact Insoles on the Plantar Stress Redistribution: A Finite Element Analysis, Clinical Biomechanics,2003, 18, 17-24.

[19] Cheung JTM, Zhang M, Leung AKL and Fan YB: Three Dimensional Finite Element Analysis of the Foot During Standing: A Material Sensitive Study, Journal of Biomechanics, 2004, 38, 1045-1054.
[20] Louis O, Willnecker J, Soykens S, Van den Winkel P and Osteaux M: Cortical Thickness Assessed by Peripheral Quantitative Computed Tomography: Accuracy Evaluated on Radius Specimens, Osteoporosis International, 1995, 5, 446-449.
[21] Ramos A and Simões J. Tetrahedral versus hexahedral finite elements in numerical modelling of the proximal femur. Medical Engineering & Physics, 2006, 28(9):916-924.
[22] Rho, JY, Tsui TY and Pharr GM: Elastic properties of human cortical and trabecular lamellar bone measured by nanoindentation, Biomaterials, 1997, 18(20), 1325- 1330.
[23] Li Z, Kim JE, Davidson JS, Etheridge BS, Alonso JE and Eberhardt AW: Biomechanical Response of the Pubic Symphysis in Lateral Pelvic Impacts: A Finite Element Study, Journal of Biomechanics 2007, 40, 2758-2766.
[24] A. Amis. Ligament Injuries and Their Treatment, chapter Biomechanics of Ligaments, 1985: 3–28.
[25] Logan S and Nowak M: Distinguishing biomechanical properties and intrinsic and extrinsic human wrist ligaments. Journal of Biomechanical Engineering, 1991, 113(1):85–93.
[26] R. Berger. The ligaments of the wrist. Hand Clinics, 1997, 13(1):63–82.
[27] Bettinger PC, Linscheid, RL, Berger RA,Cooney WP and An KN: An anatomic study of the stabilizing ligaments of the trapezium and trapeziometacarpal joint. J. Hand Surg. Am., 1999, 24(4), 786–798
[28] Berger R, Imeada T, Berglund L and An K: Constraint and material properties of the subregions of the scapholunate interosseous ligament. J Hand Surg Am, 24(5):953–62, 1999.
[29] Nowak M: Biomechanics of the Wrist Joint, chapter Material Properties of Ligaments. Springer Verlag New York, 1991.
[30] Kauer JM: The mechanism of the carpal joint, Clinical Orthopaedics and Related Research, 1986, 202, 16–26.
[31] Feipel V, Salvia P and Rooze M: A new method for measuring wrist joint ligament length changes during sagittal and frontal motion, Clinical Biomechanics, 1998, 13(2): 128-137.
[32] Grosland N, Rogge RD and Adams BD: Influence of articular geometry on prosthetic wrist stability, Clinical Orthopaedics and Related Research, 2004, 421:134-142.
[33] Fowler NK and Nicol AC: Interphalangeal Joint and Tendon Forces: Normal Model and Biomechanical Consequences of Surgical Reconstruction, Journal of Biomechanics, 2000, 33, 1055-1062.
[34] Harewood FJ and McHugh PE: Comparison of the Implicit and Explicit Finite Element Methods Using Crystal Plasticity, Computational Materials Science, 2007, 39, 481-494.

[35] Garcia-Elias M: Kinetic analysis of carpal stability during grip. Hand Clinics, 1997, 13(1):151–158.

[36] Palmer A and Werner F. Biomechanics of the distal radioulnar joint. Clin. Orthop. Rel. Res, 1984, 187:26–35.

[37] Viceconti M, Olsen S, Nolte L and Burton K: Extracting clinically relevant data from finite element simulations (editorial). Clinical Biomechanics, 2005, 20:451–454.

[38] Macleod NA, Nash DH, Stansfield BW, Bransby-Zachary M and Hems T: Cadaveric Analysis of the Wrist and Forearm Load Distribution for Finite Element Validation, Proceedings of the 6th International Hand and Wrist Biomechanics Symposium, Tainan, Taiwan 2007.

# Chapter 4

# Application of Finite Element Analysis in Implant Dentistry

B. Alper Gultekin, Pinar Gultekin and Serdar Yalcin

Additional information is available at the end of the chapter

## 1. Introduction

Since Brånemark's discovery, dental implants have become the most common restorative technique for the rehabilitation of edentulism. Many factors can impact the survival of implant-supported restorations. The most important factor for determining the long-term success of osseointegration is the state of the peri-implant bone [1-3]. Ideal biomechanical conditions directly affect bone remodeling and help to maintain the integrity of non-living structures such as the implant, abutment, and superstructures (Figures 1-7). Oral dental implant interventions involving surgical and restorative procedures for the rehabilitation of various causes of edentulism are associated with several risks. In particular, mechanical and technical risks plays a major role in implant dentistry, resulting in increased rates of repairs, unnecessary costs and lost time, and even complications that may not be easily corrected (Figures 8-10) [4-7]. Therefore, the potential mechanical and technical risks of failure or associated complications need to be evaluated before undertaking such interventions, since the application of necessary precautions may improve the survival of implant-supported restorations. Consequently, the number of biomechanical studies in the field of implant dentistry has dramatically increased in an effort to reduce failure rates.

Several methods based on photoelastic, strain-gauge, and finite element analysis (FEA)-based studies have been used to investigate stress in the peri-implant region and in the components of implant-supported restorations [8-11]. FEA is a numerical stress analysis technique that is widely used to assess engineering and biomechanical problems before they occur [12,13]. A finite element model is constructed by dividing solid objects into several elements that are connected at a common nodal point. Each element is assigned appropriate material properties corresponding to the properties of the object being modeled. The first step is to subdivide the complex object geometry into a suitable set of smaller 'elements' of 'finite' dimensions. When combined with the 'mesh' model of the investigated structures, each element can adopt a specific geometric shape (i.e., triangle, square, tetrahedron, etc.) with a specific internal strain function. Using these functions and the actual geometry of the element, the equilibrium equations between the external forces acting on the element and the displacement occuring at each node can be determined [9].

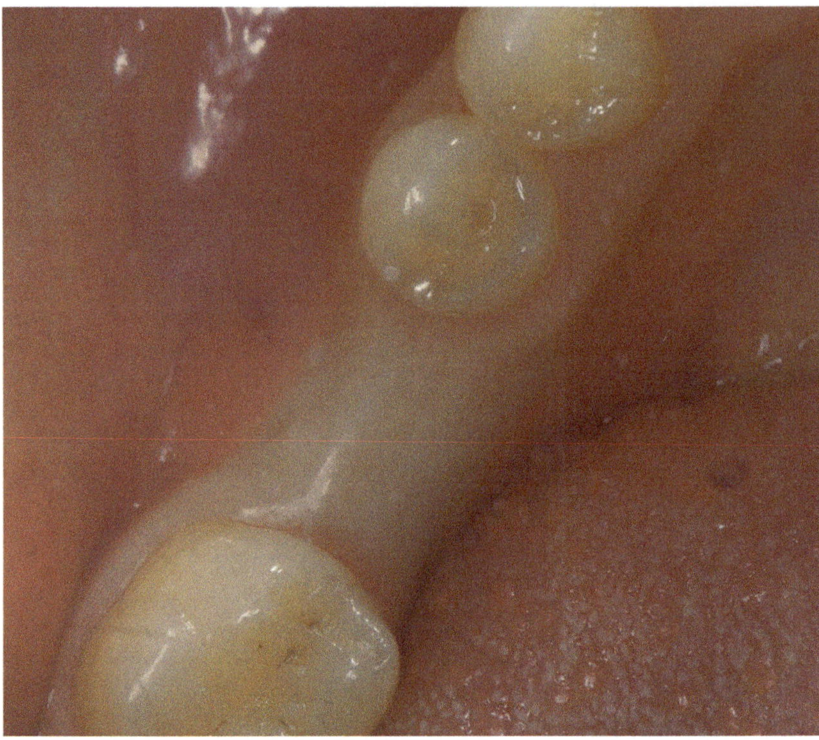

**Figure 1.** Missing molar in the mandible, to be treated with a dental implant-supported restoration

In implant dentistry literature, commonly used materials in FEA studies can be classified as either implant, peri-implant bone (cortical and cancellous bone), and restoration (Figure 11). This method allows application of simulated forces at specific points in the system and stress analysis in the peri-implant region and surrounding structures. 2-D and 3-D models can be created and models for every treatment alternative can be explored. However, 2-D models cannot simulate the behavior of 3-D structures as realistically as 3-D models, so most recent studies have focused on 3-D modeling [14-17].

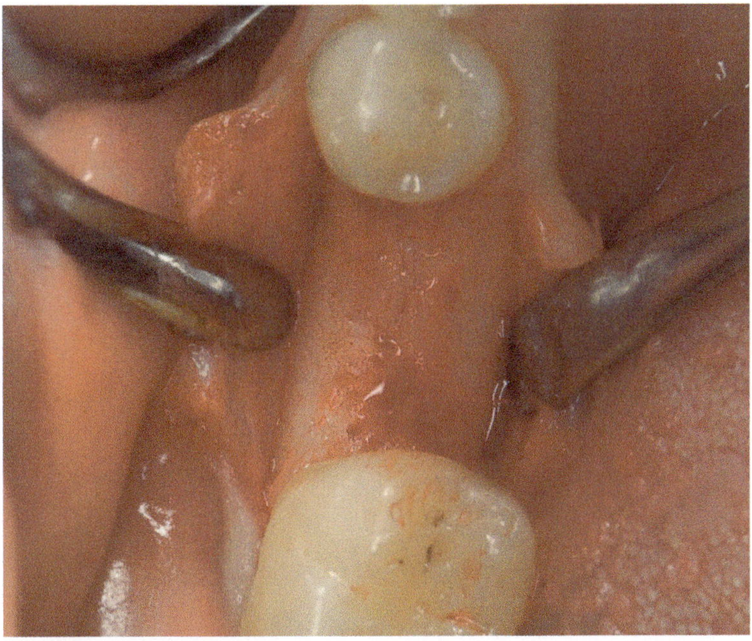

**Figure 2.** After flap elevation, the cortical bone is visible

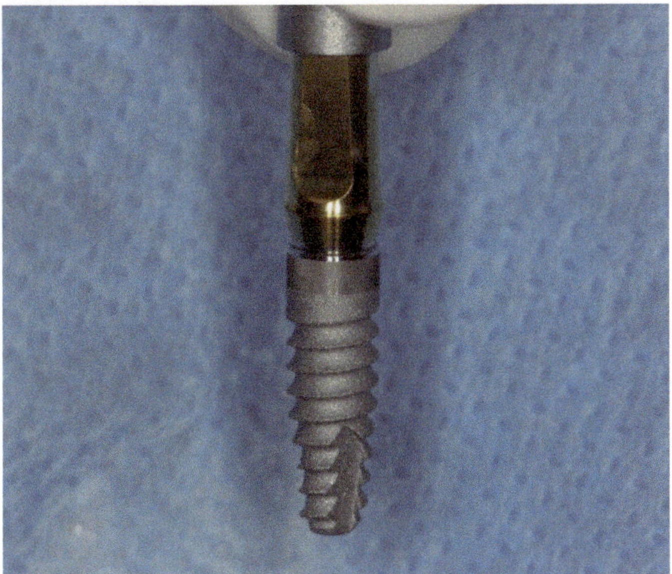

**Figure 3.** Dental implant with an abutment to be placed in the ridge created by the missing molar

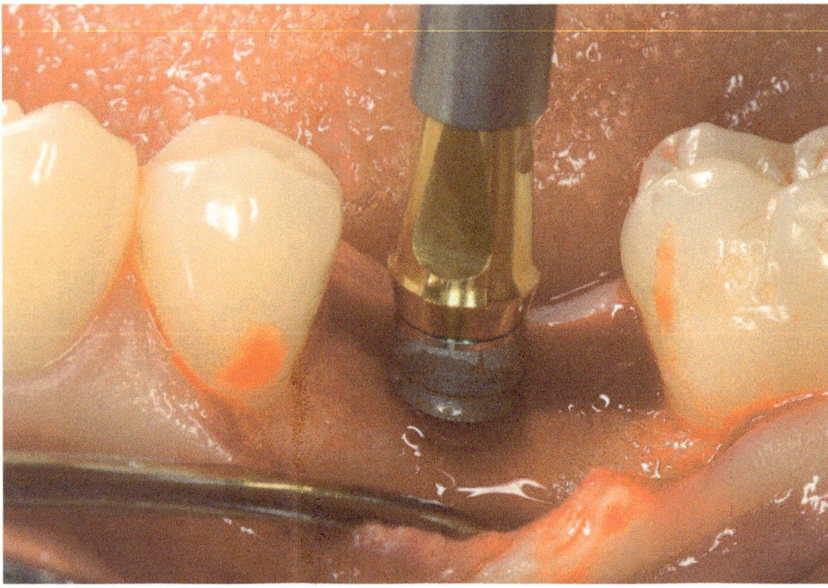

**Figure 4.** Implant is placed in the ridge

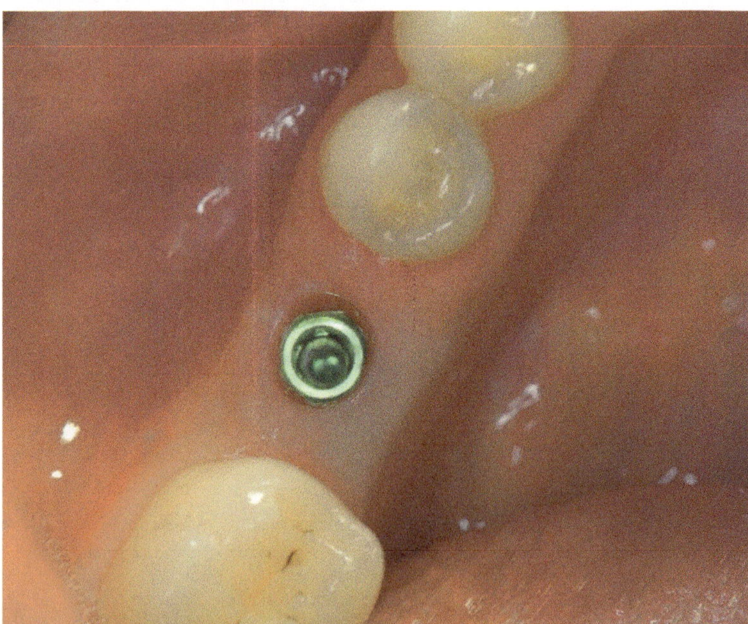

**Figure 5.** Occlusal view of the implant after 2 months of healing

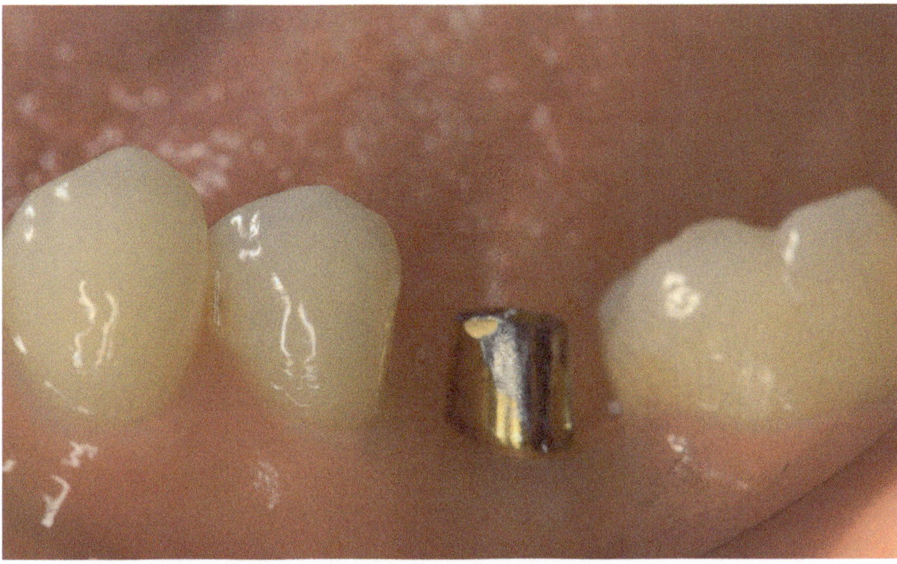

**Figure 6.** Abutment is prepared and attached to the implant

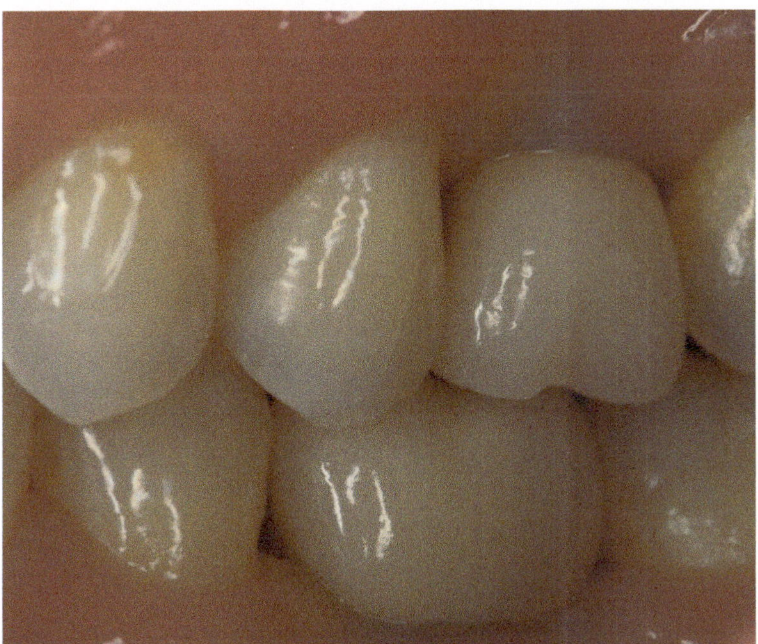

**Figure 7.** Porcelain-fused metal implant-supported restoration in use with optimum treatment planning

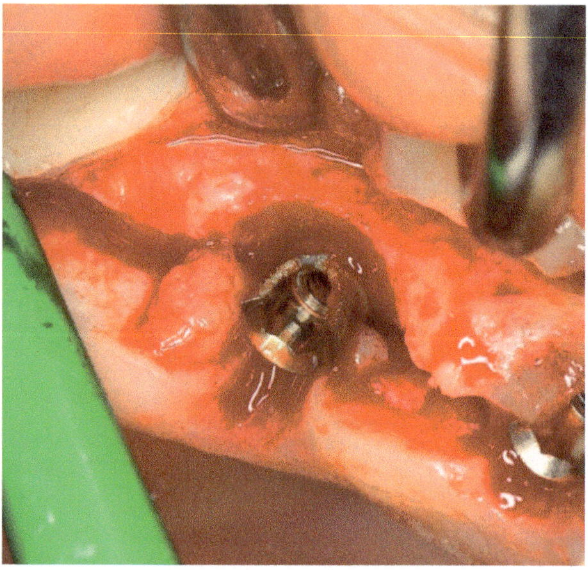

**Figure 8.** Intraoral picture of a broken implant due to excessive loading after 1 year of use

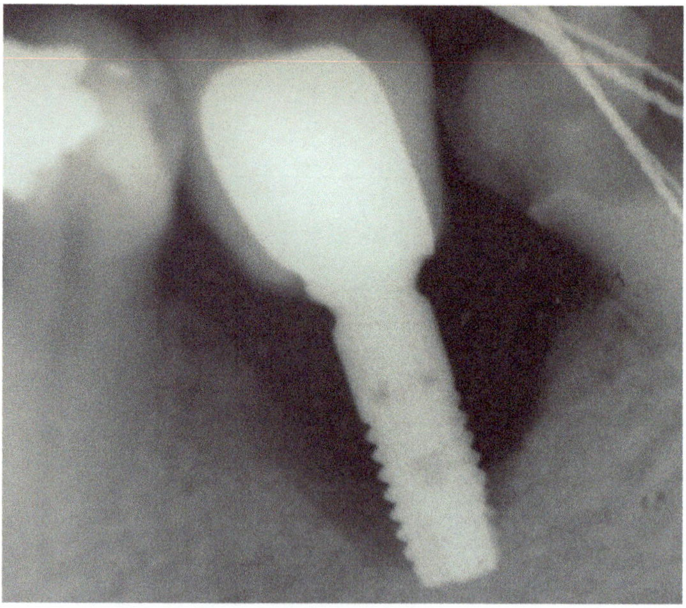

**Figure 9.** Severe bone resorption after 2 years of loading; implant and superstructure have no mechanical failure, but peri-implant bone could not resist excessive loading (biomechanical failure because of improper occlusal adjustment)

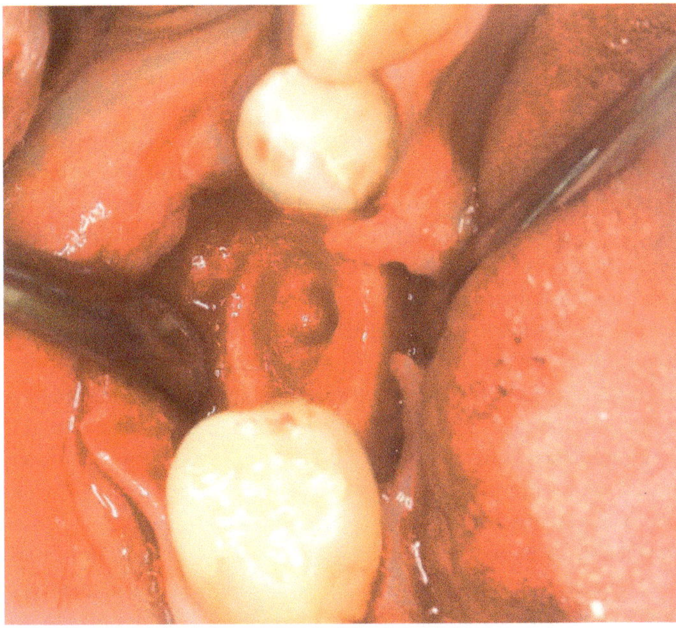

**Figure 10.** Severe bone defect is seen after implant removal; advanced bone regeneration techniques are needed to replace the implant

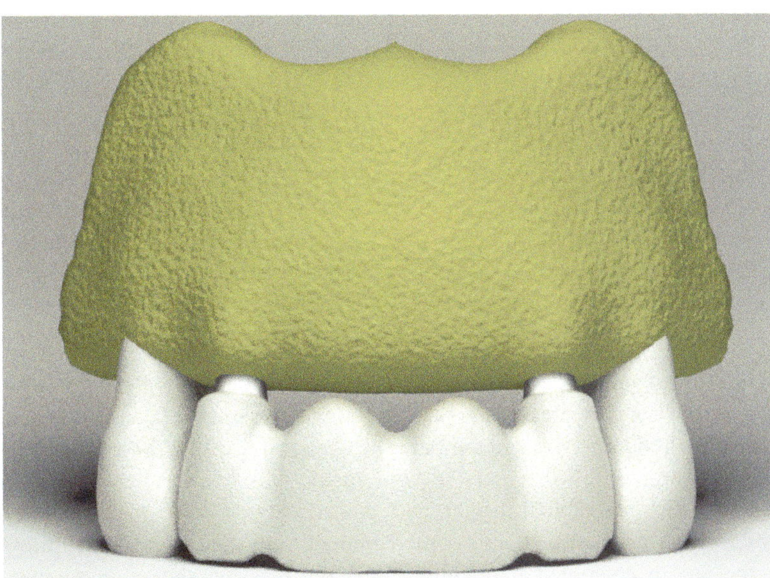

**Figure 11.** Modeling of bone, implant, abutment, and restoration

## 2. Modelization of living structure (bone)

To improve the quality of FEA research, strict attention should be paid to the modelization procedure as one of the most important part of FEA studies. The features of the model should resemble the physical properties of the actual structure as closely as possible, with respect to dimension and material properties. The most difficult and complex part of the modelization process involves capturing the detailed properties of living structures. Therefore, in general, specifications drawn from chapters of a detailed anatomy book or from tomographic scans of a jaw from a cadaveric human specimen can be used for the modeling procedure. Volumetric data obtained from tomography devices or magnetic resonance imaging are digitally reconstructed [18,19]. Then, the material properties applied to the elements can be varied according to the modeling requirements of a particular situation. Computed tomography offers another advantage for realistic modeling in not only the development of anatomic structures, but also the inclusion of material properties according to different bone density values [20,21]. In some studies, the bone is totally or partially modeled as a simple rectangle, elipsoid, or U-shape [18]. In detailed studies, especially with data obtained from scanners, bone can be modeled in a very realistical manner; however, this increased level of geometric detail will result in increased working and computing time. According to the treatment alternatives being investigated, cortical bone can be layered in milimeters or can be neglected altogether in order to simulate weak bone properties similar to those found in the posterior maxilla (Figure 12). Bone properties related to density can be calibrated to range from very soft to dense bone, according to the individual research protocol. If only a specific area and/or condition of the mandible or maxilla is being investigated, there is no need to visualize and construct a model of entire jaw. Limiting the scope or features of the model will distinctly decrease the working time and costs, as previously discussed. A region of interest can be extracted using a number of techniques, such as a Boolean process (Figures 13-15), and any implant design can be adopted for the study. Regions of interest may change according to the study protocol. Portions of the mandible or maxilla, maxillary sinus region, and temporomandibular joint are the most common anatomical areas used in studies related to implantology. In the existing literature 2-D FEA bone models are generally simplified as a rectangular shape [14]. However, recent studies have used 3-D bone modeling to better represent the realistic anatomy of these complex structures [22-24].

In a previous FEA study, the human mandible model was based on a cadaveric mandible obtained from the anatomy department [25]. The edentulous cadaver mandible was scanned using a dental volumetric computed tomography device (ILUMA, Orthocad, CBCT scanner, 3M ESPE, St. Paul, MN, USA) (Figure 16). Volumetric data were reconstructed in 0.2 mm thick sections. The mandibular height and width were at least 10 mm and 5 mm, respectively. More detailed anatomic representations could be created in future studies through the use of computed tomography scanners that can slice objects into thinner sections, but this may increase the working time and development cost of the final finite element model (FEM). In the study mentioned above, sections were digitized into the DICOM 3.0 format and visualized using 3-D Doctor software (Able Software Corp., Lexington, MA, USA). Cortical bone of 2 mm uniform thickness, and cancellous bone were also modeled (Figure 17). In this study, cortical and cancellous bone model components were considered homogenous. However, in fact, cancellous bone in particular has widely variable density properties. The non-uniform nature of the density of this anatomic structure may affect the magnitude and distribution of stress concentration after loading. These simplifications are common in studies that employ FEA and are aimed at limiting the computing difficulties associated with performance of these studies [18,26,27]. To develop more realistic models of living structures, future studies may include variable density properties obtained from bone density values measured in Hounsfield Units or from other advanced data obtained from computed tomography scans performed with individual patients (Figure 18) [28-30].

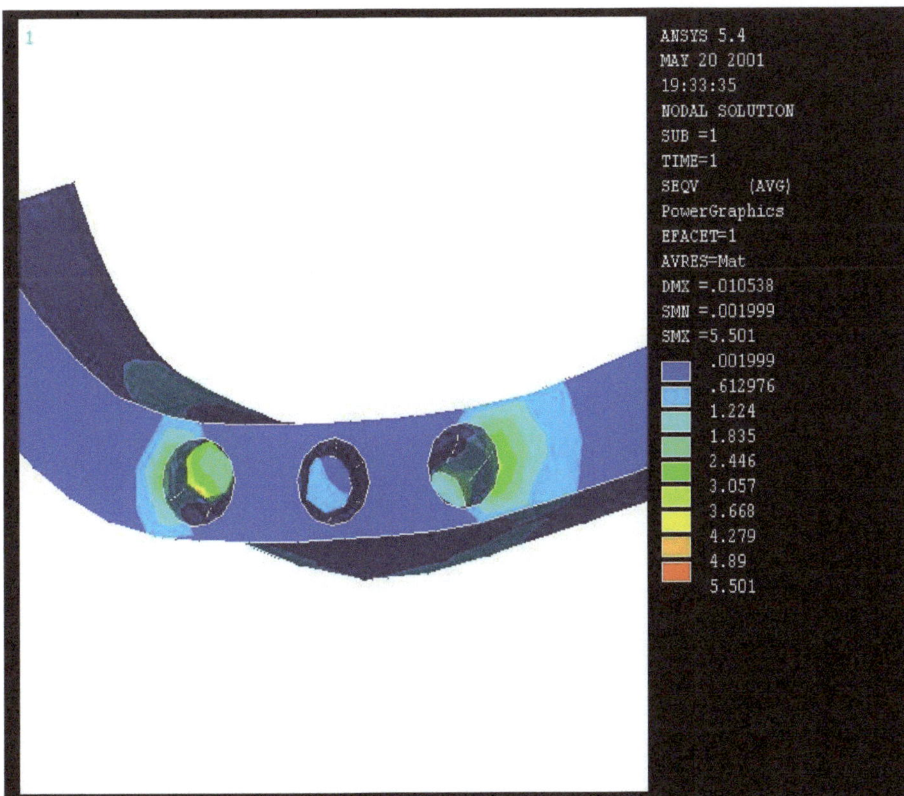


**Figure 12.** Cortical thickness of the posterior maxilla is neglected; only cancellous bone properties are modeled

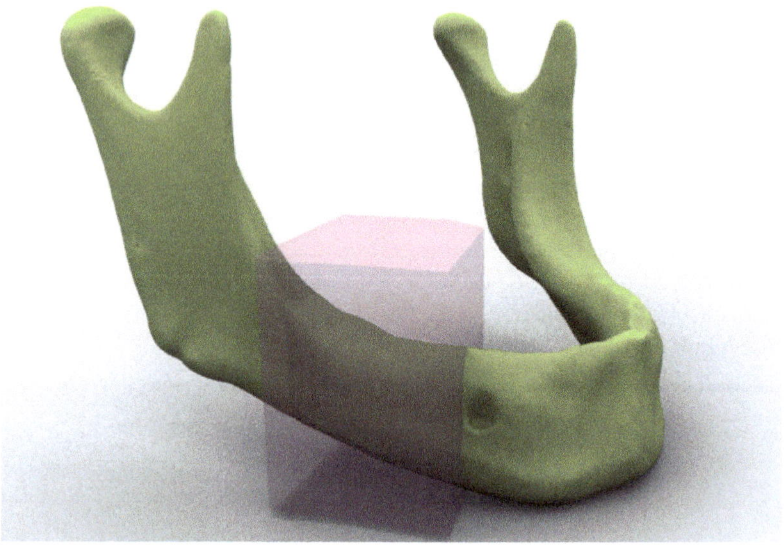

**Figure 13.** Mandible is modeled and region of interest is selected

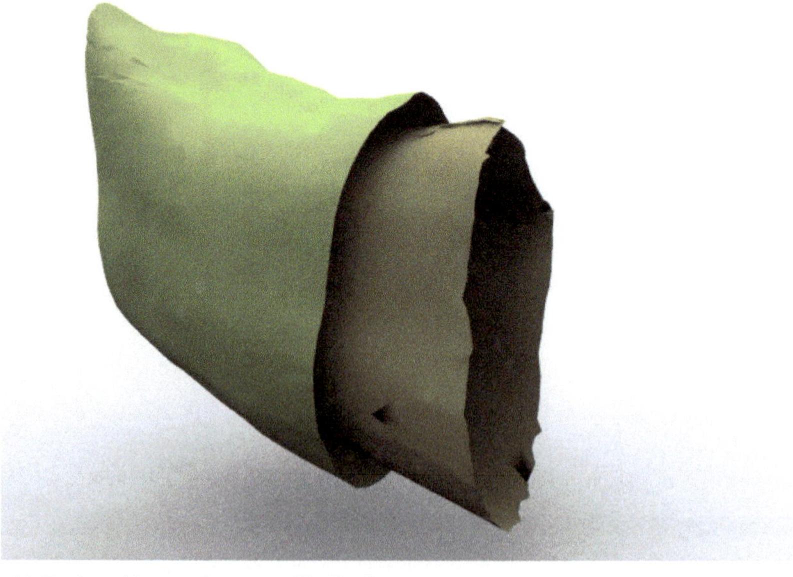

**Figure 14.** Region of interest is extracted by Boolean process

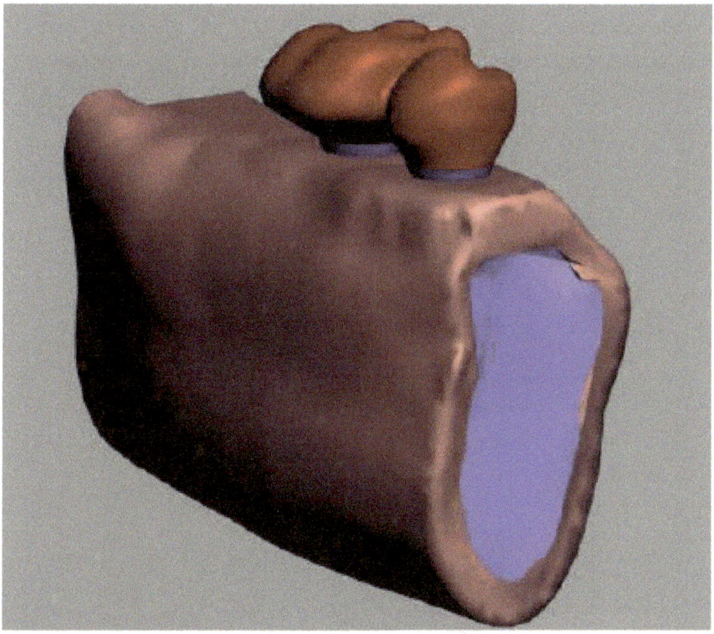

**Figure 15.** Part of the mandible modeled with superstructure, implant, and surrounding bone

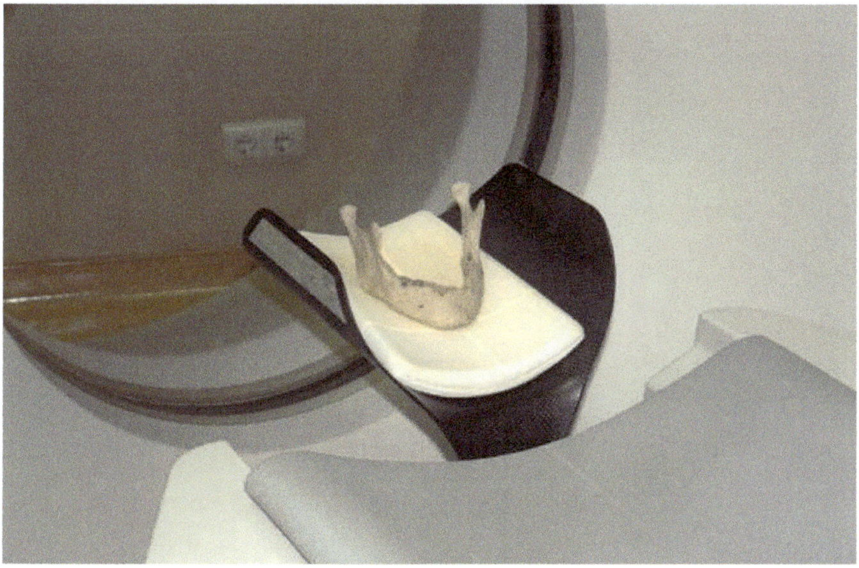

**Figure 16.** The edentulous mandible obtained from a cadaver was scanned using a dental volumetric tomography device

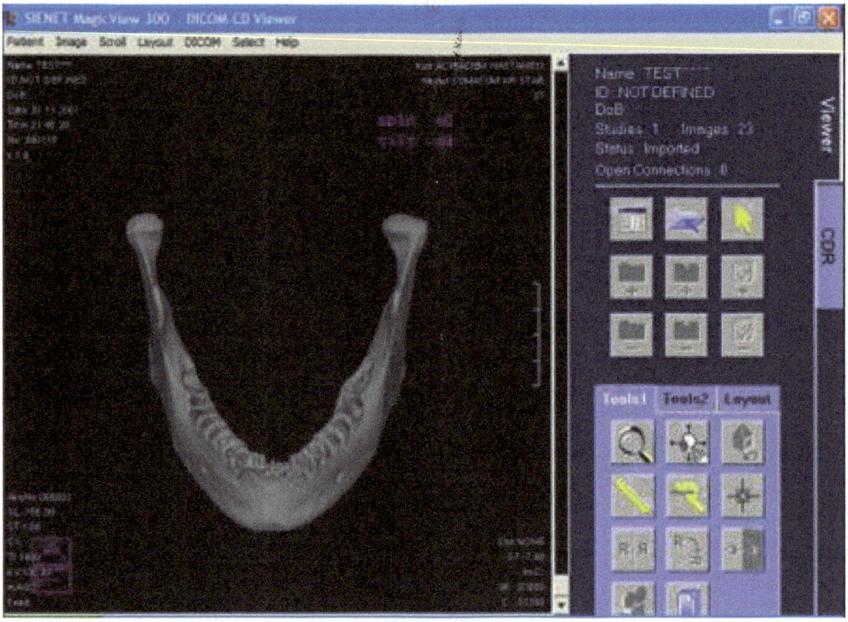


Figure 17. Volumetric data were reconstructed in 0.2 mm thick sections

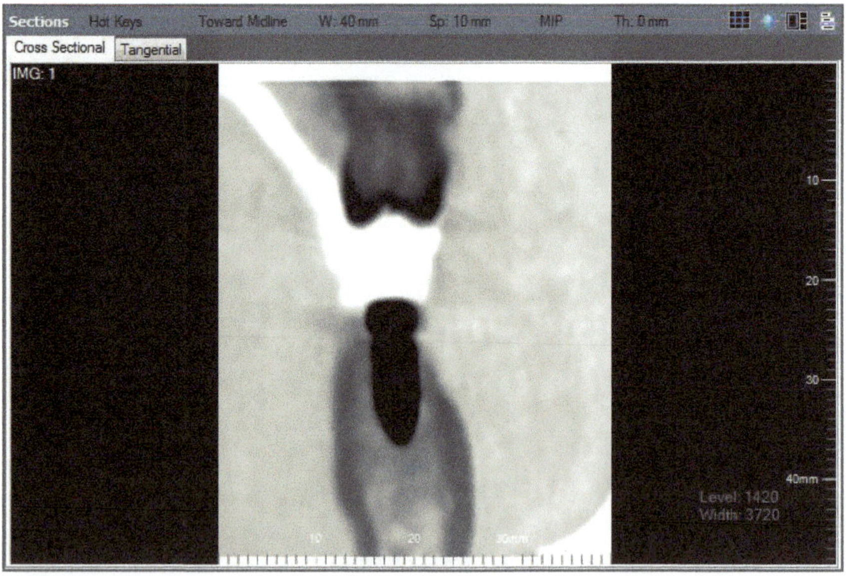


Figure 18. Bone density values can be measured according to gray scale using advanced 3-D radiographic techniques

## 3. Modelization of non-living structure (materials)

Non-living mechanical structures such as implants, abutments, and restorations can be simulated in detail and can substantially influence the calculated stress and strain values, similar to living structures. These materials can be digitally modeled in FEA studies using previously determined isotropic, transversely isotropic, orthotropic, and/or anisotropic properties [31]. In an isotropic material, the relevant material properties are the same in all directions, resulting in only 2 independent material constants, such as Young's modulus and Poisson's ratio [9,13,31]. Young's modulus (MPa), also known as the tensile modulus, is a quantity used to characterize materials and is a measure of the stiffness of an elastic material. Young's modulus is also called the elastic modulus or modulus of elasticity, because Young's modulus is the most commonly used elastic modulus [9,13,32,33]. When a sample object is stretched, Poisson's ratio is the ratio of the contraction or transverse strain (perpendicular to the applied load), to the extension or axial strain (in the direction of the applied load). When a material is compressed in 1 direction, it tends to expand in the other 2 directions perpendicular to the direction of compression. This phenomenon is called the Poisson effect. Poisson's ratio is a measure of the Poisson effect [9,13,32,33].

An anisotropic material has material properties that vary by direction [31]. Isotropic material properties are used in most FEA studies related to implant dentistry [18,25,34]. For instance, the material properties of living bone are anisotropic, and inhomogeneous. These properties of real bone greatly affect stress and strain patterns. In addition, bone density may differ among various regions of the same jaw and areas of differing densities may only be separated by milimeters. For simplification and to overcome computing difficulties, in most cases, the materials are modeled as homogenous, isotropic, and linearly elastic [35-39]. However, some studies have modeled the bone block using anisotropic properties (i.e., the material properties differ with respect to direction) [26]. The material properties of both living and non-living structures are chosen in accordance with the goal of the modeling exercise.

In some studies, implants are modeled using a screw design but without threads (Figure 19). This may simplify the computing process, but does not reflect the reality of implant geometry. If one or more study parameters are related to implant dimensions, there is little doubt that inclusion of implant threads in the model is quite important to the quality of the research. Most clinicians are interested in the magnitude and distribution of stress that may induce microdamage to the bone and result in crestal bone resorption; therefore, macro and micro threads are crucial in the modeling stage of an implant study. The implant thread design influences the induced bone stress around the implant, which contributes to crestal bone loss, and can jeopardize the maintenance of osseointegration [40-43]. In recent FEA studies, implant threads are modeled in detail (Figure 20,21). There are 2 ways to model implant and abutment materials. One way is to obtain all of the geometric information (e.g., length, diameter, macro-micro thread configuration) in milimeters from the manufacturer. The second option is to scan implants and abutment materials and digitally reconstruct them. Efficient and realistic models can be obtained by using either option. In general, for the digital preparation of crown models, an anatomy atlas of the tooth can be used as a reference to calculate the form and both mesiodistal and buccolingual dimensions [44]. The prosthetic superstructure can be simulated according to various treatment protocols. Superstructure can also be modeled as a geometric figure, such as a simple rectangular shape, but this may interfere with the realism of the model (Figure 22).

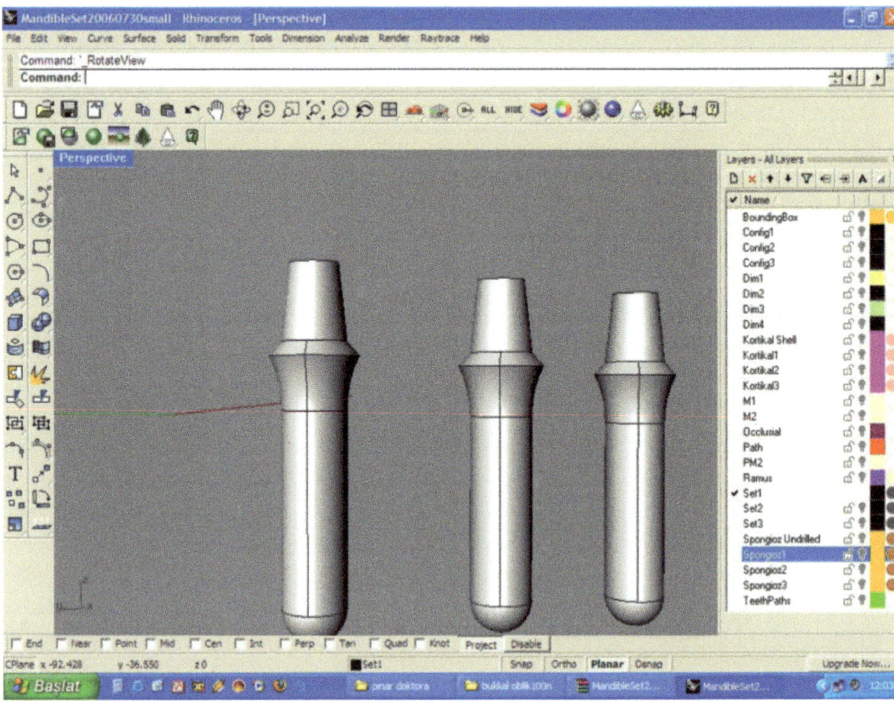


**Figure 19.** Implants are modeled without threads

In a previous study, the crown model was simulated as porcelain fused to metal restoration. To calculate the mesiodistal width of the second premolar and first molar, Wheeler's Atlas of Anatomical Natural Tooth Morphology was used (Figure 23) [44]. The atlas was used again for digital preparation of the crown models. Properties of chromium-cobalt alloy were used for the framework and feldspatic porcelain as used to simulate the second premolar and the first molar of a mandibular model. The metal thickness of the framework was 0.8 mm and the porcelain thickness was at least 2.0 mm. The thickness of porcelain changes with the creation of pits and trabeculae of the tooth surface. In most FEA studies, not only the cement thickness but also the interface between the materials is assumed to be 100%

bonded [9,18,25,31,34]. Implant, abutment, abutment screw, framework, and porcelain structures are considered to be a single unit (Figure 24). In contrast, there are some studies that use a contact condition between the abutment and implant set as a frictional coefficient [26]. In these studies, the corresponding material properties are used and modeled separately. Most studies also model the implant as rigidly anchored in the bone model along its entire interface and with total osseointegration. It is impossible to visualize these interface conditions in real life, but simplifications in interface conditions will inevitably result in considerable inaccuracy. The most common drawback of FEA from the clinical perspective is that many features that directly affect model accuracy, such as loading conditions, material properties, and interface conditions are neglected or ignored. In most cases, researchers neglect one or more features in their studies. Moreover, bias may result from interpretation of data obtained from an FEA study to that obtained from another. Within a single study, these simplifications are consistent for all the simulated models; therefore, the accuracy of the analysis from the stress distribution viewpoint is not affected, as long as the models are compared with each other in the same study [9,18,25,31,34].

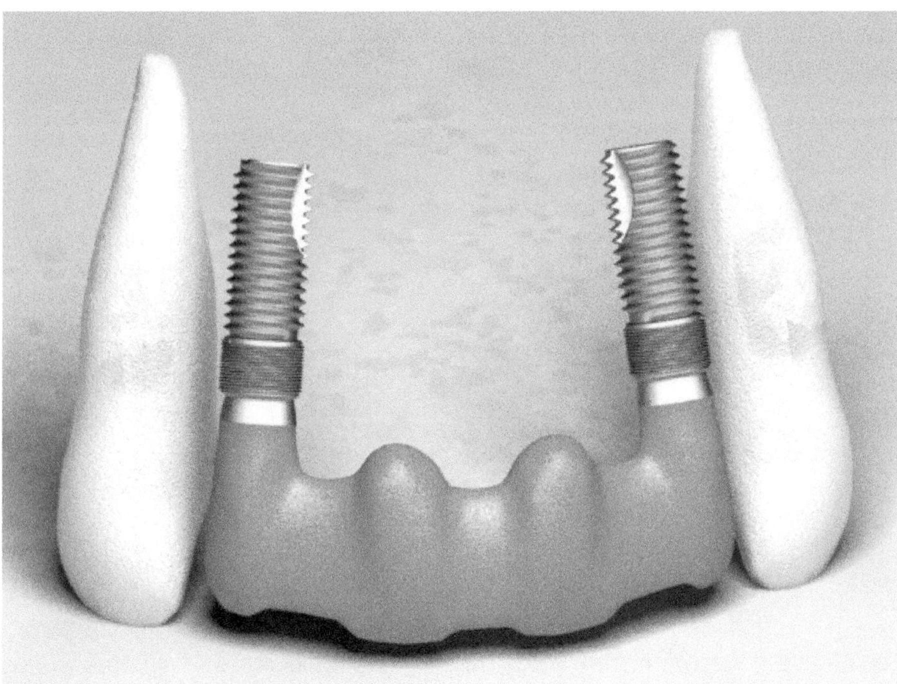

**Figure 20.** Implants are modeled with micro and macro threads

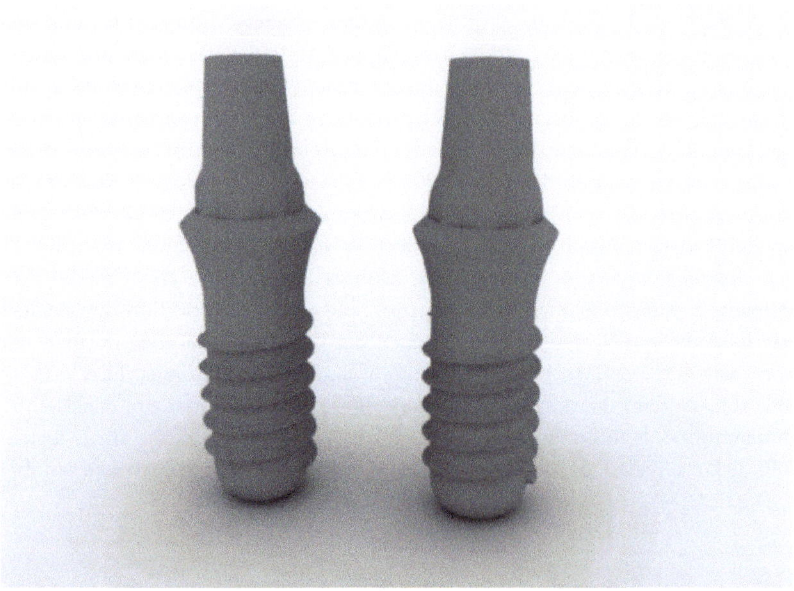

**Figure 21.** Implants are modeled with threads and abutments

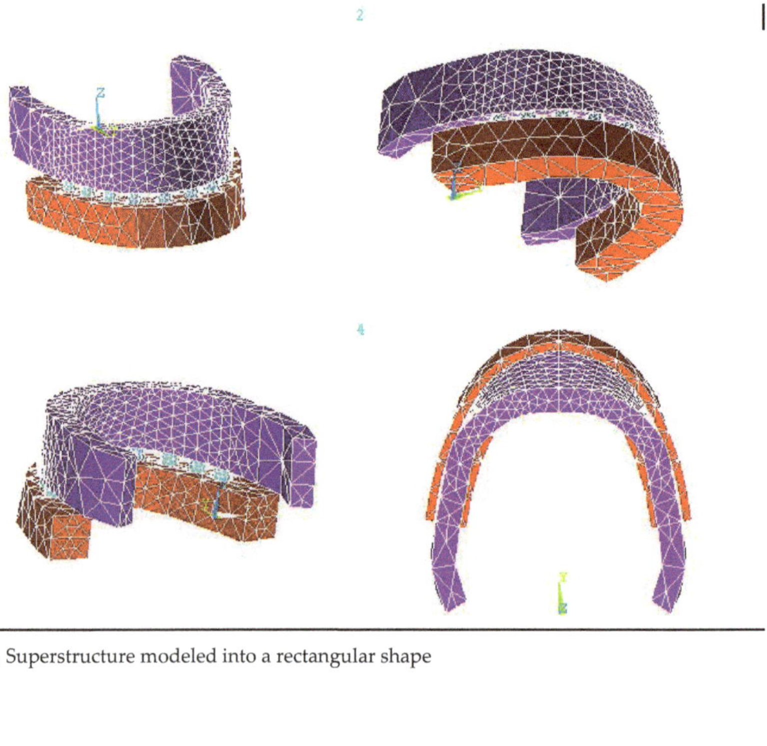

**Figure 22.** Superstructure modeled into a rectangular shape

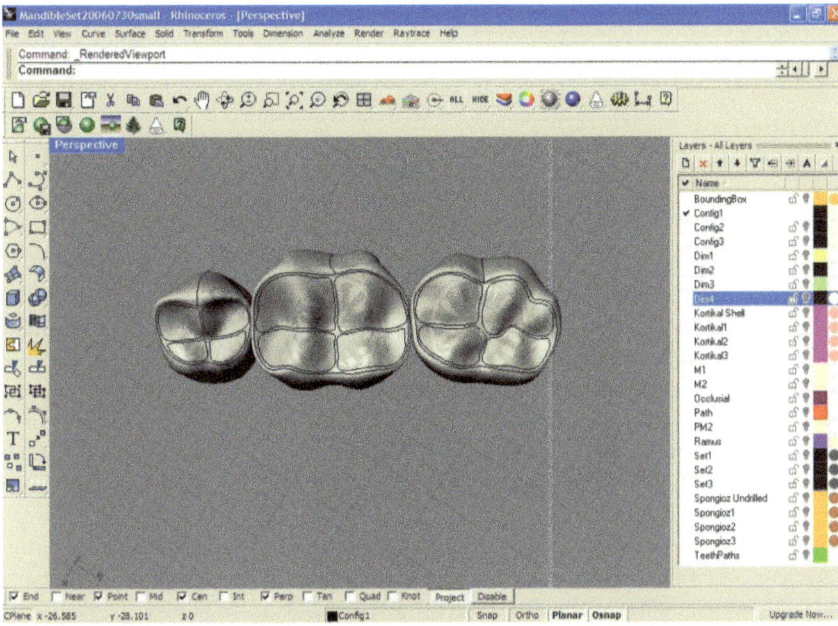


**Figure 23.** Digital preparation of crown models

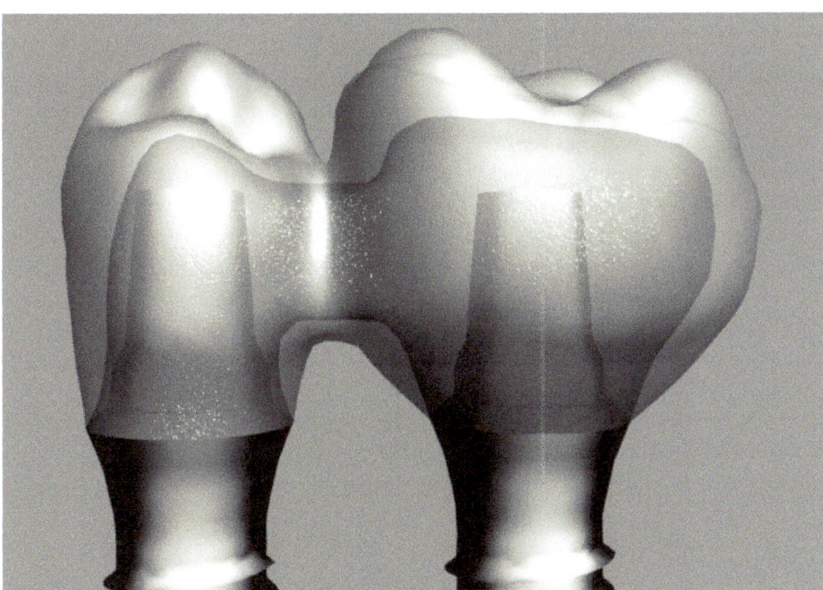

**Figure 24.** Implant, abutment, abutment screw, framework, and porcelain structure are modeled as 1 unit

Almost all of the elastic properties of selected living and non-living materials are available in the literature [9,25,31,34]. Young's modulus and Poisson's ratio are used in models to simulate reality as closely as possible. For example, alveolar bone (both cortical and cancellous portions), implant, abutment, metal framework, and porcelain can be included in the model properties.

## 4. Boundary conditions

A boundary condition is the application of force and constraint. The different ways to apply force and moment include a concentrated load (at a point or single node), force on a line or edge, a distributed load (force varying as an equation), bending moments, and torque [45]. In structural analysis, boundary conditions are applied to those regions of the model where the displacements and/or rotations are known. Such regions may be constrained to remain fixed (have zero displacement and/or rotation) during the simulation or may have specified, non-zero displacements and/or rotations. The directions in which motion is possible are called degrees of freedom (DOF). Zero-displacement constraints must be placed on some boundaries of the model to ensure an equilibrium solution. The constraints should be placed on nodes that are located far from the region of interest to prevent overlap of the stress or strain fields associated with reaction forces with the bone-implant interface. In maxillary FEA models, the nodes along the external lines of the cortical bone of the oral and nasopharyngeal cavities were fixed in all directions [46].

In most FEA studies that include models of the mandible, the boundary conditions are set as a fixed boundary [9]. Zhou et al. developed a more realistic 3-D mandibular FEA model from transversely scanned computed tomography imaging data. The functions of the muscles of mastication and the ligamentous and functional movements of the temporomandibular joints (TMJs) were simulated by means of cable elements and compressive gap elements, respectively. Using this mandibular FEA model, it was concluded that cable and gap elements could be used to set boundary conditions, improving the model mimicry and accuracy [47]. Chang et al. used a technique in which only half of the model was meshed, thus symmetry boundary conditions were prescribed at the nodes on the symmetry plane. Models were constrained in all directions at the nodes on the mesial and symmetrical distal bone surfaces [48]. Expanding the domain of the model can reduce the influence of inaccurate modeling of the boundary conditions. This, however, will be at the expense of computing and modeling time. Teixera et al. concluded that in a 3-D mandibular model, modeling the mandible at distances greater than 4.2 mm mesial or distal from the implant did not result in any significant increase in FEA accuracy [49]. Use of infinite elements is another potential method for modeling boundary conditions [9].

## 5. Loading conditions

Marginal bone loss in the peri-implant region may be the result of excessive occlusal force [50]. Extensive investigations are needed to establish and understand the correlation between marginal bone loss and occlusal forces; including the engineering principles, biomechanical relationships to living tissues, and the mechanical properties of bone surrounding implants [50]. In recent years, a greater amount of materials used for oral implantology are fabricated from titanium and titanium alloy. The Young's modulus of titanium is 5-10 times greater than that of cortical ridge bone surrounding implants [51]. The fundamental engineering principle, composite beam analysis, expresses the concept that when 2 materials of different Young's modulus are placed in direct contact with no intervening material and 1 material loaded, a stress contour will be described at the point where the 2 materials come into contact [52]. For oral implantology, these stress contours are of greater magnitude at the crestal bone. Therefore, the loading condition is another important part of FEA studies. Each component modelization stage contributes to the final analysis after loading. In other words, from the beginning to the end, all procedures and FEA stages add to the ability to extrapolate the results of bite forces surrounding the peri-implant region and prosthetic structures.

Bite forces may be defined as compressive, tensile, or shear forces. Compressive forces attempt to push materials toward each other. Tensile forces pull objects apart. Shear forces on implants cause sliding. The most detrimental forces that can increase the stress around the implant-bone interface and prosthetic assembly are tensile and shear forces. These forces tend to harm material integrity and cause stress build-up. In general, the implant-prosthetic unit can adapt to compressive forces [51]. In actual mastication, the repeated pattern of cyclic forces transmits loading via the restoration and dental implants to peri-impant bone. This generates different amounts of stress around the ridge and also in the prosthetic structure. However, randomized cyclic forces are not easily simulated. Therefore, most FEA studies use static axial and/or non-axial forces. Non-axial loads generate distinctive stress in the ridge especially in the cortical bone. The main remodeling differences between axial and non-axial loading are affected mostly by the horizontal component of the resultant stresses [53]. Therefore, for realistic simulation, combined oblique loads (axial and non-axial) are generally used. One study, comparing dynamic with static loading, revealed that dynamic loading resulted in greater stress levels than static loading [54]. Dynamic loading has consistently been found to have more osteogenic potential than static loading [55]. Sagat et al. investigated the influence of static force on peri-implant stress. In varied models, 100 N static forces were applied vertically and separately to the anterior and posterior parts of a bridge [18]. In another study, static forces of 100 N were applied at 30 degrees obliquely and separately to the lingual inclination of the buccal cusps of a crown (Figures 25,26) [25]. In another study, loading was simulated by applying an oblique load (vertical load of 100 N and horizontal load of 20 N) from buccal to palatal region at 4 different locations. An equivalent load of 200 N was applied in the vertical direction and 40 N in the buccal-palatal direction. The application point of the force was on the central and distal fossae of the crown [48]. Eskitascioglu et al. used an average occlusal force of 300 N applied to a missing second premolar implant-supported crown. Three-point vertical loads were applied to the tip of the buccal cusp (150 N) and distal fossa (150 N); the tip of the buccal cusp (100 N), distal fossa (100 N), and mesial fossa (100 N) [56].

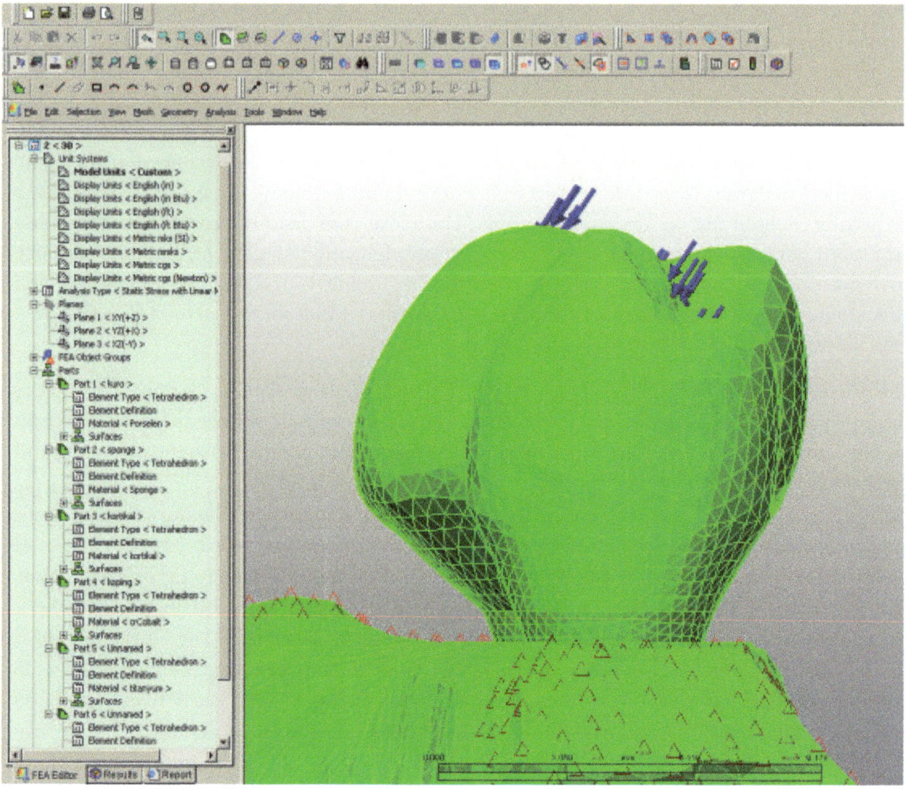

**Figure 25.** Static forces were applied at 30 degrees obliquely and separately to the lingual inclination of the buccal cusps of the crown

As mentioned before, oblique loads are more destructive to the peri-implant bone region and clinically disruptive to prosthetic structures. The magnitude of bite force may change according to age, sex, edentulism, parafunctional habits, and may differ from anterior to posterior in the same mouth [9,31]. In FEA literature, the locations for the application of bite force change according to the modeling of the restoration [9,31]. In advanced modeling studies, more realistic force application could be described including ridges of the cusp, labial or lingual surfaces of crown, occlusal surface, distal, and mesial fossa [9,27,31,57]. For realistic simulation of biting, loading forces should be applied to the restoration first, and then transmitted by the abutment to the implant and surrounding bone. Stress concentrations will then be generated, evaluated, and proper risk assessment will be considered.

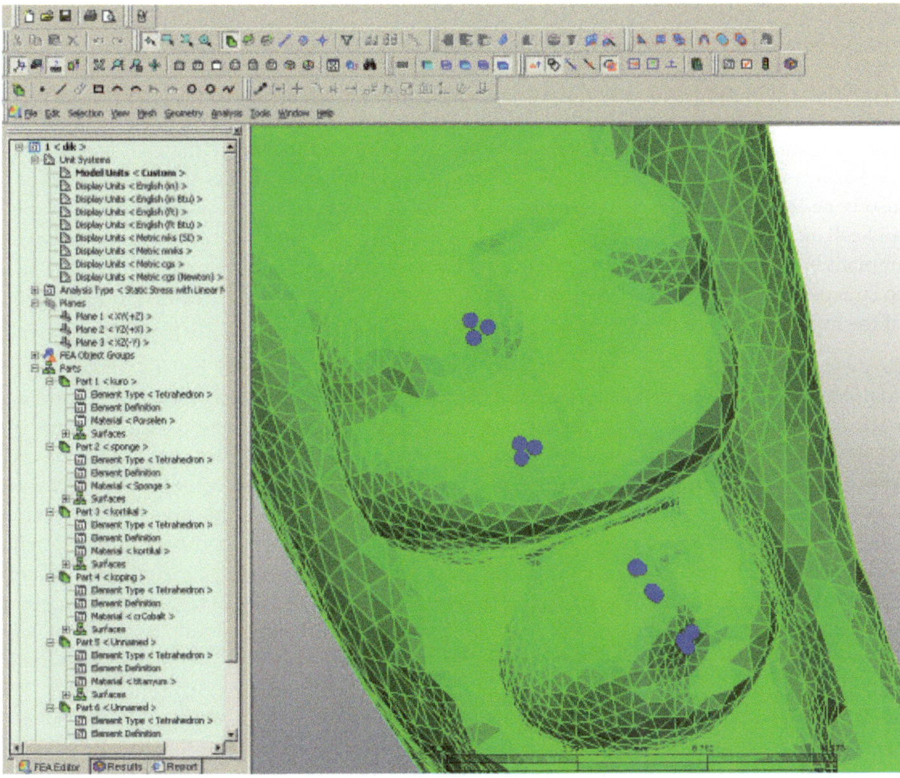

**Figure 26.** Force application to the region of restoration

## 6. Bone-implant interface

The 'osseointegration' concept was described as the direct contact between living bone and a loaded dental implant surface by Brånemark et al. [58]. The most widely used material for dental implant manufacture is pure titanium (Grade 4), titanium alloy (Grade 5), and rarely zirconia [59-62]. These materials have good biocompatibility with surrounding tissues, are resistant to deformation, and are easily manipulated for shaping as a natural tooth root forms by Computer Numerical Control (CNC) machines [59-62]. Titanium alloy has mechanical advantages over pure titanium in implant manufacture. With increases in grade number, the alloy becomes much stronger and more resistant to fractures or wearing of the components [59-62]. However biocompatability may be reduced in inverse proportion the increase in grade number. Implant companies use Grade 4 or Grade 5 titanium for the implant body and generally choose Grade 5 titanium for implant abutment manufacture. Recently, to increase the strength of implant bodies, new materials have also been introduced into the market, such as roxolid (a zirconium and titanium combination) [63]. The use of zirconium and titanium combination material as an implant body has limited scientific data and requires long-term investigations. Therefore, most FEA studies in the literature involve titanium and titanium alloys [9,18,24,31].

The most commonly used surfaces for implant bodies are rough surfaces. Different implant surface modifications (sandblasted, acid-etched, sandblasted and acid-etched, anodized, hydroxyapatite coatings, and plasma-sprayed) are proposed to change the characteristics of the surface from machined to rough, to increase the osteoblastic cell attachment level and also bone-implant contact (BIC) [64-68]. The influence of these surface modifications on BIC and cell attachment are still being investigated for a stronger osseointegration level between implant body and bone. Comparative studies show different BIC levels changing from 13% to 80% percent [69-79]. BIC values may change according to the jaw, placement of the region of the implant, healing time, implant design, and surface structure [64,69,70,72-74].

In most FEA studies, the bone-implant interface was assumed to be 100% bonded or completely osseointegrated [9,16,18,23,25]. As mentioned before, this is not proper modeling from a clinically realistic point of view. Cortical and cancellous bone also have different levels of BIC because of density and availability. Therefore, most studies use cortical bone of uniform thickness surrounding cancellous bone and proper material properties are chosen while modeling [9,16,18,23,25]. The degree of BIC distinctly affects the stress concentration value and distribution. In denser bone, there is less strain under loading compared with softer bone [80]. In some studies, BIC levels were assumed to be ≤100% for simulation of soft bone or immediate loading scenarios [9,81]. Evaluation of peri-implant stress in FEA studies is important for obtaining accurate treatment methods in implant dentistry. Implant and surrounding bone should be stressed within a certain range for dynamic physiologic remodeling. If ideal functional forces are placed on a restoration, the surrounding bone can adapt to the stresses and increase its density [82]. Overload may cause high stresses at the crest of the ridge and result in bone resorption. The direct opposite of this result is disuse atrophy of bone due to too little stress in the peri-implant region. Maintenance of bone density and stabilization is a direct result of the ideal stress distribution [80]. According to Frost studies, strains in the range of 50-1500 microstrain stimulates cortical bone mass and represents the physiological range. Strain beyond this range may cause overload and strain less than this range may not stimulate bone enough [80,83-85]. Most FEA studies, evaluate the risk assessment according to high stress values [9,16,18,23,25]. In other words, the most favorable modeling has the lowest stress values, and in contrast, the most deleterious modeling has the highest stress values [9,16,18,23,25]. However intensely lower stress values may also cause bone resorption because of inadequate bone stimulation.

## 7. Evaluation of stress

Under bite force, localized stress occur at the prosthesis structure and bone. Stress is the magnitude of the internal forces acting within a deformable body. It is a measure of the average force per unit area of a surface within the body on which internal forces act. These internal forces appear as a response to external forces directed on the body [86-88]. Internal resistance after the application of the force applied on the body is not practically measurable. Therefore an easier process is to measure the applied force to a cross-sectional area. The dimension of stress is that of pressure, the Pascal (*Pa*), which is equivalent to 1

Newton (force) per square meter (unit area), that is $N/m^2$. Stress is often reported in scientific publications as MPa. Stress is directly proportional to the force and inversely proportional to the area across which the force is applied. It is important to determine the area across which any force is applied. For example, the surface area of the occlusal pit restoration less than 4 mm. For this reason, the magnitude of stress in many restorations reaches hundreds of MPa [9,16,18,23,25,51].

When the force is applied to mass, a deformation occurs as a result of this force. A strain is a normalized measure of deformation representing the displacement between particles in the body relative to a reference length [9,16,18,23,25,51,86,87,88]. There is no measurement unit of strain. Strain can be defined as the deformation ratio of the original length.

In FEA studies related to implant dentistry, frequently von Mises stress (equivalent tensile stress), minimum principal, and maximum principal are used to evaluate the effect of loading forces on the peri-implant region or prosthesis structure [9,16,18,23,25,89]. When a specific force is applied to the body, von Mises stress is the criterion used to determine the strain energy principles. Loading forces affecting the object can be evaluated 2 or 3 dimensionally. There are 3 "Principal Stresses" that can be calculated at any point, acting in the x, y, and z directions. The von Mises criteria refer to a formula for combining these 3 stresses into an equivalent stress, which is then compared to the yield stress of the material [25,90]. The major stress values are formed when all the components of the shear are zero. When an element is in this position, the normal stresses are called principal stresses. Principal stresses are classified as maximum, intermediate, and minimum principal stresses. The maximum principal stress is a positive value indicating the highest tension. The intermediate principal stress represents intermediate values. The minimum principal stress is a negative value indicating the highest compression [9,16,18,23,25,89]. If the data obtained from the analysis are positive values, then they are considered tensile stresses, negative values indicate compression-type strains.

Frequently, different color figures are used according to the amount of stress around peri-implant regions and prosthetic structures (Figure 27). Stresses on each model are evaluated according to the stress values from low to high. In other words, the most favorable model has the lowest stress values, and in contrast, the most deleterious model has the highest stress values (Figure 28).

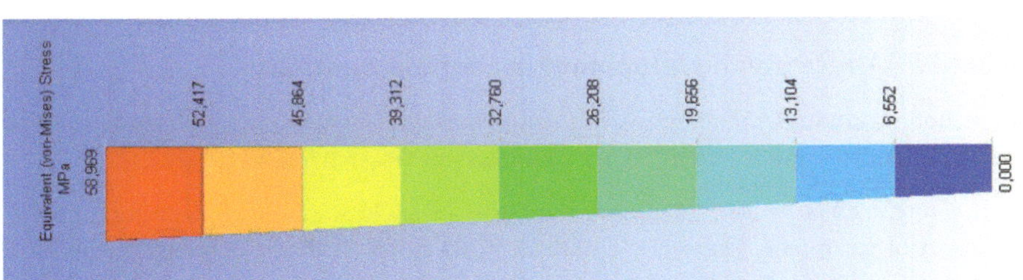


**Figure 27.** Different colors indicate the amount of stress around the peri-implant region and prosthetic structure

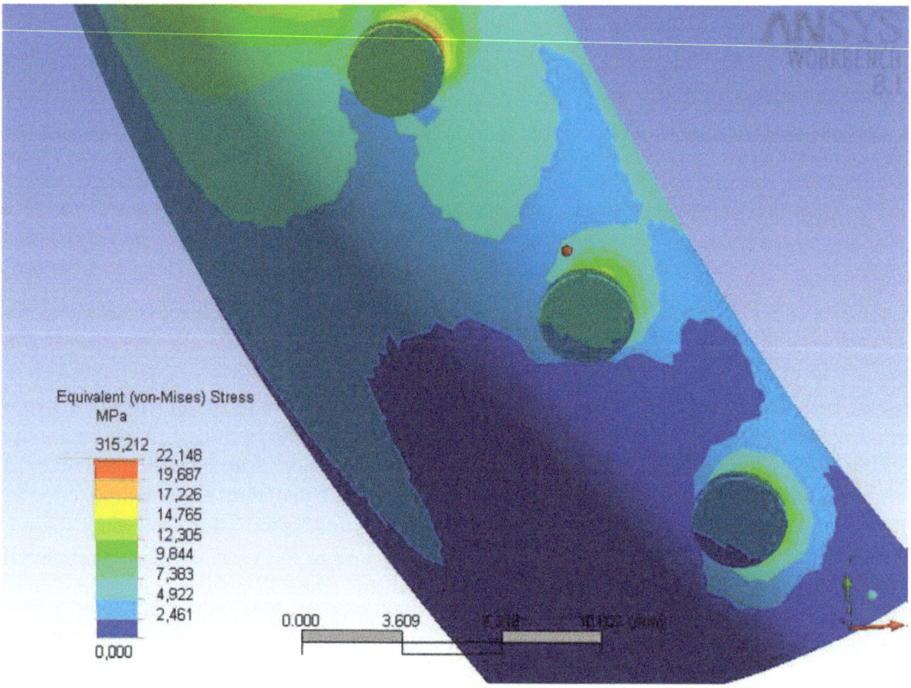


**Figure 28.** High and low stress values depicted in different colors in the models

In a previous study evaluating stress distribution, maximum von Mises (equivalent) stresses on each model are depicted around peri-implant region [18]. Eskitascioglu et al. evaluated maximum stresses (maximum von Mises) within the cortical bone surrounding the implant, framework of restoration, and occlusal surface material [56]. In a previous study, the FE model was used to calculate not only von Mises stress but also the principal stress. Authors explained their approach for this debate as follows: bone can sometimes be classified as brittle material; therefore, the principal stress was also implemented to evaluate the situation of cortical bone around implants [48].

## 8. Good FEA research development in implant dentistry

This section is provided for clinicians and researchers who want to plan FEA studies related to implant dentistry and to provide a brief summary of research methodology.

1. *Planning a scenario*: The most important part of an FEA study is planning a unique model of treatment. There are countless FEA studies in the implantology literature; therefore, at the beginning of the study, it is highly recommended that you evaluate the available literature on your subject. Implant technology is improving rapidly. There is currently no perfect dental implant design or implant-abutment connection.

Implant manufacturers change their macro design and connections according to perceived clinical benefits. The aim of these improvements are less bone resorption around peri-implant regions, less micromotion at abutments, better loading distributions at dental implant structures, and good conical sealing. These properties are commonly related to biomechanics and should be investigated not only with clinical studies but also with FEA studies. All novel designs of implants or materials can be subject to investigation and can be compared with traditional structures. Another way of instituting FEA study is investigating treatment alternatives. New and old treatment modeling can be compared, limitations, and application areas can be better understood.

2. *Computer stage*: This is the second part of FEA study. Generally clinicians have limited knowledge about modeling in computers and need help from computer engineers. It will be very wise to collaborate with friends at that field. Without a collaborator in computer engineering, too much time will be spent learning how to prepare models and developing the appropriate knowledge for the computational techniques necessary for model implementation. The clinician should manage the study and provide direction to the engineer. If the engineer does not have knowledge of the field of implant dentistry, seminars can be given to introduce the basic concepts of implantology. The seminars can include concepts such as indications for dental implants, dental implant parts, bone physiology, biting forces, connections of implants with bone, and the logic of implantology. As mentioned before, the shape of the materials can be scanned and converted digitally. Dental volumetric or computed tomography are good alternatives to scan and build bone structures. Devices used for routine treatments, can be found easily and are not expensive. For modeling of implant parts and superstructure, there are many sources, including manufacturers guidelines, scanning (advanced engineering 3-D scanning needed), and tooth atlas. The clinician should make every effort to maintain contact with their colleagues to allow frequent and efficient model evaluation and adaptation. The number of elements and nodes, can be increased to achieve more detailed modeling. However, this may be quiet time-consuming and may implicate computing complications. Therefore, the engineer should clearly understand the aim of the research. Boundaries, limitations can be applied at modeling and element numbers can be increased only at the region of interest. These applications should not directly affect the results achieved. In the literature there are many software packages available for FEA study. The computer engineer can aid clinicians in choosing the appropriate software package for the specific application. In general, von Mises (equivalent stress), minimum, and maximum principal stress values are being used in FEA studies related to implant dentistry. These stress values are evaluated from low to high, and assessments are made according to these values. Higher values are considered more destructive and involve greater risk than low values. The most common material properties used in FEA studies of implant dentistry are listed in Table 1 [9,27,48,56,57,91-109].

| Material | Young Modulus (MPa) | Poisson Ratio | Ref. No. |
|---|---|---|---|
| *Ti-6Al-4V* | 110,000<br>110,000<br>100,000 | 0.35<br>0.33<br>0.35 | 27, 48, 57, 91 |
| *Pure titanium* | 117,000 | 0.3 | 9, 92, 93 |
| *Type 3 gold alloy* | 90,000<br>100,000<br>80,000 | 0.3<br>0.3<br>0.33 | 48, 94, 95 |
| *Cortical bone* | 13,700<br>13,400<br>10,000<br>15,000 | 0.3<br>0.3<br>0.3<br>0.3 | 27, 56, 57, 96, 97 |
| *Trabecular bone* | 1,370<br>1,500<br>1,370<br>150,000<br>250,000<br>790,000 | 0.3<br>0.3<br>0.31<br>0.3<br>0.3<br>0.3 | 27, 56, 57, 98, 99 |
| *Periodontal ligament* | 170 | 0.45 | 108 |
| *Ni-Cr alloy* | 204,000 | 0.3 | 108 |
| *Dentin* | 18,600 | 0.31 | 108 |
| *Porcelain* | 66,900<br>67,700 | 0.29<br>0.28 | 31, 48, 109 |
| *Co-Cr alloy* | 218,000 | 0.33 | 56 |
| *Feldspathic porcelain* | 82,800 | 0.35 | 56 |
| *Enamel* | 41,400<br>46,890<br>82,500<br>84,000 | 0.3<br>0.3<br>0.33<br>0.33 | 97, 100-102 |
| *Mucosa* | 10 | 0.40 | 103 |
| *Ag-Pd alloy* | 95,000<br>80,000 | 0.33<br>0.33 | 109 |
| *Resin* | 2,700 | 0.35 | 31 |
| *Resin composite* | 7,000 | 0.2 | 31 |
| *Gold alloy screw* | 100,000 | 0.3 | 93 |
| *Titanium abutment* | 110,000 | 0.28 | 109 |
| *Titanium abutment screw* | 110,000 | 0.28 | 109 |
| *Zirconia implant* | 200,000 | 0.31 | 105, 107 |
| *Zirconia abutment* | 200,000 | 0.31 | 105, 107 |
| *Zirconia core* | 200,000 | 0.31 | 105, 107 |
| *Zirconia veneer* | 80,000 | 0.265 | 106, 107 |

**Table 1.** Material properties used in finite element analysis studies of implant dentistry

3. *Interpretation of results:* FEA studies have several advantages over clinical, pre-clinical, and in vitro studies. Most importantly, patients will not be harmed by the application of new materials and treatment modalities that have not been previously tested. Animals will not suffer from these biomechanical studies. However, clinicians should be aware that all of these applications are being performed on a computer, with critical limitations and assumptions that will clearly affect the applicability of the results to a real scenario. In the application of FEA studies, the most common drawback is overemphasis of the results. Simplifications are made for all simulated models; therefore, the models should be compared with each other within the same study. Other studies may use varied material properties and different planning scenarios. Confirming the FEA results with mechanical tests, conventional clinical model analysis, and preclinical tests are essential. It should not be forgotten that FEA studies are helpful for clinical trials but the results achieved from these studies are not valuable as clinical study results. However, before beginning biomechanical clinical trials, it will be wise to refer to FEA studies.

## 9. Conclusion

FEA is a numerical stress analysis technique and is extensively used in implant dentistry to evaluate the risk factors from a biomechanical point of view. Simplifications and assumptions are the limitations of FEA studies. Although advanced computer technology is used to obtain results from simulated models, many factors affecting clinical features such as implant macro and micro design, material properties, loading conditions, and boundary conditions are neglected or ignored. Therefore, correlating FEA results with preclinical and long-term clinical studies may help to validate research models.

## Author details

B. Alper Gultekin and Serdar Yalcin
*Istanbul University Faculty of Dentistry, Department of Oral Implantology, Istanbul, Turkey*

Pinar Gultekin*
*Istanbul University Faculty of Dentistry, Department of Prosthodontics, Istanbul, Turkey*

## 10. References

[1] Malevez C, Hermans M, Daelemans P (1996) Marginal bone levels at Brånemark system implants used for single tooth restoration. The influence of implant design and anatomical region. Clin Oral Implants Res Jun;7(2):162-9.
[2] Hermann F, Lerner H, Palti A (2007) Factors influencing the preservation of the periimplant marginal bone. Implant Dent. Jun;16(2):165-75.

* Corresponding Author

[3] Bateli M, Att W, Strub JR (2011) Implant neck configurations for preservation of marginal bone level: a systematic review. Int J Oral Maxillofac Implants. Mar-Apr;26(2):290-303.

[4] Baqain ZH, Moqbel WY, Sawair FA (2012)Early dental implant failure: risk factors. Br J Oral Maxillofac Surg. Apr;50(3):239-43.

[5] Real-Osuna J, Almendros-Marqués N, Gay-Escoda C (2012)Prevalence of complications after the oral rehabilitation with implant-supported hybrid prostheses. Med Oral Patol Oral Cir BucalJan 1;17(1):e116-21.

[6] Karabuda C, Yaltirik M, Bayraktar M (2008) A clinical comparison of prosthetic complications of implant-supported overdentures with different attachment systems. Implant Dent. Mar;17(1):74-81.

[7] Rieger MR, Adams WK, Kinzel GL, Brose MO (1989) Finite element analysis of bone-adapted and bone-bonded endosseous implants. J Prosthet Dent. Oct;62(4):436-40.

[8] Shen WL, Chen CS, Hsu ML (2010) Influence of implant collar design on stress and strain distribution in the crestal compact bone: a three-dimensional finite element analysis. Int J Oral Maxillofac Implants. Sep-Oct;25(5):901-10.

[9] Geng JP, Tan KB, Liu GR (2001) Application of finite element analysis in implant dentistry: a review of the literature. J Prosthet Dent. Jun;85(6):585-98.

[10] Assunção WG, Barão VA, Tabata LF, Gomes EA, Delben JA, dos Santos PH (2009) Biomechanics studies in dentistry: bioengineering applied in oral implantology. J Craniofac Surg. Jul;20(4):1173-7.

[11] Srinivasan M, Padmanabhan TV (2008) Intrusion in implant-tooth-supported fixed prosthesis: an in vitro photoelastic stress analysis. Indian J Dent Res. Jan-Mar;19(1):6-11.

[12] Rieger MR, Fareed K, Adams WK, Tanquist RA (1989) Bone stress distribution for three endosseous implants. J Prosthet Dent. Feb;61(2):223-8.

[13] DeTolla DH, Andreana S, Patra A, Buhite R, Comella B (2000) Role of the finite element model in dental implants. J Oral Implantol. 26(2):77-81.

[14] Georgiopoulos B, Kalioras K, Provatidis C, Manda M, Koidis P (2007) The effects of implant length and diameter prior to and after osseointegration: a 2-D finite element analysis. J Oral Implantol.33(5):243-56.

[15] Juodzbalys G, Kubilius R, Eidukynas V, Raustia AM (2005) Stress distribution in bone: single-unit implant prostheses veneered with porcelain or a new composite material. Implant Dent. Jun;14(2):166-75.

[16] Meriç G, Erkmen E, Kurt A, Eser A, Ozden AU (2012) Biomechanical comparison of two different collar structured implants supporting 3-unit fixed partial denture: a 3-D FEM study. Acta Odontol Scand. Jan;70(1):61-71.

[17] Erkmen E, Meriç G, Kurt A, Tunç Y, Eser A (2011) Biomechanical comparison of implant retained fixed partial dentures with fiber reinforced composite versus conventional metal frameworks: a 3D FEA study. J Mech Behav Biomed Mater. Jan;4(1):107-16.

[18] Sagat G, Yalcin S, Gultekin BA, Mijiritsky E (2010) Influence of arch shape and implant position on stress distribution around implants supporting fixed full-arch prosthesis in edentulous maxilla. Implant Dent. Dec;19(6):498-508.

[19] Gröning F, Fagan M, O'higgins P (2012) Modeling the Human Mandible Under Masticatory Loads: Which Input Variables are Important? Anat Rec (Hoboken). Mar 30. doi: 10.1002/ar.22455.

[20] Keyak JH, Meagher JM, Skinner HB, Mote CD Jr (1990) Automated three-dimensional finite element modelling of bone: a new method. J Biomed Eng. Sep;12(5):389-97.

[21] Cahoon P, Hannam AG (1994) Interactive modeling environment for craniofacial reconstruction. Visual data exploration and analysis. SPIE Proc. 2178:206-15.

[22] Ormianer Z, Palti A, Demiralp B, Heller G, Lewinstein I, Khayat PG (2012) Implant-supported first molar restorations: correlation of finite element analysis with clinical outcomes. Int J Oral Maxillofac Implants. Jan-Feb;27(1):e1-12.

[23] Canullo L, Pace F, Coelho P, Sciubba E, Vozza I (2011) The influence of platform switching on the biomechanical aspects of the implant-abutment system. A three dimensional finite element study. Med Oral Patol Oral Cir Bucal. Sep 1;16(6):e852-6.

[24] Tabata LF, Rocha EP, Barão VA, Assunção WG (2011) Platform switching: biomechanical evaluation using three-dimensional finite element analysis. Int J Oral Maxillofac Implants. May-Jun;26(3):482-91.

[25] Bayraktar M (2011) The influence of crown-implant ratio and dental implant parameters on implant and periimplant bone: A finite element analysis. Istanbul University, Institute of Health Science, Department of Prosthetic Dentistry. PhD Thesis.

[26] Chu CM, Huang HL, Hsu JT, Fuh LJ (2012) Influences of internal tapered abutment designs on bone stresses around a dental implant: three-dimensional finite element method with statistical evaluation. J Periodontol. Jan;83(1):111-8.

[27] Pessoa RS, Vaz LG, Marcantonio E Jr, Vander Sloten J, Duyck J, Jaecques SV (2010) Biomechanical evaluation of platform switching in different implant protocols: computed tomography-based three-dimensional finite element analysis. Int J Oral Maxillofac Implants. Sep-Oct;25(5):911-9.

[28] Arisan V, Karabuda ZC, Avsever H, Ozdemir T (2012) Conventional Multi-Slice Computed Tomography (CT) and Cone-Beam CT (CBCT) for Computer-Assisted Implant Placement. Part I: Relationship of Radiographic Gray Density and Implant Stability. Clin Implant Dent Relat Res. Jan 17. doi: 10.1111/j.1708-8208.2011.00436.

[29] Lee CY, Prasad HS, Suzuki JB, Stover JD, Rohrer MD (2011) The correlation of bone mineral density and histologic data in the early grafted maxillary sinus: a preliminary report. Implant Dent. Jun;20(3):202-14.

[30] Turkyilmaz I, Ozan O, Yilmaz B, Ersoy AE (2008) Determination of bone quality of 372 implant recipient sites using Hounsfield unit from computerized tomography: a clinical study. Clin Implant Dent Relat Res. Dec;10(4):238-44.

[31] Geng J, Yan W, Xu W (2008) Application of the finite element method in implant dentistry. Springer. pp. 81-89.

[32] AD McNaught, A. Wilkinson (1997) Compendium of Chemical Terminology, 2nd ed. (the "Gold Book"). Blackwell Scientific Publications, Oxford. pp. 22-45.

[33] H. Gercek (2007) Poisson's ratio values for rocks. Int. Journal of Rock Mec. and Min. Sci.; Elsevier; January; 44 (1): pp. 1–13.

[34] Ozgen M (2011) Evaluation of streses around implants that were placed in anterior maxillary vertical defect region: A finite element analysis study. Istanbul University, Institute of Health Science, Department of Prosthetic Dentistry. PhD Thesis.
[35] Papavasiliou G, Kamposiora P, Bayne SC, Felton DA (1996) Three-dimensional finite element analysis of stress-distribution around single tooth implants as a function of bony support, prosthesis type, and loading during function. J Prosthet Dent. Dec;76(6):633-40.
[36] Meyer U, Vollmer D, Runte C, Bourauel C, Joos U (2001) Bone loading pattern around implants in average and atrophic edentulous maxillae: a finite-element analysis. J Craniomaxillofac Surg. Apr;29(2):100-5.
[37] Holmes DC, Loftus JT (1997) Influence of bone quality on stress distribution for endosseous implants. J Oral Implantol. 23(3):104-11.
[38] Tada S, Stegaroiu R, Kitamura E, Miyakawa O, Kusakari H (2003) Influence of implant design and bone quality on stress/strain distribution in bone around implants: a 3-dimensional finite element analysis. Int J Oral Maxillofac Implants. May-Jun;18(3):357-68.
[39] Papavasiliou G, Kamposiora P, Bayne SC, Felton DA (1997) 3D-FEA of osseointegration percentages and patterns on implant-bone interfacial stresses. J Dent. Nov;25(6):485-91.
[40] Frost HM (2004) A 2003 update of bone physiology and Wolff's Law for clinicians. Angle Orthod. Feb;74(1):3-15.
[41] Brunski JB, Puleo DA, Nanci A (2000) Biomaterials and biomechanics of oral and maxillofacial implants: current status and future developments. Int J Oral Maxillofac Implants. Jan-Feb;15(1):15-46.
[42] Huang HL, Hsu JT, Fuh LJ, Lin DJ, Chen MY (2010) Biomechanical simulation of various surface roughnesses and geometric designs on an immediately loaded dental implant. Comput Biol Med. May;40(5):525-32.
[43] Chun HJ, Cheong SY, Han JH, Heo SJ, Chung JP, Rhyu IC, Choi YC, Baik HK, Ku Y, Kim MH (2002) Evaluation of design parameters of osseointegrated dental implants using finite element analysis. J Oral Rehabil. Jun;29(6):565-74.
[44] Wheeler RC (1963) An atlas of tooth form. Philadelphia and London. WB Saunders 26 p.
[45] Desai SR, Shinde HH (2012) Finite Element Analysis: Basics And Its Applications In Dentistry. Indian J Dent Sci. 4(1):60-5.
[46] Saab XE, Griggs JA, Powers JM, Engelmeier RL (2007) Effect of abutment angulation on the strain on the bone around an implant in the anterior maxilla: a finite element study. J Prosthet Dent. Feb;97(2):85-92.
[47] Zhou X, Zhao Z, Zhao M, Fan Y (1999) The boundary design of mandibular model by means of the three-dimensional finite element method. Hua Xi Kou Qiang Yi Xue Za Zhi.17(1):29-32.
[48] Chang CL, Chen CS, Hsu ML (2010) Biomechanical effect of platform switching in implant dentistry: a three-dimensional finite element analysis. Int J Oral Maxillofac Implants. 25(2):295-304.

[49] Teixeira ER, Sato Y, Akagawa Y, Shindoi N (1998) A comparative evaluation of mandibular finite element models with different lengths and elements for implant biomechanics. J Oral Rehabil. 25(4):299-303.
[50] Misch CE, Suzuki JB, Misch-Dietsh FM, Bidez MW (2005) A positive correlation between occlusal trauma and peri-implant bone loss: literature support. Implant Dent. Jun;14(2):108-16.
[51] Misch CE (2008) Comtemporary implant dentistry. Mosby, Elsevier. pp. 68-88 and 544-546.
[52] Baumeister T, Avallone EA (1978) Marks' standard handbook of mechanical engineers, ed 8, New York, McGraw-Hill.
[53] Barbier L, Vander Sloten J, Krzesinski G, Schepers E, Van der Perre G (1998) Finite element analysis of non-axial versus axial loading of oral implants in the mandible of the dog. J Oral Rehabil. Nov;25(11):847-58.
[54] Zhang JK, Chen ZQ (1998) The study of effects of changes of the elastic modulus of the materials substitute to human hard tissues on the mechanical state in the implant-bone interface by three-dimensional anisotropic finite element analysis. West China J Stomatol 1998;16:274-8.
[55] Akuz E, Braun TJ, Brown NA (2006) Static versus dynamic loading in the mechanical modulation of vertebral growth. Spine 31:952-958.
[56] Eskitascioglu G, Usumez A, Sevimay M, Soykan E, Unsal E (2004) The influence of occlusal loading location on stresses transferred to implant-supported prostheses and supporting bone: A three-dimensional finite element study. J Prosthet Dent. Feb;91(2):144-50.
[57] Hsu ML, Chen FC, Kao HC, Cheng CK (2007) Influence of off-axis loading of an anterior maxillary implant: a 3-dimensional finite element analysis. Int J Oral Maxillofac Implants. Mar-Apr;22(2):301-9.
[58] Brånemark PI, Adell R, Breine U, Hansson BO, Lindström J, Ohlsson A (1969) Intra-osseous anchorage of dental prostheses. I. Experimental studies. Scand J Plast Reconstr Surg.3(2):81-100.
[59] Lautenschlager EP, Monaghan P (1993) Titanium and titanium alloys as dental materials. Int Dent J. Jun;43(3):245-53.
[60] Lincks J, Boyan BD, Blanchard CR, Lohmann CH, Liu Y, Cochran DL, Dean DD, Schwartz Z (1998) Response of MG63 osteoblast-like cells to titanium and titanium alloy is dependent on surface roughness and composition. Biomaterials. Dec;19(23):2219-32.
[61] Eisenbarth E, Meyle J, Nachtigall W, Breme J (1996) Influence of the surface structure of titanium materials on the adhesion of fibroblasts. Biomaterials. Jul;17(14):1399-403.
[62] Thompson GJ, Puleo DA (1996) Ti-6Al-4V ion solution inhibition of osteogenic cell phenotype as a function of differentiation timecourse in vitro. Biomaterials. Oct;17(20):1949-54.
[63] Barter S, Stone P, Brägger U (2011) A pilot study to evaluate the success and survival rate of titanium-zirconium implants in partially edentulous patients: results after 24 months of follow-up. Clin Oral Implants Res. Jun 24. doi: 10.1111/j.1600-0501.2011.02231.

[64] Wennerberg A, Albrektsson T, Johansson C, Andersson B (1996) Experimental study of turned and grit-blasted screw-shaped implants with special emphasis on effects of blasting material and surface topography. Biomaterials. Jan;17(1):15-22.
[65] Abron A, Hopfensperger M, Thompson J, Cooper LF (2001) Evaluation of a predictive model for implant surface topography effects on early osseointegration in the rat tibia model. J Prosthet Dent. Jan;85(1):40-6.
[66] Blumenthal NC, Cosma V (1989) Inhibition of apatite formation by titanium and vanadium ions. J Biomed Mater Res. Apr;23(A1 Suppl):13-22.
[67] Klokkevold PR, Johnson P, Dadgostari S, Caputo A, Davies JE, Nishimura RD (2001) Early endosseous integration enhanced by dual acid etching of titanium: a torque removal study in the rabbit. Clin Oral Implants Res. Aug;12(4):350-7.
[68] Weng D, Hoffmeyer M, Hürzeler MB, Richter EJ (2003) Osseotite vs. machined surface in poor bone quality. A study in dogs. Clin Oral Implants Res. Dec;14(6):703-8.
[69] Piattelli A, Degidi M, Paolantonio M, Mangano C, Scarano A (2003) Residual aluminum oxide on the surface of titanium implants has no effect on osseointegration. Biomaterials. Oct;24(22):4081-9.
[70] Wennerberg A, Albrektsson T, Andersson B, Krol JJ (1995) A histomorphometric and removal torque study of screw-shaped titanium implants with three different surface topographies. Clin Oral Implants Res. Mar;6(1):24-30.
[71] Gotfredsen K, Nimb L, Hjörting-Hansen E, Jensen JS, Holmén A (1992) Histomorphometric and removal torque analysis for TiO2-blasted titanium implants. An experimental study on dogs. Clin Oral Implants Res. Jun;3(2):77-84.
[72] Ivanoff CJ, Hallgren C, Widmark G, Sennerby L, Wennerberg A (2001) Histologic evaluation of the bone integration of TiO(2) blasted and turned titanium microimplants in humans. Clin Oral Implants Res. Apr;12(2):128-34.
[73] van Steenberghe D, De Mars G, Quirynen M, Jacobs R, Naert I (2000) A prospective split-mouth comparative study of two screw-shaped self-tapping pure titanium implant systems. Clin Oral Implants Res. Jun;11(3):202-9.
[74] Kohal RJ, Weng D, Bächle M, Strub JR (2004) Loaded custom-made zirconia and titanium implants show similar osseointegration: an animal experiment. J Periodontol. Sep;75(9):1262-8.
[75] Xue W, Liu X, Zheng X, Ding C (2005) In vivo evaluation of plasma-sprayed titanium coating after alkali modification. Biomaterials. Jun;26(16):3029-37.
[76] Piattelli A, Corigliano M, Scarano A, Costigliola G, Paolantonio M (1998) Immediate loading of titanium plasma-sprayed implants: an histologic analysis in monkeys. J Periodontol. Mar;69(3):321-7.
[77] Scarano A, Iezzi G, Petrone G, Marinho VC, Corigliano M, Piattelli A (2000) Immediate postextraction implants: a histologic and histometric analysis in monkeys. J Oral Implantol. 26(3):163-9.
[78] Galli C, Guizzardi S, Passeri G, Martini D, Tinti A, Mauro G, Macaluso GM (2005) Comparison of human mandibular osteoblasts grown on two commercially available titanium implant surfaces. J Periodontol. Mar;76(3):364-72.

[79] Ivanoff CJ, Widmark G, Johansson C, Wennerberg A (2003) Histologic evaluation of bone response to oxidized and turned titanium micro-implants in human jawbone. Int J Oral Maxillofac Implants. May-Jun;18(3):341-8.
[80] Frost HM (1987) Bone 'mass' and the 'mechonastat': a proposal, Anar Rec. 219 pp. 1-9.
[81] Ding X, Zhu XH, Liao SH, Zhang XH, Chen H (2009) Implant-bone interface stress distribution in immediately loaded implants of different diameters: a three-dimensional finite element analysis. J Prosthodont. Jul;18(5):393-402.
[82] Roberts WE, Garetto LP, DeCastro RA (1989) Remodeling of devitalized bone threatens periosteal margin integrity of endosseous titanium implants with threaded or smooth surfaces: indications for provisional loading and axially directed occlusion. J Indiana Dent Assoc. Jul-Aug;68(4):19-24.
[83] Cowin SC, Hegedus DH (1976) Bone remodeling I: theory of adaptive elasticity. J Elast. 6:313-326.
[84] Cowin SC, Hegedus DH (1976) Bone remodeling II: small strain adaptive elasticity. J Elast 6:337-352.
[85] Cowin SC, Nachlinger RR (1978) Bone remodeling II: uniqueness and stability in adaptive elasticity theory. J Elast. 8:285-295.
[86] Atanackovic, Teodor M.; Guran, Ardéshir (2000). Theory of elasticity for scientists and engineers. Springer. pp. 1–46.
[87] Ameen, Mohammed (2005). Computational elasticity: theory of elasticity and finite and boundary element methods. Alpha Science Int'l Ltd.. pp. 33–66.
[88] Chakrabarty, J. (2006). Theory of plasticity (3 ed.). Butterworth-Heinemann. pp. 17–32.
[89] Turkoglu P (2006) Finite element stress analysis of in-line and staggered placement of mandibular dental implants. Istanbul University, Institute of Health Science, Department of Prosthetic Dentistry. PhD Thesis.
[90] Franklin FE (1998) Stress analysis. Mechanical Engineers' Handbook. Wiley Interscience pp. 191-245.
[91] Colling EW (1984) The physical metallurgy of titanium alloys. Metal Park (OH): American Society for Metals.
[92] Ronald LS, Svenn EB (1995) Nonlinear contact analysis of preload in dental implant screws. Int J Oral Maxillofac Implants.10:295-302.
[93] Lang LA, Kang B, Wang RF, Lang BR (2003) Finite element analysis to determine implant preload. J Prosthet Dent. Dec;90(6):539-46.
[94] Sakaguchi RL, Borgersen SE (1995) Nonlinear contact analysis of preload in dental implant screws. Int J Oral Maxillofac Implants.10(3):295-302.
[95] Lewinstein I, Banks-Sills L, Eliasi R (1995) Finite element analysis of a new system (IL) for supporting an implant-retained cantilever prosthesis. Int J Oral Maxillofac Implants.10(3):355-66.
[96] Cook SD, Klawitter JJ, Weinstein AM (1982) A model for the implant-bone interface characteristics of porous dental implants. J Dent Res.61(8):1006-9.
[97] Farah JW, Craig RG, Meroueh KA (1989) Finite element analysis of three- and four-unit bridges. J Oral Rehabil.16(6):603-11.

[98] MacGregor AR, Miller TP, Farah JW (1980) Stress analysis of mandibular partial dentures with bounded and free-end saddles. J Dent.8(1):27-34.

[99] Knoell AC (1977) A mathematical model of an in vitro human mandible. J Biomech.10(3):159-66.

[100] Davy DT, Dilley GL, Krejci RF (1981) Determination of stress patterns in root-filled teeth incorporating various dowel designs. J Dent Res. 60(7):1301-10.

[101] Wright KW, Yettram AL (1979)Reactive force distributions for teeth when loaded singly and when used as fixed partial denture abutments. J Prosthet Dent. 42(4):411-6.

[102] Farah JW, Hood JA, Craig RG (1975) Effects of cement bases on the stresses in amalgam restorations. J Dent Res. 54(1):10-5.

[103] Maeda Y, Wood WW (1989) Finite element method simulation of bone resorption beneath a complete denture. J Dent Res. 68(9):1370-3.

[104] Ronald LS, Svenn EB (1995) Nonlinear contact analysis of preload in dental implant screws. Int J Oral Maxillofac Implants. 10:295-302.

[105] Kohal RJ, Papavasiliou G, Kamposiora P, Tripodakis A, Strub JR (2002) Three-dimensional computerized stress analysis of commercially pure titanium and yttrium-partially stabilized zirconia implants. Int J Prosthodont. Mar-Apr;15(2):189-94.

[106] White SN, Miklus VG, McLaren EA, Lang LA, Caputo AA (2005) Flexural strength of a layered zirconia and porcelain dental all-ceramic system. J Prosthet Dent. 94(2):125-31.

[107] Caglar A, Bal BT, Karakoca S, Aydın C, Yılmaz H, Sarısoy S (2011) Three-dimensional finite element analysis of titanium and yttrium-stabilized zirconium dioxide abutments and implants. Int J Oral Maxillofac Implants. 26(5):961-9.

[108] Lanza MDS, Seraidarian PI, Jansen WC, Lanza MD (2011) Stress analysis of a fixed implant-supported denture by the finite element method (FEM) when varying the number of teeth used as abutments. J Appl Oral Sci. 19(6):655-61.

[109] Quaresma SET, Cury PR, Sendyk WR, Sendyk C (2008) A finite element analysis of two different dental implants: Stress distribution in the prosthesis, abutment, implant, and supporting bone. Journal of Imp. 34(1):1-6.

Section 2

# Recent Advances of Finite Element Analysis in "Electrical Engineering"

# Chapter 5

# Finite Element Analysis of the Direct Drive PMLOM

Govindaraj Thangavel

Additional information is available at the end of the chapter

## 1. Introduction

The **finite element method (FEM)** (its practical application often known as **finite element analysis** (FEA)) is a numerical technique for finding approximate solutions of partial differential equations (PDE) as well as integral equations. The solution approach is based either on eliminating the differential equation completely (steady state problems), or rendering the PDE into an approximating system of ordinary differential equations, which are then numerically integrated using standard techniques such as Euler's method, Runge-Kutta, etc. In solving partial differential equations, the primary challenge is to create an equation that approximates the equation to be studied, but is numerically stable, meaning that errors in the input and intermediate calculations do not accumulate and cause the resulting output to be meaningless. There are many ways of doing this, all with advantages and disadvantages. The finite element method is a good choice for solving partial differential equations over complicated domains (like cars and oil pipelines), when the domain changes (as during a solid state reaction with a moving boundary), when the desired precision varies over the entire domain, or when the solution lacks smoothness. For instance, in a frontal crash simulation it is possible to increase prediction accuracy in "important" areas like the front of the car and reduce it in its rear (thus reducing cost of the simulation). Another example would be in Numerical weather prediction, where it is more important to have accurate predictions over developing highly-nonlinear phenomena (such as tropical cyclones in the atmosphere, or eddies in the ocean) rather than relatively calm areas.

The finite difference method (FDM) is an alternative way of approximating solutions of PDEs. The differences between FEM and FDM are:

- The most attractive feature of the FEM is its ability to handle complicated geometries (and boundaries) with relative ease. While FDM in its basic form is restricted to handle rectangular shapes and simple alterations thereof, the handling of geometries in FEM is theoretically straightforward.

- The most attractive feature of finite differences is that it can be very easy to implement.
- There are several ways one could consider the FDM a special case of the FEM approach. E.g., first order FEM is identical to FDM for Poisson's equation, if the problem is discretized by a regular rectangular mesh with each rectangle divided into two triangles.
- There are reasons to consider the mathematical foundation of the finite element approximation more sound, for instance, because the quality of the approximation between grid points is poor in FDM.
- The quality of a FEM approximation is often higher than in the corresponding FDM approach, but this is extremely problem-dependent and several examples to the contrary can be provided.

Generally, FEM is the method of choice in all types of analysis in structural mechanics (i.e. solving for deformation and stresses in solid bodies or dynamics of structures) while computational fluid dynamics (CFD) tends to use FDM or other methods like finite volume method (FVM). CFD problems usually require discretization of the problem into a large number of cells/gridpoints (millions and more), therefore cost of the solution favors simpler, lower order approximation within each cell. This is especially true for 'external flow' problems, like air flow around the car or airplane, or weather simulation.A variety of specializations under the umbrella of the mechanical engineering discipline (such as aeronautical, biomechanical, and automotive industries) commonly use integrated FEM in design and development of their products. Several modern FEM packages include specific components such as thermal, electromagnetic, fluid, and structural working environments. In a structural simulation, FEM helps tremendously in producing stiffness and strength visualizations and also in minimizing weight, materials, and costs.FEM allows detailed visualization of where structures bend or twist, and indicates the distribution of stresses and displacements. FEM software provides a wide range of simulation options for controlling the complexity of both modeling and analysis of a system. Similarly, the desired level of accuracy required and associated computational time requirements can be managed simultaneously to address most engineering applications. FEM allows entire designs to be constructed, refined, and optimized before the design is manufactured.

The 3-D finite element method (FEM) involves important computational methods. Many efforts have been undertaken in order to use 3-D FEM (FEMLAB6.2 WITH MATHWORKS). Analytical and Numerical Analysis have been developed for the analysis of the end zones of electrical machine.This paper presents different methodologies based on 3-D geometries using analytical solutions, This method has been implemented in conjunction with various geometry optimization techniques as it provides very fast solutions and has exhibited very good convergence with gradient free algorithms. Interior permanent magnet motors are widely applied to the industry because of many advantages. Also the characteristics of magnetic materials are important to the performance and efficiency of electrical devices.

Tradeoffs between accuracy, robustness, and speed are central issues in numerical analysis, and here they receive careful consideration. The principal purpose of the work is to evaluate the performances of the PMLOM models when implemented in the FEM analysis of electrical machines. The developed methods are applied in an in-house FEM code, specialized for the design and analysis of electrical machines. The FEM simulations and the analysis on axial flux PMLOM, and the numerical results are validated experimentally. The techniques developed for the calculation of integral parameters involve particular assumptions and simplifications and present specific advantages.

LINEAR motors are finding increasing applications in different specific areas like high-speed transport, electric hammers, looms, reciprocating pumps, heart pumps etc. [1]-[7]. They are also well suited for manufacturing automation applications. Therefore, design of energy efficient and high force to weight ratio motors and its performance assessment has become a research topic for quite a few years. The Permanent Magnet Linear Oscillating Motors (PMLOMs) are one of the derivatives of the linear motors in the low power applications having the advantages of higher efficiency. They can be supplied with dc or ac voltages [4]-[7] of which, the dc motors are having better efficiency due to the absence of the core losses.

The motor designed and analyzed in this paper finds the suitability of application in the loads having low frequency and short stroke requirements. One such application is the heart pump, where frequency of oscillation is to be adjusted between 0.5 to 1.5 Hz, with the requirement of variable thrust depending on the condition of the heart under treatment. For analysis of such motors the main task is to determine the essential equivalent circuit parameters, which are its resistances and inductances. The resistances, for the machine, though vary with operating conditions due to temperature, do not affect much on its performance assessment. However, the inductances for these machines are mover position dependent and mostly affect the machine performance. Therefore, determination of these parameters is essentially required for analyzing the machine model. There are several works [6], [9] executed which assumes the machine inductance to be constant for simplicity of the model although different other works [4], [7]and [8] dynamically estimate the inductance through FEM and field analysis and control[10-15] for getting correct results. In this paper, the machine under consideration is an axial flux machine and the mover is having a non-magnetic structure, which is aluminium. Also the rare earth permanent magnets used in the mover are having a relative permeability nearly equal to unity and therefore the magnetic circuit under consideration will be unsaturated due to major presence of air in the flux path. Hence, consideration of constant inductance is quite errorless for such kind of machines, which also conforms to the experimental data shown later. Finally the machine is analyzed with the help of the field equations and solved for forces and resultant flux densities through FEMLAB6.2 WITH MATHWORKS backed by suitable experimental results. A controller using PIC16F877A microcontroller has been developed for its speed and thrust control for successful implementation in the proposed application.

## 2. Machine construction

The construction of the prototype PMLOM is shown in Fig.1 below. Also the dimensional details of the motor are shown in Fig.2. There are two concentric coils on the surface of the stators connected in such polarities that the fluxes for both the coils aid each other to form the poles in the iron parts. The formation of the N and the S poles of the electromagnet of the stator are shown in the Fig.2.

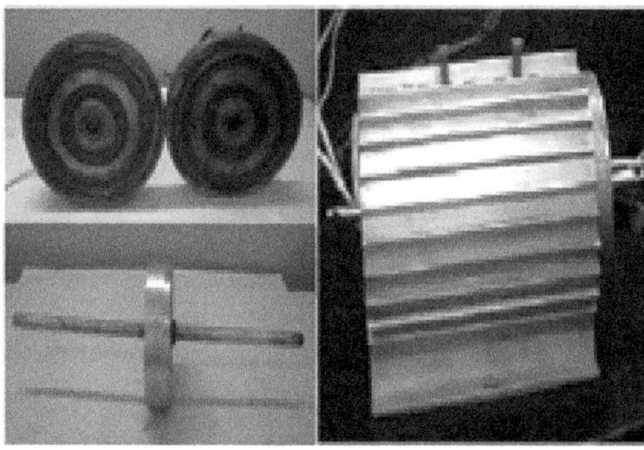

**Figure 1.** Construction details of the developed PMLOM (i) Stators to be mounted on both sides of the mover and (ii) the mover (iii) the PMLOM machine

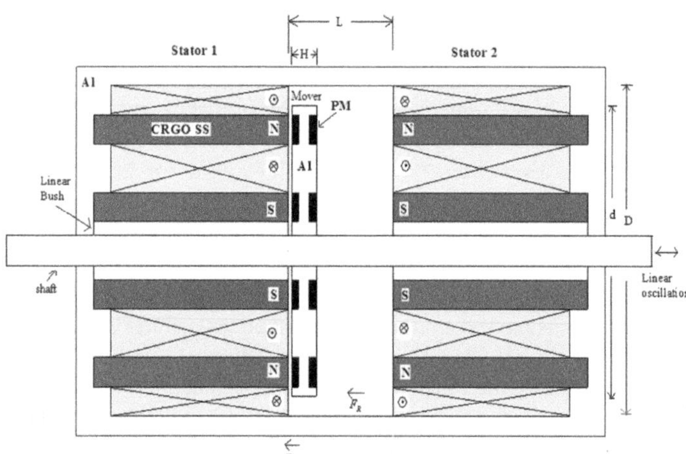


Al – Aluminium material PM-N42 Permanent Magnet

Attraction Force $F_A$ and Repulsion Force $F_R$

Coil 1 – aa′ and bb′ Coil 2 – cc′ and dd′

**Figure 2.** Dimensional details of the developed PMLOM

## 3. Simulation and experimental results

The proposed scheme is simulated under FEMLAB6.2 WITH MATHWORKS environment, which provides a finite element analysis. The machine specification used for both simulation and experiment is given in Table-1.

The classical description of static magnetic fields are provided by Maxwell's equations

$$\nabla \times H = J \tag{1}$$

$$\nabla \cdot B = 0 \tag{2}$$

Where $H$ is magnetic field intensity, $B$ is magnetic flux density and $J$ is the current density of the magnetic field.

Subject to a constitutive relationship between B and H for each material:

$$B = \mu H \tag{3}$$

Where $\mu$ denotes material permeability. Boundary conditions that must be satisfied at the interface between two materials having finite conductivities are,

$$\hat{n} \times (H_1 - H_2) = 0 \tag{4}$$

$$\hat{n} \cdot (B_1 - B_2) = 0 \tag{5}$$

Since the divergence of the curl of any vector must always be zero, it follows from (2) that there exists a so-called magnetic vector potential A such that,

$$B = \nabla \times A \tag{6}$$

Substituting (3) and (6) into (1) and taking a curl on both sides yields

$$\nabla \times \left( \frac{1}{\mu} \nabla \times A \right) = J \tag{7}$$

If

$$J = J\hat{z} \tag{8}$$

Then,

$$A = A\hat{z} \tag{9}$$

Thus, (7) reduces to,

$$-\nabla \cdot \left( \frac{1}{\mu} \nabla A \right) = J \tag{10}$$

The above equation (10) may be written in the expanded form as,

$$\frac{\partial}{\partial x}\left(\frac{1}{\mu}\frac{\partial A}{\partial x}\right)+\frac{\partial}{\partial y}\left(\frac{1}{\mu}\frac{\partial A}{\partial y}\right)=-J \tag{11}$$

This equation (11) represents the *scalar Poisson equation.*

The mover consists of aluminium structure embedded with rare earth permanent magnets with the polarities as shown. The force developed will be attractive on one side and simultaneously repulsive on the other side. These two forces act in the same direction to enhance the total force on the mover, assisting the linear oscillation of the mover cyclically.

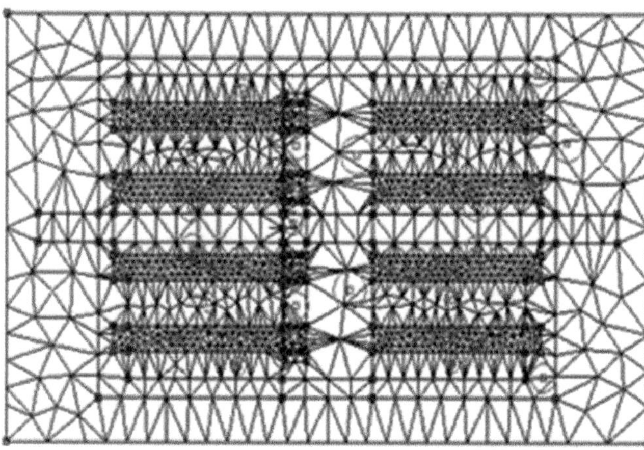

(a)

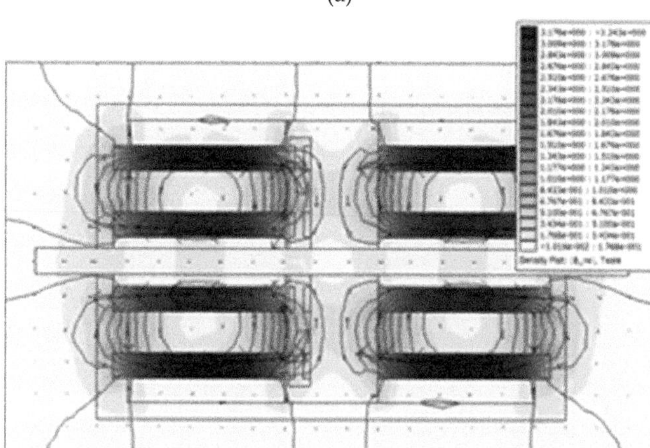

(b)

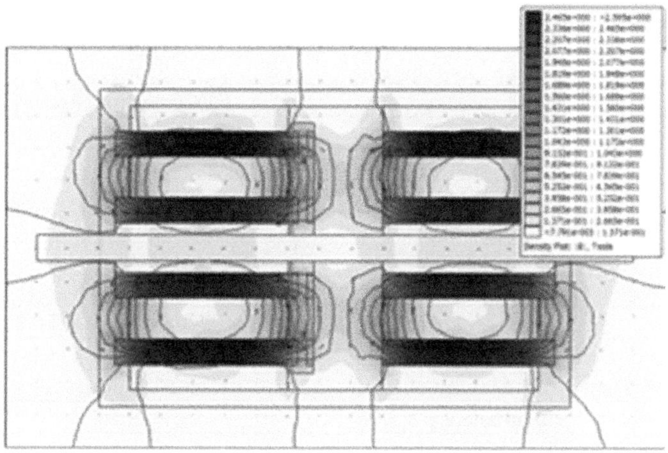

(c)

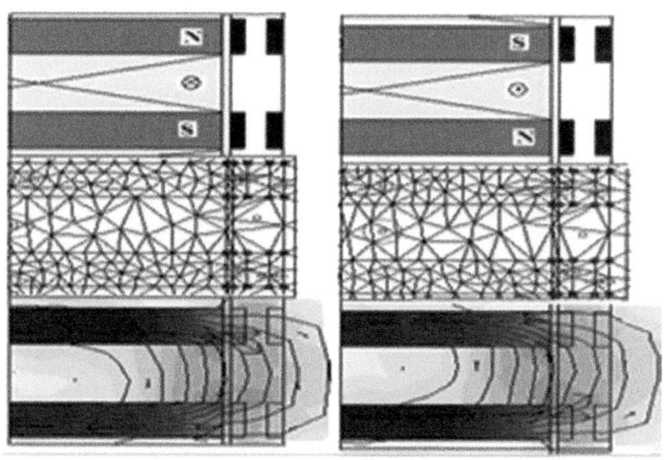

(d)

**Figure 3.** (a) Finite element mesh of PMLOM while mover is oscillating with in Stator 1. (b) Magnetic flux plotting of PMLOM while mover is oscillating with in Stator 1, at 1 Hz, 4Amps, (c) Magnetic flux plotting of PMLOM while mover is oscillating with in Stator 1, at 0 Hz. (d) Finite element Magnetic flux plotting at upper and lower part of the airgap while mover oscillates within stator 1. Now Mover is attracted to the Stator 1

| | |
|---|---|
| Rated Input Voltage | 70V |
| Rated input power | 200 watts |
| Stroke length | 10 mm |
| Outer Diameter (Stator) | 85 mm |
| Stator core type | CRGO Silicon Steel |
| Thickness of lamination | 0.27 mm |
| Stator length | 60 mm |
| Number of turns in Coil aa',cc' | 800 |
| Number of turns in Coil bb',dd' | 400 |
| Coil resistance | 17.8 ohms |
| Slot depth | 45 mm |
| Permanent Magnet Type | Rare Earth N42, Nd-Fe-B |
| Permanent Magnet Length | 2 mm |
| Coercivity | 925000 A/m |
| Remanence | 1.3 T |
| Outer diameter (Mover) | 65 mm |
| Shaft diameter | 8 mm |
| Coil Inductance | 0.18 Henry |

**Table 1.** PMLOM Design Parameters

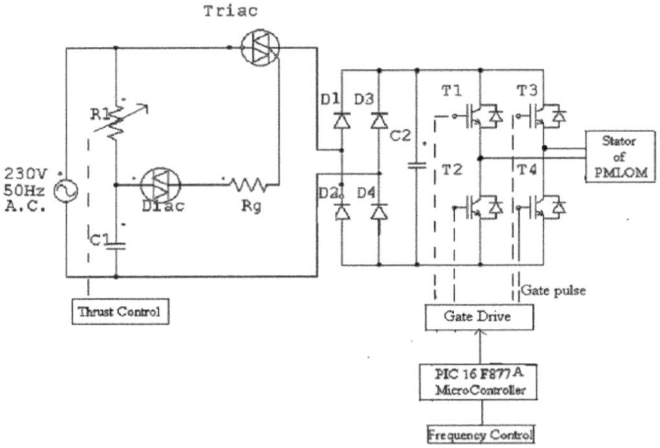


**Figure 4.** Power Circuit of PMLOM

Figure 3(a), shows the FEM mesh configuration for the PMLOM Prototype. Figure 3(b) shows the Magnetic flux plotting of PMLOM while mover is oscillating within Stator 1, at 1 Hz, 4Amps.Figure 3(c) is the corresponding flux plotting of the machine while mover is oscillating within Stator 1, at 0 Hz. Figure 3(d) illustrates the finite element analysis of the PMLOM at the axial airgap. Thus, the geometries of the mover and stator have been accurately discretized with fine meshes. Symmetry was exploited to reduce the problem domain to half of the axial cross section of the overall motor.

The halved longitudinal cross section of the motor has created the calculation area, with Dirichlet boundary conditions (Fig. 3(*d*)). Thus, the magnetic field has been analyzed. For the calculations, material linearity of the NdFeB permanent magnet ( $\mu_r$ =1.048) was supposed. Its coercive force was assumed to be $H_c$ = 925 KA/m and the magnetization vector direction were adopted for the calculations. Very small air gaps compared with the main motor dimensions between permanent magnets and ferromagnetic rings were neglected due to very small magnetic permeability of the permanent magnets, it is acceptable.

The finite-element mesh (Fig. 3(*d*)) is dense in the air gap between stator cylinder sleeve and the mover. The fine mesh is also used near the edges of ferromagnetic parts where the magnetic field is expected to vary rapidly (Fig. 3(*d*)). In order to predict the integral parameters of the PMLOM, it is necessary to analyze the magnetic field distribution in the stator and mover. Obviously, it is possible to optimize the construction by making changes in the stator and mover geometries. The improvements of the structure result from knowledge of the magnetic field distribution. The presented results have been obtained for one variant of the motor construction.

The control block diagram along with the experimental set-up power electronic control circuit is shown in Fig.4. Here the thrust control is provided with the help of phase controlled ac supply which can vary the input voltage. The frequency control is provided with the help of a low cost and commercially available microcontroller PIC16F877A.

The set up is reliable and provides a scope for portability to any remote place. Fig. 5 shows the plot of the input voltage and current of the machine at 5 Hz. From which the assumption of constant inductance for the machine can be well validated. Fig. 6 shows the characteristics plot of input power, voltage and force as a function of current for the machine taken at a frequency of 1Hz. Figure 7 shows Force at different axial airgap length. Force observed by measurement is compared with the theoretical Force value and shown in fig. 8.

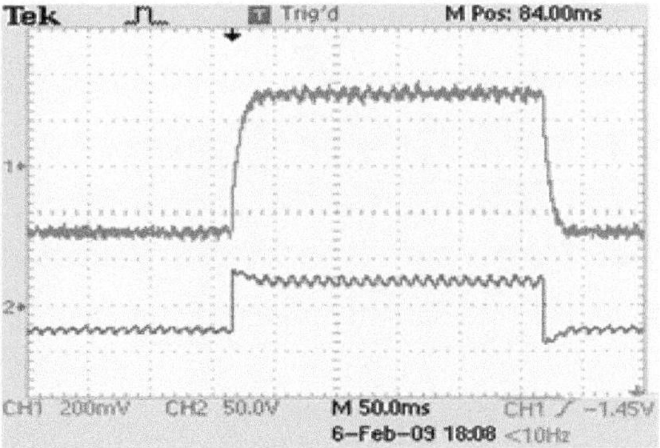


**Figure 5.** Current waveform of PMLOM taken from Tektronix make Storage Oscilloscope

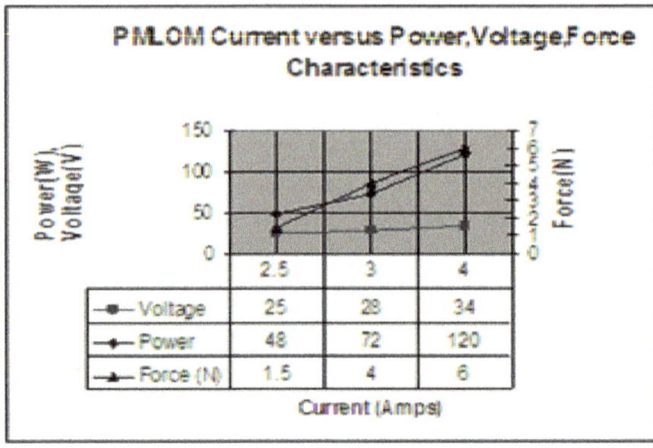


**Figure 6.** Measured Coil current versus Power( (W).Voltage(V),Force(N) Characteristics of PMLOM

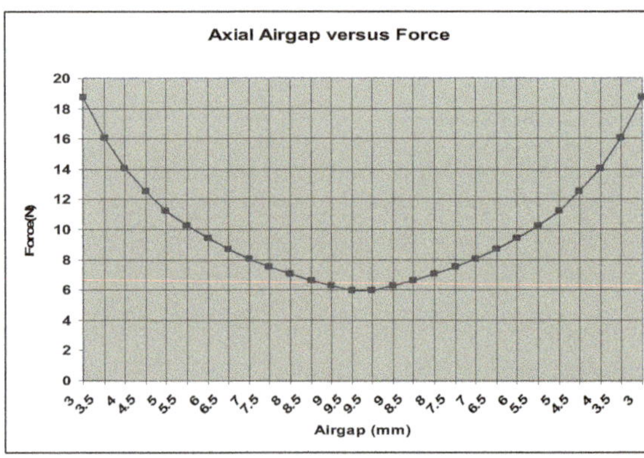


**Figure 7.** Axial Airgap Length versus Force

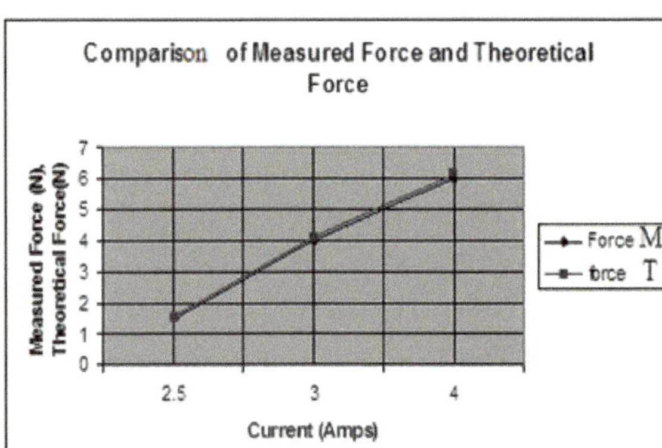


**Figure 8.** Comparison of measured Force Versus Theoretical Force

## 4. Conclusions

A simple control method along with the development of an axial flux PMLOM suitable for low frequency and short stroke application is presented. Analytical solution to the forces and determination method of the integral parameters of a PMLOM are shown. Finite element method with FEMLAB6.2 WITH MATHWORKS is used for the field analysis of the different values of the exciting current and for variable mover position. Computer simulations for the magnetic field distribution, forces are given. To obtain experimentally the field distribution and its integral parameters, a physical model of the motor together with its electronic controller system has been developed and tested. The Prototype has been operated in the oscillatory mode with small loads at low frequency up to 5 Hz. The theoretically calculated results are compared with the measured ones and found a good conformity.

## Author details

Govindaraj Thangavel
*Department of Electrical and Electronics Engineering, Muthayammal Engineering College, India*

## 5. References

[1] Kou Baoquan, Li Liyi, and Zhang Chengming, "Analysis and Optimization of Thrust Characteristics of Tubular Linear Electromagnetic Launcher for Space-Use,"*IEEE Trans.Magn., vol.. 45, no. 1,* pp. 250- 255,Jan.2009

[2] Ge Baoming, Aníbal T. de Almeida, and Fernando J. T. E.Ferreira, "Design of Transverse Flux Linear Switched Reluctance Motor,"*IEEE Trans. Magn., vol 45, no.1,* pp.113-119 Jan 2009

[3] H. D. Chai, *Electromechanical Motion Devices*. Upper Saddle River, NJ: Prentice Hall, 1998.

[4] G. Kang, J. Hong, and G. Kim, "Design and analysis of slotless-type permanent- magnet linear brushless motor by using equivalent magnetizing current," *IEEE Trans. Ind. Appl.*, vol. 37, no. 5, pp. 1241–1247,2001.

[5] S. A. Nasar and I. Boldea, *Linear Electric Motors*..Englewood Cliffs, NJ: Prentice-Hall, 1987.

[6] B. Tomczuk and M. Sobol, "Influence of the supply voltage on the dynamics of the one-phase tubular motor with reversal motion," in *Proc. 39th Int. Symp. Electrical Machines—SME'2003*, Gdansk/Jurata, Poland, Jun. 9–11, 2003, pp. 417–426

[7] N. Sadowski, R. Carlson, A. M. Beckert, and J. P. A.Bastos, "Dynamic modeling of a newly designed linear Actuator using 3D edge elements analysis," *IEEE Trans. Magn.*, vol. 32, no. 3, pp. 1633–1636, May 1996.

[8] D. G. Taylor and N. Chayopitak, "Time-optimal position control of electric motors with steady-state temperature constraints," in *Proc.IEEE Int. Symp. Industrial Electronics*, Montreal, QC, Canada, Jul 2006, pp. 3142– 3146.

[9] S. Vaez-Zadeh and A. Isfahani, "Multi-objective design optimization of air-core linear permanent-magnet synchronous motors for improved thrust and low magnet consumption," *IEEE Trans. Magn.*, vol.42, no. 3, pp. 446–452, 2006.

[10] Govindaraj T, Debashis Chatterjee, and Ashoke K. Ganguli, "A Permanent Magnet Linear Oscillating Motor for Short Strokes,"in *Proc.Int .Conf. Electrical Energy Systems & Power Electronics in Emerging Economies- ICEESPEEE 09, SRM University, India,* Apl. 16-18, 2009, pp. 351- 355

[11] Govindaraj T, Debashis Chatterjee, and Ashoke K. Ganguli,"Development, Finite Element Analysis and Electronic Controlled Axial Flux Permanent Magnet Linear Oscillating Motor direct drive suitable for short Strokes" in *Proc. Int. Conf. Control,Automation, Communication and Energy Conservation -INCACEC 2009, Kongu Engineering College, India* Jun.4–6, 2009, pp. 479–483.

[12] Govindaraj T, Debashis Chatterjee, and Ashoke K. Ganguli,"FE Magnetic Field Analysis Simulation Models based Design, Development, Control and Testing of An Axial Flux Permanent Magnet Linear Oscillating Motor" in *Proc. Int. Conf. Electrical andElectronics Engineering,ICEEE2009, Int. Association of Engineers,World Congress on Engineering 2009vol 1,* London, United Kingdom, July 1-3, 2009

[13] Govindaraj T, Debashis Chatterjee, and Ashoke K. Ganguli, "Development, Analysis and Control of an Axial Flux Permanent Magnet Linear Oscillating Motor suitable for Short Strokes" in *Proc. 2009 IEEE Int. Sym. on Industrial Electronics,IEEE ISIE 2009,* Seoul Olympic Parktel, Seoul, Korea, July 5-8, 2009, 29-34

[14] Govindaraj T, Debashis Chatterjee, and Ashoke K. Ganguli,"Development, Control and Testing of a New Axial Flux Permanent Magnet Linear Oscillating Motor using FE Magnetic Field Analysis Simulation Models", *Proc. 2009 Int. Conf. on Mechanical and Electronics Engineering, ICMEE2009,Int. Association of Computer Science and Information Technology, IACSIT,* Chennai, India, July 24-26, 2009, pp 191 - 195

[15] D.G. Holmes, T. A. Lipo, B. P. McGrath, and W.Y. Kong," Optimized Design of Stationary Frame Three phase AC Current Regulators"*IEEE Trans. Power Electronics, vol. 24, no. 11,* pp 2417-2426, Nov 2009.

Chapter 6

# Finite Element Analysis of Stationary Magnetic Field

Elena Otilia Virjoghe, Diana Enescu, Mihail-Florin Stan and Marcel Ionel

Additional information is available at the end of the chapter

## 1. Introduction

Computer-aided analysis of field distribution for evaluating electromagnetic device or component performance has become the most advantageous way of design. Analytical methods have limited uses and experimental methods are time intensive and expensive (Morozionkov et al., 2008).

The problems of magnetic fields calculation are aimed at determining the value of one or more unknown functions for the field considered, such as magnetic field intensity, magnetic flux density, magnetic scalar potential and magnetic vector potential. As the field has infinite points, the function values are in infinite number.

Physical phenomena of electromagnetic nature are described by Maxwell's equations from the mathematical point of view. These are differential equations with the given boundary conditions. By means of them, the exact solution of the problem is obtained. In this way, the value of function or functions in any point of the studied range is calculated. This represents the analytical way for solving the problems.

Analytical methods (conformable representation method, method of separation of the variables, Green function method) are applied to solve relatively simple problems. Problems which occur in practice are often complex concerning the geometric construction, material heterogeneity, loading conditions, boundary conditions, so that the integration of differential equations is difficult or sometimes impossible. In this case, the analytical solution can be carried out only by creating a simplified model so that the integration of differential equations is possible. Therefore, an exact solution for a simplified model can be obtained (Gârbea, 1990).

It is sometimes preferable to obtain, instead of the exact solution of the simplified model, an approximate solution of the real problem. Approximate solutions which are obtained by numerical methods reflect better the reality than exact solutions of a simplified model.

The software package ANSYS can be used for investigation of the magnetic field distribution (the magnetic flux density, the magnetic field intensity and the magnetic vector potential) and basic electromagnetic characteristics (inductance and electromagnetic force). A typical magnetic field problem is described by defining the geometry, material properties, currents, boundary conditions, and the field system equations. The computer requires the input dates, the numerical solution of the field equation and output of desired parameters. If the values are found unsatisfactory, the design modified and parameters are recalculated. The process is repeated until optimum values for the design parameters are obtained.

The ANSYS program is based on the finite element method (FEM) for solving Maxwell's equations and can be used for electromagnetic field modeling, where the field is electrostatics, magnetostatics, eddy currents, time-invariant or time-harmonic and permanent magnets (ANSYS Documentation).

The finite elements method assures sufficient accuracy of electromagnetic field computation and very good flexibility when geometry is modeled and field sources are loaded.

## 2. The fundamental relations of the stationary magnetic field

In this section, we discuss the particular forms of the electromagnetic field theory laws for the magnetic stationary field. We consider the models of the magnetic induction versus magnetic field intensity (*B-H*) relation, passing conditions through discontinuity surfaces, the enunciation of stationary magnetic field (the sources of the field, boundary conditions), the enunciation of scalar magnetic potential - magnetostatic field problems (Dirichlet conditions, Neumann conditions) and the enunciations using the magnetic vector potential (stationary magnetic field problems). The general formulation of the uniqueness conditions gets particular forms, adapted to some geometrical configurations (plane-parallel fields, with rotation symmetry, etc.).

Depending on the relation between the magnetic induction and the intensity of the magnetic field, a few types of materials are distinguished, the most important being linear and isotropic materials, linear and non-isotropic materials, linear and non-isotropic materials, non-linear and isotropic materials, without permanent magnetization, non-linear and non-isotropic materials, materials with hysteresis.

Non-linear and isotropic materials, without permanent magnetization, are ferromagnetic materials, which are frequently used in the production of electric equipment.

### 2.1. Particular forms of the electromagnetic field theory laws for the stationary magnetic field

The stationary magnetic field is established by non-moving, permanently magnetized bodies and by non-moving connecting wires crossed by direct current (Mocanu, 1981). Fundamental magnetic field relationships result by customizing the general laws and material laws of the electromagnetic field in the following conditions: bodies are non-moving $\vec{v}=0$ and the electric and magnetic quantities are invariable in time, $\frac{\partial(.)}{\partial t}=0$. A stationary magnetic field in a conducting domain satisfies the following system of equations:

- the magnetic circuit law (Ampère's theorem)

$$rot\,\vec{H}=\vec{J} \quad (1)$$

- the magnetic flux law (local form)

$$div\,\vec{B}=0 \quad (2)$$

- the connection law in magnetic field

$$\vec{B}=\mu_0\left(\vec{H}+\vec{M}\right) \quad (3)$$

- the temporary magnetization law

$$\vec{M}_t=\vec{M}_t\left(\vec{H}\right). \quad (4)$$

Here, $\vec{H}$ is the magnetic field intensity, $\vec{J}$ is the total current density, $\vec{B}$ is the magnetic induction, quantity $\vec{M}_t$ is called the temporary magnetization of the material and $\mu_0=4\pi\cdot10^{-7}$ $\left[H\cdot m^{-1}\right]$ is the vacuum permeability.

Magnetostatics is the branch in electromagnetism that studies the stationary magnetic states that do not accompany the conduction electric currents. This magnetic field is produced by permanent magnets ($\vec{J}=0\,;\vec{v}=0$).

## 2.2. Models for the *B-H* relation

Depending on the relation between the magnetic induction $\vec{B}$ and the magnetic field intensity $\vec{H}$, a few types of materials are distinguished (Andrei et al. 2012).

### 2.2.1. *Linear and isotropic materials*

The most important type of materials consists of the linear and isotropic materials, in which:

$$\vec{B}=\mu\vec{H}+\mu_0\vec{M}_p=\mu\vec{H}+\vec{I}_p \quad (5)$$

where $\vec{M}_p$ is called the permanent magnetization and $\vec{I}_p$ is the magnetic polarization. This category includes the materials for which the temporary magnetization law is (Răduleţ, 1975) :

$$\vec{M}_t=\chi_m\vec{H} \quad (6)$$

where $\chi_m$ is called the magnetic susceptibility, representing a dimensionless and constant scalar quantity.

In the absence of permanent magnetization ($\vec{M}_p$ =0, $\vec{I}_p$ =0) the relation becomes:

$$\vec{B} = \mu\vec{H} \tag{7}$$

$$\vec{H} = \nu\vec{B}\,. \tag{8}$$

The quantity $\mu$ is the magnetic permeability and $\nu = \dfrac{1}{\mu}$ is called the reluctivity. The $\vec{B}$ and $\vec{H}$ vectors are collinear.

### 2.2.2. *Linear and non-isotropic materials*

In these materials, the $\vec{B}$ and $\vec{H}$ vectors are not, generally, collinear, but the connection between them remains linear.

For some crystalline materials, the dependence between $\vec{M}_t$ and $\vec{H}$ is linear, but each component of the temporary magnetization depends on all components of the magnetic field.

The relation between them can be written, in the absence of permanent magnetization, under the form:

$$\vec{B} = \overline{\overline{\mu}}\vec{H} \tag{9}$$

where $\overline{\overline{\mu}}$ is a tensor. In Cartesian coordinates, the relation becomes:

$$\begin{bmatrix} B_x \\ B_y \\ B_z \end{bmatrix} = \begin{bmatrix} \mu_{xx} & \mu_{xy} & \mu_{xz} \\ \mu_{yx} & \mu_{yy} & \mu_{yz} \\ \mu_{zx} & \mu_{zy} & \mu_{zz} \end{bmatrix} \cdot \begin{bmatrix} H_x \\ H_y \\ H_z \end{bmatrix} \tag{10}$$

The permittivity matrix is symmetrical ($\mu_{ij} = \mu_{ji}$) and positively defined. In these conditions, there are three orthogonal directions, called main directions, with respect to which the relation between $\overline{B}$ and $\overline{H}$ becomes (Hănţilă, 2004):

$$\begin{bmatrix} B_1 \\ B_2 \\ B_3 \end{bmatrix} = \begin{bmatrix} \mu_1 & 0 & 0 \\ 0 & \mu_2 & 0 \\ 0 & 0 & \mu_3 \end{bmatrix} \cdot \begin{bmatrix} H_1 \\ H_2 \\ H_3 \end{bmatrix} \tag{11}$$

### 2.2.3. *Non-linear and isotropic materials, without permanent magnetization*

In these materials, the $\vec{B}$ and $\vec{H}$ vectors are collinear, but the relation between them is non-linear:

$$B = f(H) \quad f : R^3 \to R^3\,. \tag{12}$$

This is, usually, the behavior of ferromagnetic materials, which are frequently used in the production of electric equipment.

### 2.2.4. *Hysteresis materials*

In hysteresis materials, the instantaneous value of the magnetic induction depends not only on the value of the intensity of the magnetic field, but also on the previous evolution of these quantities.

Assume that the magnetic field intensity is gradually reduced after following the first magnetization curve *OA*, corresponding to a value $+H_{max}$ (Figure 1).

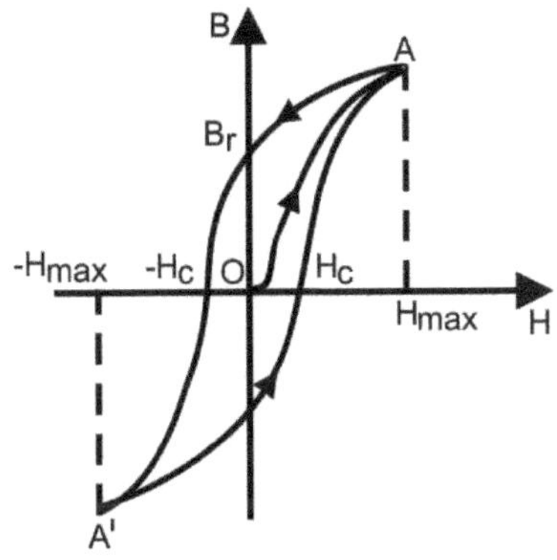


**Figure 1.** The *B-H* relation for a hysteresis material

The curve obtained during the magnetic field intensity reduction differs from the first magnetization curve. When *H* is null, the magnetic induction has a value different from zero called the residual magnetic induction:

$$B_r = \mu_0 \cdot M_r \quad (13)$$

where $M_r$ is the residual magnetization.

For further reduction of the magnetic induction, the sense of magnetic field intensity is changed (as well as the sense of magnetization current), with respect to the initial one.

The magnetic field intensity necessary to compensate the magnetic induction is called the coercitive field $H_c$. Increasing the field in the contrary sense to $-H_{max}$ and then returning to the values of *H* up to $H_{max}$, the hysteresis cycle is obtained. By repeating several times the magnetization cycle between the limits $+H_{max}$ and $-H_{max}$, a closed curve and a stabilized cycle are obtained, with the reversal points *A* and *A′* symmetrical with respect to the origin of the coordinate system.

It is important to mention that in the case of a periodic magnetization, the existence of the hysteresis cycle leads to energy losses that occur in the ferromagnetic core as heat. These energy losses are called the hysteresis iron losses (Şora, 1982).

*2.2.5. Non-linear and non-isotropic materials*

In these materials, the $\vec{B}$ and $\vec{H}$ vectors are not, generally, collinear and the relation between them is non-linear.

## 2.3. The magnetic vector potential

The magnetic vector potential is a vector field, which does not have a specific physical meaning. Its utilization allows simplification of the mathematical approach of many physical problems.

The condition $div\,\vec{B}=0$, which expresses the continuity of the magnetic flux, is identically satisfied if the $\vec{B}$ vector is expressed under the form of an auxiliary vector $\vec{A}$, called the magnetic potential vector:

$$\vec{B} = rot\,\vec{A} \tag{14}$$

The potential vector $\vec{A}$ is univocally determined only after $div\,\vec{A}=0$ is chosen. The choice of the value for $\vec{B}$ is called the calibration of the vector potential and the respective condition is called calibration condition. Depending on the context, one can adopt convenient *calibration conditions*. One of the most used calibration conditions is the Coulomb condition:

$$div\,\vec{A} = 0 \tag{15}$$

If the calculation of the magnetic flux through an open surface is expressed by means of magnetic induction, then the magnetic vector potential must be taken into account by the Stokes' theorem (Moraru, 2002).

The *magnetic flux* through a surface $S_\Gamma$ bounded by a contour Γ can be computed as a contour integral of the vector potential:

$$\Phi_{S_\Gamma} = \int_{S_\Gamma} \vec{B}\,d\vec{A} = \int_{S_\Gamma} rot\,\vec{A}\,d\vec{A} = \oint_\Gamma \vec{A}\,d\vec{l} \tag{16}$$

The magnetic flux through the surface $S_\Gamma$ is equal to the line integral of the magnetic vector potential along the contour $\Gamma$ on which this surface is supported. Equation (16) relieves the fact that the value of a magnetic flux does not depend on the surface shape, as it is computed only by considering the contour on which that surface is supported. Let us consider a material with linear magnetic properties and without permanent magnetization, for which $\vec{B}=\mu\vec{H}$. From the magnetic circuit law, it results that (Hănţilă, 2004):

$$rot\,\vec{H} = rot\left[\frac{1}{\mu} rot\,\vec{A}\right] = \vec{J} \tag{17}$$

In linear and homogeneous mediums, where μ is constant:

$$rot\,rot\,\vec{A} = \mu\vec{J} = grad\,div\,\vec{A} - \Delta\vec{A} \tag{18}$$

the magnetic vector potential verifies the *Poisson's* vector equation:

$$\Delta\vec{A} = -\mu\,\vec{J} \tag{19}$$

and if $\vec{J} = 0$, it verifies the *Laplace's* equation:

$$\vec{A} = 0 \tag{20}$$

Solving Equations (19) and (20) requires the boundary condition to be known. The vector equations are divided after the Cartesian coordinates in scalar equations of Poisson type

$$\Delta A_x = -\mu\,J_x \qquad \Delta A_y = -\mu\,J_y \qquad \Delta A_z = -\mu\,J_z \tag{21}$$

respectively, scalar equations of Laplace type

$$\Delta A_x = 0 \qquad \Delta A_y = 0 \qquad \Delta A_z = 0\,. \tag{22}$$

The integral of Equation (19) in all space is determined by using the scalar forms (21).

The magnetic vector potential of the filiform circuit with current *i* is expressed as:

$$\vec{\mathbf{A}} = \frac{\mu \cdot i}{4\pi}\oint_{\Gamma}\frac{\mathrm{d}\vec{l}}{R}, \tag{23}$$

$d\vec{l}$ is line unit vector.

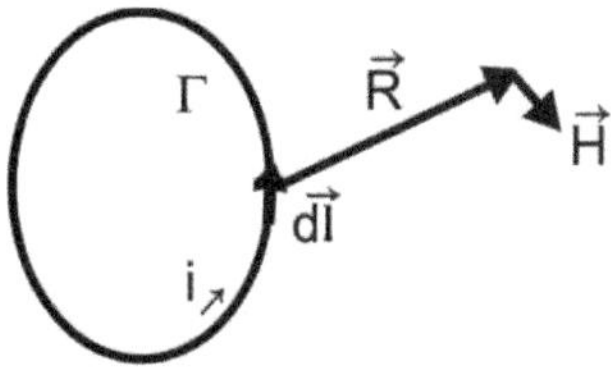


**Figure 2.** Biot–Savart–Laplace relation for filiform conductors

The magnetic field intensity is (Figure 2):

$$\vec{\mathbf{H}} = \frac{\mathrm{rot}\,\vec{\mathbf{A}}}{\mu} = \frac{i}{4\pi}\oint_{\Gamma}\frac{\mathrm{d}\vec{l}\times\vec{\mathbf{R}}}{R^3}. \tag{24}$$

The Biot-Savart-Laplace relation becomes:

$$\vec{B} = \frac{\mu}{4\pi}\int_{D_\infty}\frac{\vec{J}\times\vec{R}}{R^3} \tag{25}$$

## 2.4. The scalar magnetic potential

The magnetic field is not irrotational for a circuit with current flow, therefore this can be deduced fom a scalar potential. But the rotor of magnetic field intensity is equal to zero if there is no current which flows, $\vec{J}=0$. If there is no conductor in the considered space, the following equation is available (Mocanu, 1981):

$$rot\,\vec{H}=0 \tag{26}$$

Therefore, $\vec{H}$ can be deduced by a scalar potential:

$$\vec{H}=-grad\,V_m \tag{27}$$

where is $V_m$ is the scalar magnetic potential.

In the presence of some conductors crossed by electric current, the scalar magnetic potential is not uniform:

$$\int_1^2 \vec{H}d\vec{l}=i=-\int_1^2 \nabla V_m dl=-\int_1^2 dV=V_{m1}-V_{m2}\neq 0 \tag{28}$$

For uniformity, a cut can be introduced an arbitrary surface bounded by the contour crossed by current (Figure 3).

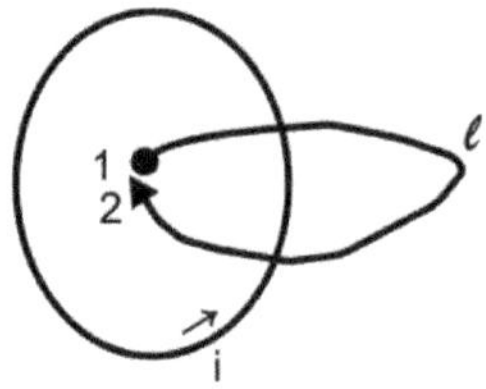


**Figure 3.** Cut in order to uniform the scalar magnetic potential

In the absence of permanently magnetized bodies, the partial derivative equation of the scalar magnetic potential is deducted from the magnetic flux law:

$$\vec{B}=-\mu\nabla V_m \qquad div\vec{B}=0\rightarrow -div\left(\mu\nabla V_m\right)=0 \tag{29}$$

In homogeneous materials (where μ=cost), the Laplace equation is obtained:

$$\nabla V_m=0 \tag{30}$$

This expression is used to determine the scalar magnetic potential of the magnetic field produced by a filiform circuit crossed by electric current.

Applying the magnetic circuit law for a closed curve which surrounds the conductor, the scalar magnetic potential is written as:

$$V_m = \frac{i}{4\pi}\Omega \tag{31}$$

where Ω is the solid angle under which the Γ contour is seen from the point where the field is calculated:

$$\Omega = \int_{S_\Gamma} \frac{R}{R^3} dS \tag{32}$$

In this case, the magnetic field intensity $\vec{H}$ is:

$$\vec{H} = -\frac{1}{4\pi} i \nabla \Omega \tag{33}$$

## 2.5. Energy of the stationary magnetic field

Magnetic energy is located in a magnetic field with a volume density $w_m$ whose expression is (Şora, 1982):

$$w_m = \int_0^B \vec{H} d\vec{B} \tag{34}$$

If the medium is linear (μ=constant), then:

$$HdB = Hd(\mu H) = d\left(\frac{\mu H^2}{2}\right) \tag{35}$$

In this case, the following expressions are obtained:

$$w_m = \frac{\mu H^2}{2} = \frac{B^2}{2\mu} = \frac{\vec{H}\vec{B}}{2} \tag{36}$$

$$W_m = \frac{1}{2}\int_v \vec{H}\vec{B} dv \tag{37}$$

Assuming that inside the field limited by a closed surface Σ and considering an isotropic medium, the magnetic energy can be written as:

$$div\left(\vec{A} \cdot \vec{H}\right) = \vec{H} rot \vec{A} - \vec{A} rot \vec{H} \tag{38}$$

taking into account the vector operation.

In other words, the magnetic energy is:

$$W_m = \frac{1}{2}\int_v \vec{H} rot \vec{A} dv = \frac{1}{2}\int_v div\left(\vec{A} \cdot \vec{H}\right) dv + \frac{1}{2}\int_v \vec{A}\vec{J} dv \tag{39}$$

Applying the Gauss-Ostrogradski's theorem to the first term on the right-hand side, the following expression is obtained:

$$W_m = \frac{1}{2}\int_{\Sigma}(\vec{A}\cdot\vec{H})d\vec{s} + \frac{1}{2}\int_{v}\vec{A}\vec{J}dv \tag{40}$$

## 2.6. Generalized forces in the stationary magnetic field

In the case of the stationary magnetic field, the general expressions of the generalized forces $X_k$ associated to a generalized coordinate $x_k$ are given by one of the two generalized forces theorems in a stationary magnetic field (Timotin, 1970):

$$X_k = -\left[\frac{\partial W_{em}}{\partial x_k}\right]_{\Phi=ct} \tag{41}$$

$$X_k = \left[\frac{\partial W_{em}}{\partial x_k}\right]_{i=ct} \tag{42}$$

in which the transformations are supposed to be done at constant fluxes on any surface, respectively to constant currents through any conducting contour.

## 2.7. Uniqueness theorems of the solutions of the equations of stationary and magnetostatic magnetic fields

### 2.7.1. *The enunciation of stationary and magnetostatic magnetic field*

In stationary magnetic field problems, the electric currents distribution (the $\vec{J}$ field) is supposed to be known (for example, by solving a stationary electrokinetic stationary regime, in the case of massive conductors, or by indicating the value of the current through the coils in the domain that is being studied).

In magnetostatic field problems, the sources of the field are represented by the distribution of the permanent magnetization ( $\vec{M}_p$ or the permanent magnetic polarization), which are supposed to be known (Andrei et al, 2012).

In conformity to the general uniqueness theorem of the solutions of the stationary magnetic fields equations, the solution of the electromagnetic field equations in a domain $D_\Sigma$ bounded by closed surface $\Sigma = S_H \cup S_B$ is uniquely determined by the following uniqueness conditions:

The electric currents distribution in the domain:

$$J(P);\quad \forall P \in D_\Sigma \tag{43}$$

The distribution of the permanent magnetization:

$$M_p(P);\quad \forall P \in D_\Sigma \tag{44}$$

Boundary conditions, that can be of the following types (Figure 4):

- tangent component of the magnetic field intensity on the surface $S_H$:

$$H_t(P) = f(p); \quad \forall P \in S_\Sigma \tag{45}$$

- normal component of the magnetic induction on the surface $S_B$:

$$B_n(P) = g(p); \quad \forall P \in S_\Sigma \tag{46}$$

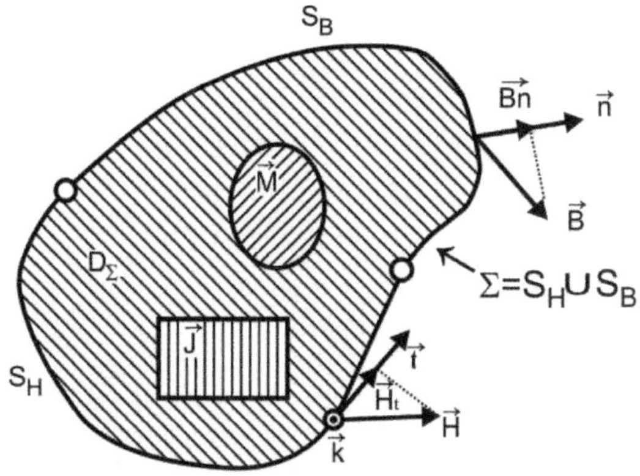


**Figure 4.** The uniqueness theorem for the stationary magnetic field (Andrei et al. 2012)

The theorem stands for linear materials, or for non-linear materials, but having *B-H* monotone magnetization characteristics. The case of materials which have hysteresis is not included. Particular forms can be deducted from this general formulation, expressed by field potentials.

### *2.7.2. The enouncing by scalar magnetic potential (magnetostatic field problems)*

In magnetostatic regime problems, the sources of the magnetic field are represented by the permanent magnetization of the bodies (Andrei et al., 2012).

The boundary conditions, expressed by the scalar magnetic potential, are of the following types:

a. Dirichlet Conditions, which consist of imposing values for the scalar magnetic potential at the points on the surface, denoted by $S_D$:

$$V(P) = f(P); \quad \forall P \subset S_D \tag{47}$$

These conditions imply knowing the value of the tangent component of the intensity of the magnetic field in the respective points, which is equal to the derivative by the tangent direction of the scalar magnetic potential (Flueraşu & Flueraşu, 2007).

b. Neumann Conditions, which consist of imposing the values of the derivative of the scalar magnetic potential iny the direction of the normal to the surface, denoted by $S_N$. Practically, this type of conditions imposes the normal component of the magnetic induction in the respective points on the surface:

$$B_n = -\mu \frac{dV}{dn} = g(P); \quad \forall P \subset S_N \tag{48}$$

c. mixed conditions, that consist of imposing a condition in the form of a linear combination between the two above condition types, on a portion $S_M$ of the surface.

$$\alpha V + \beta \frac{dV}{dn} = h(P); \quad \forall P \subset S_M \tag{49}$$

## 3. Finite element analysis

### 3.1. Triangular finite elements

The first step in solving the problems using the Finite Element Method (FEM) begins by dividing the analysis area in finite elements, as well as the choice of the finite element type. Currently, a wide range of finite elements is used, but their classification, their description, as well as their criteria presentation for choosing adequate finite element types does not represent the subject of this chapter. In the presented application, the triangular finite element with three nodes is used. At the same time with the choice of finite element type, the shape functions are chosen, so that the description of finite elements is followed by the associated shape function presentation. Concerning the shape functions, the interpolating polynomials are mainly used due to the facility in their derivation and their integration. The interpolation on a triangle supposes a shape or interpolating function which links the nodal values (triangle vertices). An approximation of the solution of the magnetic vector potential $\vec{A}$ is allowed at the level of each triangular element „e" (Figure 5), according to the following interpolating polynomial (Stammberg, 1995):

$$\alpha_k = a_k + b_k \cdot x + c_k \cdot y \tag{50}$$

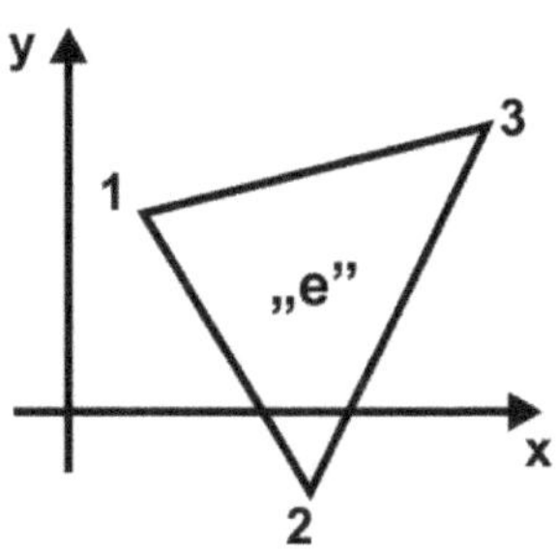


**Figure 5.** The triangular element „e"

The shape function coefficients $a_k$, $b_k$, and $c_k$ are called the generalized coordinates. These coefficients are constant because they depend on the constant coordinates of the nodes only.

The values of the shape functions vary between 0 and 1. They are equal to 1 in node $k$ and liniarly decrease in the elements adjacent to this node, being null in the rest of the nodes and elements (Figure 6). Thus:

$$\begin{aligned} &\alpha_k\left(x = x_k, y = y_k\right) = 1 \qquad k = 1,2,3 \\ &\alpha_k\left(x = x_i, y = y_i\right) = 0 \qquad i \neq k \end{aligned} \tag{51}$$

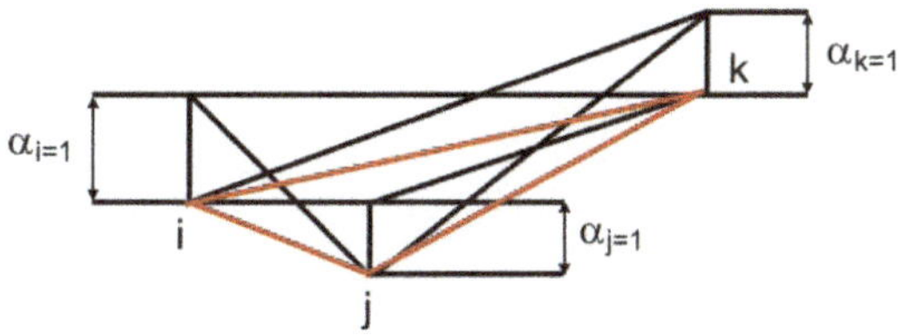


**Figure 6.** Graphical representation of the shape function

The shape function $\alpha_1$ for node 1 results from the formula (Stammberg, 1995):

$$\begin{pmatrix} 1 & x_1 & y_1 \\ 1 & x_2 & y_2 \\ 1 & x_3 & y_3 \end{pmatrix} \cdot \begin{pmatrix} a_1 \\ b_1 \\ c_1 \end{pmatrix} = \begin{pmatrix} 1 \\ 0 \\ 0 \end{pmatrix}. \tag{52}$$

The shape function coefficients are:

$$\begin{aligned} a_k &= \frac{x_{i+1} \cdot y_{i+2} - x_{i+2} \cdot y_{i+1}}{D} \\ b_k &= \frac{y_{i+1} - y_{i+2}}{D} \\ c_k &= \frac{x_{i+2} - x_{i+1}}{D} \end{aligned} \tag{53}$$

where:

$$D = x_2 y_3 - x_3 y_2 - x_1 y_3 + x_1 y_2 + x_3 y_1 - x_2 y_1 \tag{54}$$

and the index $i$ takes the values by circular permutations in nodal set of an element „e", in clockwise order. Writing with *Se*, the area of element „e":

$$S_e = \frac{1}{2} \begin{vmatrix} 1 & x_1 & y_1 \\ 1 & x_2 & y_2 \\ 1 & x_3 & y_3 \end{vmatrix} \tag{55}$$

Considering:

$$D = 2S_e \tag{56}$$

The magnetic vector potential in an arbitrary point (*x,y,z*) is obtained with the following equation:

$$A(x,y,z) = \sum_{k=1}^{n(e)} \alpha_k(x,y,z) A_k \tag{57}$$

where:

$\alpha_k$ - shape function

$n(e)$ - number of nodes on element

$A_k$ - magnetic vector potential of node $k$

### 3.2. FEM application for two-dimensional problems of stationary magnetic field

Finite element methods (FEM) use most of the times a variation principle. According to the variation computation, solving a differential equation in a field and under certain boundary conditions is equivalent with minimizing, in that field, a functional corresponding to the differential equation with its boundary conditions. A functional integral is an integral expression, a function that depends on the unknown functions. The functional integral has a finite value.

The problem concerning solving the system of differential equations of the electromagnetic field with some boundary conditions is equivalent with the problem of finding a function which gives the integral minimum by which the energy system is expressed.

Let's consider the energy functional associated with the arbitrary three-dimensional field D:

$$\Im = \int_{\mathcal{D}} \left\{ \left( \int_0^{\vec{E}} \vec{D} d\vec{E} - \int_0^{\vec{B}} \vec{H} d\vec{B} \right) + \left( \vec{J}\vec{A} - \rho_V V \right) \right\} d\mathcal{D} \tag{58}$$

where $\vec{D}, \vec{E}, \vec{B}, \vec{H}$ are the vectors associated with electric and magnetic fields, $\vec{A}$ is the magnetic vector potential, $V$ is the scalar electric potential, $\vec{J}$ is the density vector of conduction electric current and is the volume density of electric charge (Silvester & Ferrari, 1996).

The first parenthesis of the integrand represents the difference between the volume density of the electric and magnetic energy. The second parenthesis represents the difference between the volume density of interaction energies between the conduction current and magnetic field, as well as between the electric charge and the electric field. The interaction energies are equal to the work done by the field forces in order to bring the current density, respectively electric charge, from infinity, where the potentials are considered $(\vec{A} = 0, V = 0)$,, to the states characterized by the values $\vec{A}$ and $V$.

The electromagnetic potentials $\vec{A}$ and $V$ define the vectors $\vec{E}$ and $\vec{B}$ of the electromagnetic field:

$$\vec{E} = -grad\,V \tag{59}$$

and

$$\vec{B} = rot\,\vec{A} \tag{60}$$

The energy functional associated with the stationary magnetic field produced by the direct currents and in case by permanent magnets is expressed as:

$$\Im(\vec{A}) = \int_{\mathcal{D}} \left( \int_0^{\vec{B}} \vec{H} d\vec{B} - \vec{J}\vec{A} \right) d\mathcal{D} \tag{61}$$

Two-dimensional problems of stationary magnetic field are by definition problems in which the unknown, the vector magnetic potential $\vec{A}$, is orientated as one axis of the coordinate system and depends on the other two coordinates of the system (Stammberg, 1995).

2D problems in Cartesian coordinates (x,y,z) are called parallel-plane. The current density $\vec{J}$ is oriented by the axis Oz and the magnetic vector potential has the structure $\vec{A} = A\vec{k}$ and its orientation is also by the axis Oz. The magnetic induction is written as:

$$\vec{B} = rot\vec{A} = \begin{vmatrix} \vec{i} & \vec{j} & \vec{k} \\ \frac{\partial}{\partial x} & \frac{\partial}{\partial y} & \frac{\partial}{\partial z} \\ 0 & 0 & A(x,y) \end{vmatrix} = \vec{i}\frac{\partial A}{\partial y} - \vec{j}\frac{\partial A}{\partial x} = \vec{k} \times (-gradA) \tag{62}$$

The square of magnetic induction is:

$$\vec{B}^2 = \left(rot\,\vec{A}\right)^2 = \begin{pmatrix} \frac{\partial A}{\partial y} \\ -\frac{\partial A}{\partial x} \\ 0 \end{pmatrix} = \left(\frac{\partial A}{\partial y}\right)^2 + \left(\frac{\partial A}{\partial x}\right)^2 = \left(grad\,A\right)^2. \tag{63}$$

The magnetic vector potential $\vec{A}$ is obtained by minimizing the functional:

$$\delta\Im(\vec{A}) = 0 \tag{64}$$

In the case of stationary magnetic field, the functional is related to the physical size of a known issue, such as the total energy of the magnetic field inside the domain *D*:

$$W_m = \frac{1}{2}\int_V \vec{H} \cdot \vec{B} dv \tag{65}$$

The magnetic energy is located in the magnetic field with the volume density $w_m$:

$$w_m = \frac{\vec{H} \cdot \vec{B}}{2} = \frac{\vec{B}^2}{2\mu} \tag{66}$$

Eckhardt describes in detail the solution to this problem. The magnetic field density is reduced to the following scalar equation (Stammberg, 1995; Eckhardt, 1978):

$$w_m = \frac{1}{2\mu} B^2 - J \cdot A = \frac{1}{2\mu}\left[\left(\frac{\partial A}{\partial x}\right)^2 + \left(\frac{\partial A}{\partial y}\right)^2\right] - J \cdot A \tag{67}$$

In the case of the parallel-plane fileds, the boundary conditions are:

$$A(x,y) = f(P), \quad P \in C_D \tag{68}$$

$$-\frac{1}{\mu}\frac{dA}{dn}\bigg|_{C_N} = g(P), \quad P \in C_N \tag{69}$$

where Eq. (68) represents the Dirichlet boundary conditions on the boundary $C_D$ and Eq. (69) represents the Neumann boundary conditions on the boundary $C_N$. The unknown function $A(x,y)$ is the solution of the Poisson's equation in a two-dimensional domain, the boundary $\Gamma$ being composed of two disjoint parts $C_D$ and $C_N$ where the Neumann and Dirichlet conditions are described (Figure 7):

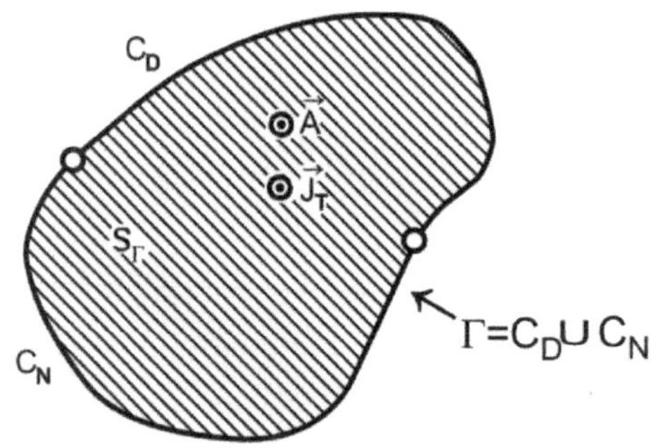


**Figure 7.** The domain of computation for a two-dimensional problem (Andrei et al. 2012)

$$\ell = \int_S \left\{ \frac{1}{2\mu}\left[\left(\frac{\partial A}{\partial x}\right)^2 + \left(\frac{\partial A}{\partial y}\right)^2\right] - J \cdot A \right\} dxdy \tag{70}$$

where the function $\ell$ represents the magnetic energy computed on the surface $S_r$ bounded by the curve $\Gamma$. The unknown function $A(x,y)$ for node $i$ is determined by minimizing the functional:

$$\frac{\partial \ell}{\partial A_i}=\int_S \left\{ \frac{1}{\mu}\left[ \frac{\partial A}{\partial x}\cdot\frac{\partial}{\partial A_i}\left(\frac{\partial A}{\partial x}\right)+\frac{\partial A}{\partial y}\cdot\frac{\partial}{\partial A_i}\left(\frac{\partial A}{\partial y}\right)^2 \right]-J\frac{\partial A}{\partial A_i}\right\}dxdy=0 \tag{71}$$

Therefore, the following equations system is obtained:

$$f_k=\sum_{z=1}^{q(i)}\frac{\partial \ell}{\partial A_i}=0 \tag{72}$$

$A_i$ is the magnetic vector potential of the node $i$ of the element $z$.

The differentials involved in Eq. (71) are written according to the shape functions and have the following expressions (Stammberg, 1995; Eckhardt, 1978):

$$\frac{\partial A}{\partial x}=\sum_{k=1}^{n(e)}A_k\frac{\partial \alpha_k}{\partial x}=\sum_{k=1}^{n(e)}A_k\cdot b_k \tag{73}$$

$$\frac{\partial A}{\partial y}=\sum_{k=1}^{n(e)}A_k\frac{\partial \alpha_k}{\partial y}=\sum_{k=1}^{n(e)}A_k\cdot c_k \tag{74}$$

$$\frac{\partial}{\partial A_i}\left(\frac{\partial A}{\partial x}\right)=\frac{\partial \alpha_i}{\partial x}=b_i \tag{75}$$

$$\frac{\partial}{\partial A_i}\left(\frac{\partial A}{\partial y}\right)=\frac{\partial \alpha_i}{\partial y}=c_i \tag{76}$$

$$\frac{\partial A}{\partial A_i}=\frac{\partial}{\partial A_i}\left(\sum_{k=1}^{n(e)}\alpha_k A_k\right)=\alpha_i=a_i+b_i\cdot x+c_i\cdot y \tag{77}$$

$$\frac{\partial \ell_e}{\partial A_i}=\int_S \left\{ \frac{1}{\mu}\left[ \left(\sum_{k=1}^{n(e)}b_k\cdot A_k\right)\cdot b_i+\left(\sum_{k=1}^{n(e)}c_k\cdot A_k\right)\cdot c_i \right]-J\alpha_i\right\}dxdy=0 \tag{78}$$

A homogeneous medium is considered, thus the magnetic permittivity $\mu$ is constant for each finite element, being independent of the $x$ and $y$ coordinates:

$$\frac{\partial \ell_e}{\partial A_i}=\frac{D}{2\mu}\left[A_1\left(b_1b_i+c_1c_i\right)+A_2\left(b_2b_i+c_2c_i\right)+A_3\left(b_3b_i+c_3c_i\right)\right]-J\frac{D}{6}=0 \tag{79}$$

where D is the determinant computed according to Eq.(56).

Eq. (79) can be written under matrix form as:

$$\frac{\partial \ell_e}{\partial A_i} = \frac{D}{2\mu}\begin{pmatrix} b_1 b_i + c_1 c_i & b_2 b_i + c_2 c_i & b_3 b_i + c_3 c_i \end{pmatrix} \cdot \begin{pmatrix} A_1 \\ A_2 \\ A_3 \end{pmatrix} - J\frac{D}{6} = 0 \tag{80}$$

The functional is sum of contributions other than the „*ne*" finite elements. Stationarization of the functional requires (Stammberg, 1995; Silvester & Ferrari, 1996):

$$\sum_{e=1}^{ne} \frac{\partial \ell_e}{\partial A_i} = 0 \tag{81}$$

where „*ne*" is the total number of finite elements.

$$\{F\} = [K] \cdot \{A\} - [P] + [Q] \tag{82}$$

The term $[P]$ is used for the case of the „source" type elements where the current density $\bar{J}$ is non-null, and the term $[Q]$ is used for the case of the elements which have in one of the sizes a non-homogeneous Neumann boundary. By assembling the„*ne*" equations in Eq. (81), a linear system of equations of magnetic vector potential values in the mesh nodes is obtained (Ioan, 1993).

## 4. Applications

Direct current (DC), which was one of the main means of distributing electric power, is still widespread today in the electrical plants supplying particular industrial applications. The advantages in terms of settings, offered by the applicants of DC motors and by supply through a single line, make direct current supply a good solution for railway and underground systems, trams, lifts and other transport means. Current-limiting circuit breakers play an important role in electrical low-voltage circuits. Due to the high short-circuit currents it is necessary a very short time to switch off the faulted branch. For this reason the current limiting circuit breakers are conceived as elaborated solutions especially for the arc quenching system, meaning the path of current and the arcing chamber (Vîrjoghe, 2010).

This section presents the calculation of the magnetic field in the arcing chamber of a current-limiting d.c. circuit breaker of 1250 A, 750 V, and in a DC circuit breaker-separator of 3150 A, 1000 V. The authors present few optimization solutions of some quenching systems which will lead to more performing constructive choices. Two-dimensional (2D) and three-dimensional (3D) problems of stationary magnetic field are addressed.

The finite element software package ANSYS is used for calculation of the magnetic field components. This tool includes three stages: preprocessor, solver and postprocessor. The procedure for carrying out a static magnetic analysis consists of following main steps: create the physics environment, build and mesh the model and assign physics attributes to each region within the model, apply boundary conditions and loads (excitation), obtain the solution, and review the results (ANSYS Documentation).

A typical magnetic field problem is described by defining its geometry, material properties, currents, boundary conditions, and the field system equations. The computer requires the input data and provides the numerical solution of the field equation and the output of desired parameters. If the values are found unsatisfactory, the design is modified and parameters are recalculated. The process is repeated until optimum values for the design parameters are obtained.

In order to define the physics environment for an analysis, it is necessary to enter in the ANSYS preprocessor (PREP7) and to establish a mathematical simulation model of the physical problem. In order to this, the following steps are presented below: set GUI Preferences, define the analysis title, define the element types and options, define the element coordinate systems, set real constants and define a system of units, and define the material properties (ANSYS Documentation).

The Global Cartesian coordinate system is the default. A different coordinate system can be specified by the user by indicating its origin location and orientation angles. The coordinate system types are Cartesian, cylindrical (circular or elliptical), spherical, and toroidal.

Some materials with magnetic properties are defined in the ANSYS material library. The materials can be modified to correspond more closely to the analysed problem and to be loaded in the ANSYS database. The copper property shows temperature which depends on resistivity and relative permeability. All other properties are described in terms of *B-H* curves. Most of the materials included in ANSYS are used for modeling the electromagnetic phenomenon. The element types are used to establish the physics of the problem domain. Some element types and options are defined to represent the different regions in the model. If some laminated materials are aligned in an arbitrary form, the element coordinate system or systems have to be identified and used. The applications presented in this chapter use the PLANE53 element in the two-dimensional problem and the SOLID97 element for the three-dimensional problem.

In order to obtain the magnetic field values, the Maxwell's equations are solved by using the imput data. The nodal values of the magnetic vector potential are considered as main or primary unknows. Their derivatives (e.g., flux density) are the secondary unknows. After this, it is possible to choose the type of solver to be used. The available options include Sparse solver (default), Frontal solver, Jacobi Conjugate Gradient (JCG) solver, JCG out-of-memory solver, Incomplete Cholesky Conjugate Gradient (ICCG) solver, Preconditioned Conjugate Gradient solver (PCG), and PCG out-of-memory solver (ANSYS Documentation).

The results of the calculations are shown in the postprocessing phase, which is a graphical program. Here, it can be observed if the applied loads affect the design, if the finite element mesh is good, and so on. The resulting fields in the form of contour and density plots are displayed by this praphical program. The analysis of the field at arbitrary points, the evaluation of a number of different integrals, and the plot of some quantities along pre-defined contours are also made with this program. The plotted results are saved in the Extended Metafile (EMF) format.

### 4.1. Numerical modelling of stationary magnetic field in area slope-slider-ferromagnetic profile of arc chamber in case of a current-limiting DC circuit breaker – 2D application

The problem of magnetic field distribution in the arc chamber of DC a circuit breaker with rated current 1250A was numerically solved. The conductor where a current of 1250 A flows is located in the immediate vicinity of a ferromagnetic profile. This has the role of enhancing and orienting the magnetic field in the arc-quenching chamber for obtaining a strong force that moves the arc up inside the extinction chamber (Vîrjoghe, 2004).

It is considered the plane parallel model, whose cross section is shown in Figure 8. This model is an *I* shaped ferromagnetic profile, with cross section in the vertical plane and the dimmensions of 60x3 $mm^2$. The cross section of copper conductor is 5x15 $mm^2$. The conductor is surrounded by a slider with *U* shaped cross section and a thickness of 1mm.

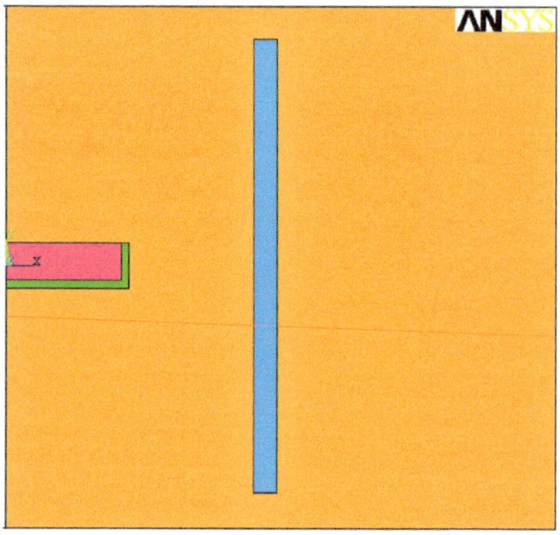


**Figure 8.** The physical model in the area with ramp, slider and ferromagnetic profile

For numerical computation the *PLANE53* element was chosen, which allows two-dimensional modeling of the magnetic field in plane parallel and axisymmetric problems. This element is based on the magnetic vector potential formulation with Coulomb calibration. This element is also applicable to the stationary magnetic field with the possibility of modeling the magnetic nonlinearities. The material used for other two ferromagnetic profiles is a steel chosen from the ANSYS library and having the properties in the *emagM3.SI_MPL* folder. The material is M3 steel and its magnetization curve is shown in Figure 9. This domain was discretized in a number of 2436 triangular finite elements uniformly distributed (Vîrjoghe, 2004).

The boundary conditions and loads are applied to a 2-D static magnetic analysis either on the solid model (key points, lines, and areas) or on the finite element model (nodes and elements). The loads applied to the solid model to the mesh during solution are automatically transferred by ANSYS (ANSYS Documentation).

To observe the influence of the ferromagnetic slider and of the ferromagnetic plate on the magnetic fied distribution, the magnetic induction is computed and the magnetic flux lines are drawn. The slider and the ferromagnetic plate case were studied independently of each other.

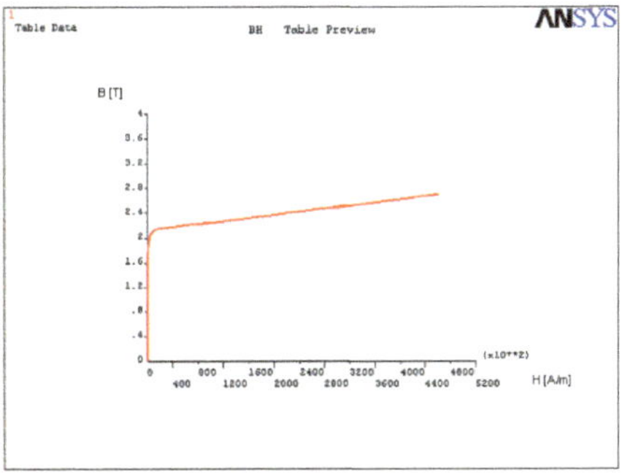


**Figure 9.** The magnetization characteristic for the M3 steel

Figure 10 and Figure 11 respectively show the magnetic induction spectrum and the magnetic field lines only, for the case of ferromagnetic slider.

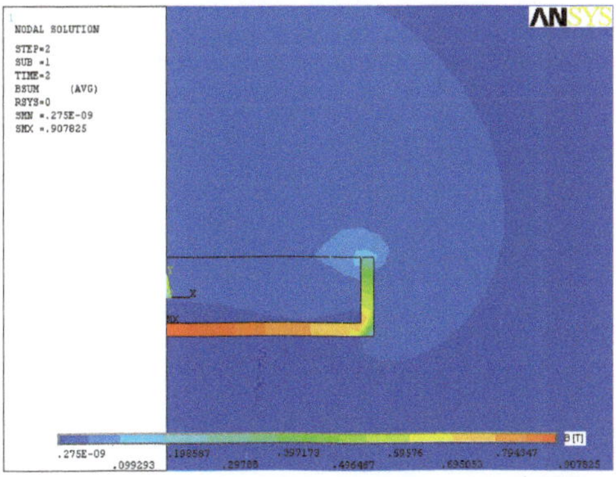


**Figure 10.** The magnetic induction spectrum in the presence of the ferromagnetic slider

Figure 12 and Figure 13 respectively show the magnetic induction spectrum and the magnetic field lines only, for the case of *I* shaped ferromagnetic profile. Figure 14 and Figure 15 respectively show the magnetic induction spectrum and the magnetic field lines when using the ferromagnetic profile and the ferromagnetic slider (Vîrjoghe, 2004).

Analyzing these simulations, a strong influence of the ferromagnetic slider on the orientation of the magnetic field was observed. When using only the ferromagnetic slider, a shielding of the field lines is observed, and the maximum values of the magnetic induction is 0.907 T. The *I* shaped ferromagnetic profile makes a good shielding of the field lines obtaining the maximum values of magnetic induction of 0.153 T. If both methods of magnetic field orientation are used together then a maximum value of magnetic induction of 0.947 T is obtained. To obtain the system optimization in this area, the simulations for a thickness of 2 mm have been repeated.

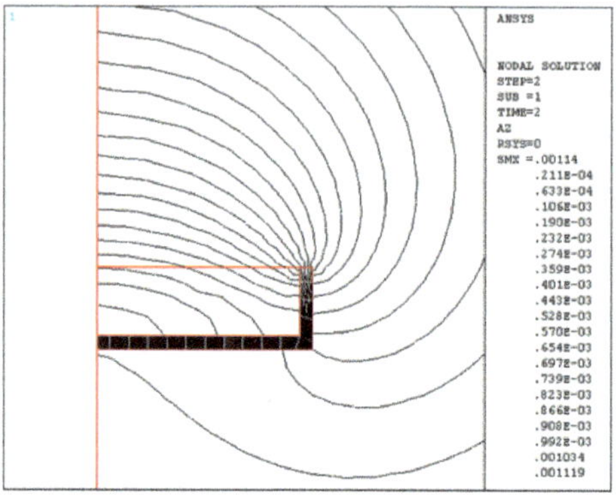


**Figure 11.** The magnetic equipotential lines in the presence of the ferromagnetic slider

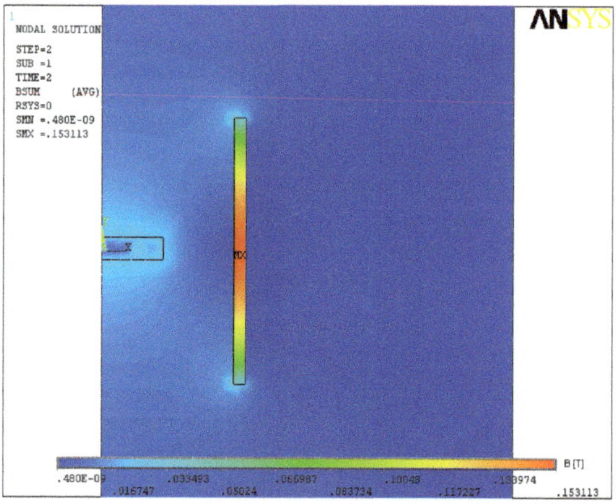


**Figure 12.** The magnetic induction spectrum in the case with the ferromagnetic profile

Thus, the obtained results are plotted in comparison with those presented for the slider of 1 mm (Figure 16). For the slider with the thickness of 2 mm, the values of magnetic induction are lower (up to 0.5 T). Using the slider of 1 mm thickness a better orientation of the field lines, as well as a better arc transmission toward arc-quenching chamber are observed. The path for the displayed charts is chosen between two points placed symmetrically one from another in the middle of a figure which contains the conductor, slider and ferromagnetic profile (Vîrjoghe, 2004).

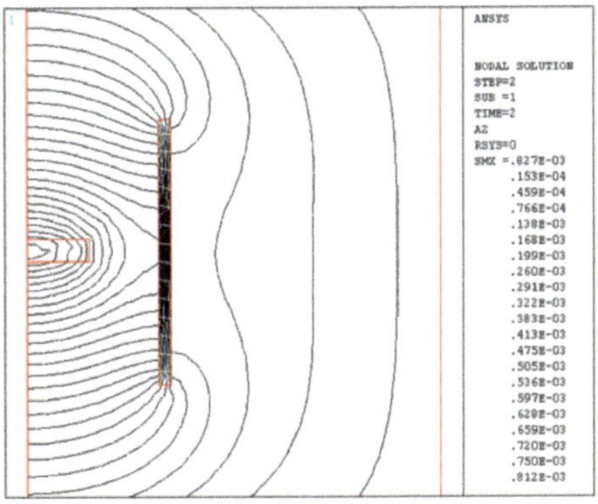


**Figure 13.** The magnetic equipotential lines in the presence of the ferromagnetic profile

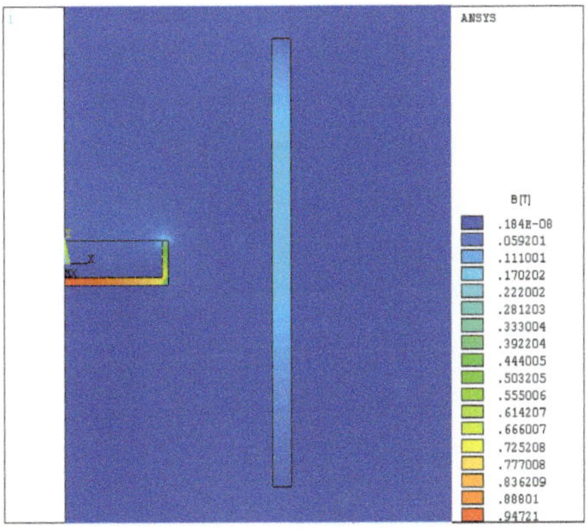


**Figure 14.** The magnetic induction spectrum in the area with ramp, slider and ferromagnetic profile.

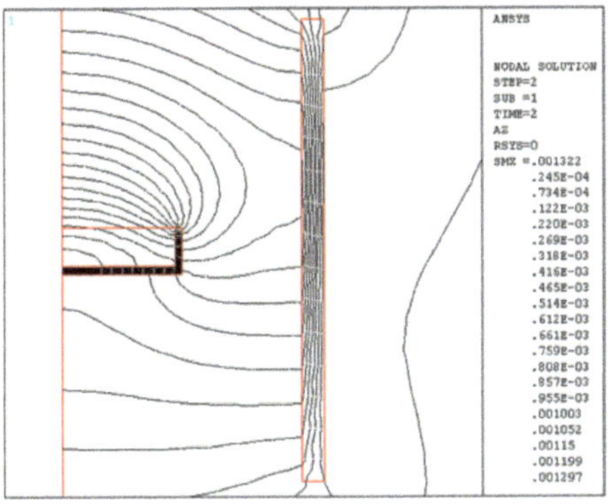


**Figure 15.** The magnetic equipotential lines in the area with ramp, slider and ferromagnetic profile.

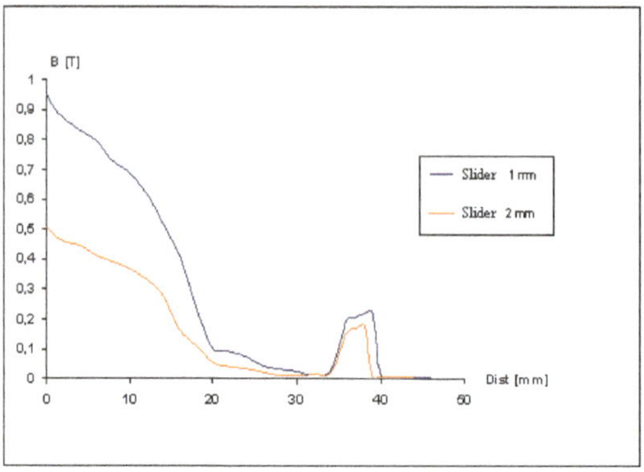


**Figure 16.** Magnetic induction variation depending on the thickness of the slider.

## 4.2. Numerical modelling of stationary magnetic field of arc-quenching chamber in case of a DC circuit breaker-separator – 3D application

The physical model of the arc-quenching chamber in case of a DC circuit breaker-separator of 3125 A having the ramps-ferromagnetic profiles is shown in Figure 17. In this model, two profiles composed of a ferromagnetic material are presented. The magnetization curve of the two profiles is shown in Figure 9.

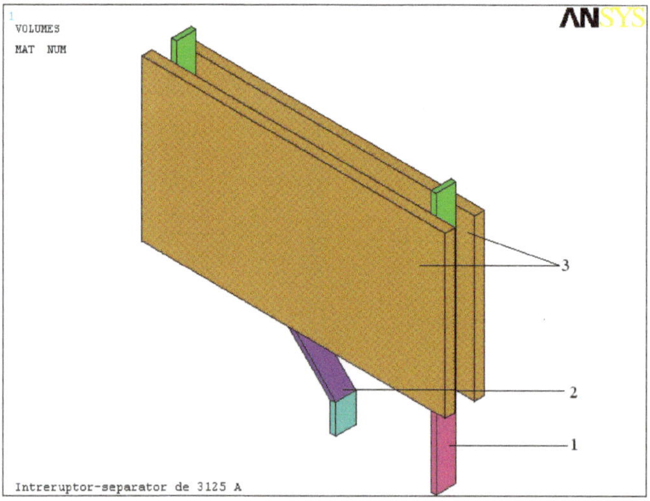


**Figure 17.** The DC circuit breaker-separator model with a current of 3125A.

These profiles form a rectangular prism with length of 150 mm, height of 100 mm and thickness of 5 mm. The electric arc ramps 1 and 2 are made of copper and have width of 10mm and thickness of 2.5 mm. The left ramp is inclined to the vertical with an angle of 45º.

The arc chamber model together with ramps 1 and 2 are incorporated in a boundary volume, where the air is defined as material. In order to achieve the circuit continuity, two ramps have been unified with a bar 3 having the same dimensions of the ramps (Figure 18).

In the preprocessing phase, the materials are defined and chosen. For the current path, consisting of two ramps and the connecting bar, the copper was chosen. For the two ferromagnetic plates, from ANSYS library a *M3* steel is chosen, having the properties contained in the *emagM3.SI_MPL* file (ANSYS Documentation). The next step in the preprocessor phase is the mesh generation and load application upon the elements (Figure 19). In this application, for modeling the three-dimensional stationary magnetic field a *SOLID97* element is chosen. For the numerical computation of the stationary magnetic field, the model of the DC circuit-breaker together with the boundary volume is discretized in a number of 1268 nodes and 3623 triangular elements (Vîrjoghe, 2004).

In the postprocessing phase is also applied the load on elements and boundary conditions.

The load on elements is represented by the conduction current density. For *3D* analysis, a positive value indicates current flowing in the +*Z* direction in the plan case and the -*Z* (loop) direction in the asymmetrical case. The current density is directly applied on the finite elements which form the conductors and its value is $125 \cdot 10^6$ A/m$^2$. As boudary conditions the Dirichlet condition, $A$=0, is applied.

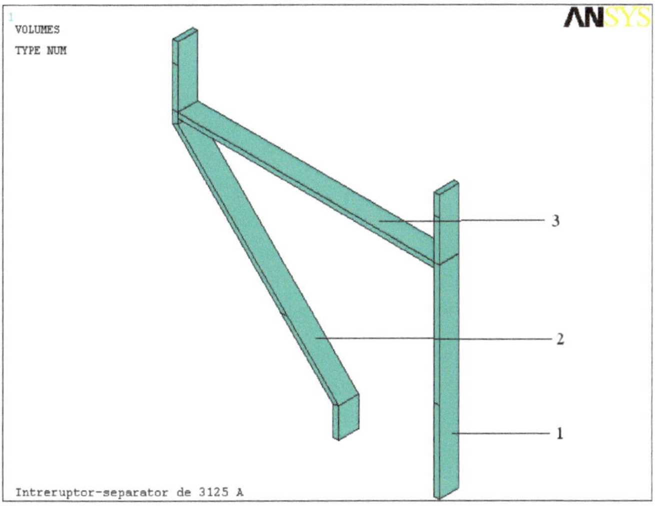


**Figure 18.** The current path for DC circuit breaker – separator of 3125 A

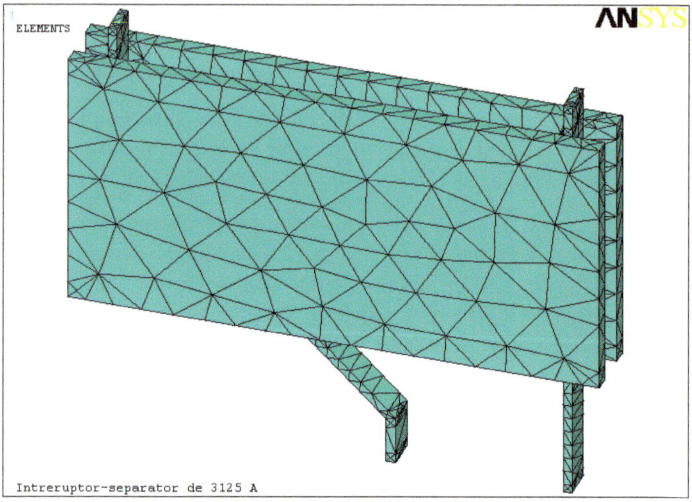


**Figure 19.** The discretized model of the DC circuit breaker-separator of 3125 A

The Maxwell's equations solver is based on the Finite Element Method (FEM). The results are the nodal values of the primary unknowns (magnetic vector potential) and derivatives of these values for obtaining the secondary unknowns (magnetic induction).

In the postprocessing phase, the tool allows visualization of magnetic induction spectrum, determination of magnetic sizes in arbitrarily chosen points, as well as the evaluation of the different charts. Figure 20 shows the magnetic induction spectrum in the arc-quenching chamber of DC circuit breaker-separator of 3125 A. A maximum value of magnetic induction 2.149 T is obtained.

The DC circuit breaker-separator is designed as a particularly elaborated solution for the current path and the arc-quenching chamber. It is widely known that in electromechanic design of a switching device, the arc-quenching chamber together with current paths and contacts represent the essential element due to their switching performances in normal operating conditions and in abnormal conditions. An optimization criterion of this arc-quenching chamber concerns the ferromagnetic material used in the construction of the ferromagnetic profiles. Simulation was performed for three different steels. For the two ferromagnetic plates, three different steels from ANSYS library are chosen, namely (ANSYS Documentation):

- carpenter (silicon) steel with material properties contained in *emagSilicon.SI_MPL* file;
- iron cobalt vanadium steel with material properties contained in the *emagVanad.SI_MPL file*;
- SA1010 steel with material properties contained in the *emagSa1010.SI_MPL* file.

It was established that in the case of steel EmagSilicon utilization, the maximum value of magnetic induction is 1.883 T, in the case of EmagVanad the maximum value of magnetic induction is 1.975 T and in the case of EmagSa1010 the maximum value of magnetic induction is 1.44 T. Hence, the optimal material for construction of these profiles is EmagVanad. Although the steel with vanadium is an expensive material, it assures an optimal value of magnetic induction. The high price is compensated by improving the arc-quenching chamber performance and thus increase the breaking capacity of the device (Vîrjoghe, 2004).

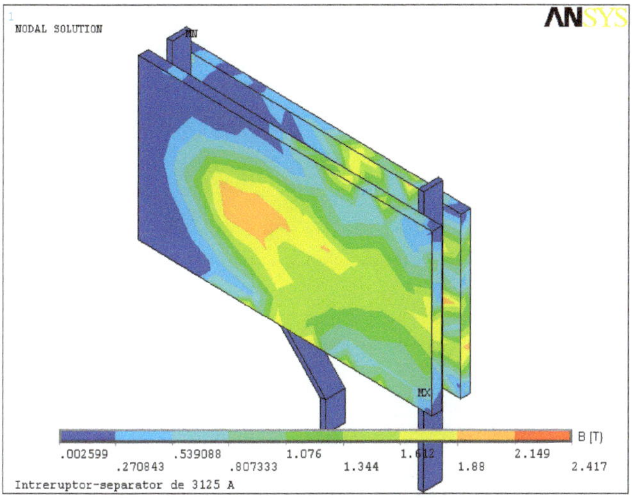


**Figure 20.** The magnetic induction spectrum in arc-quenching chamber of DC circuit breaker-separator of 3125 A

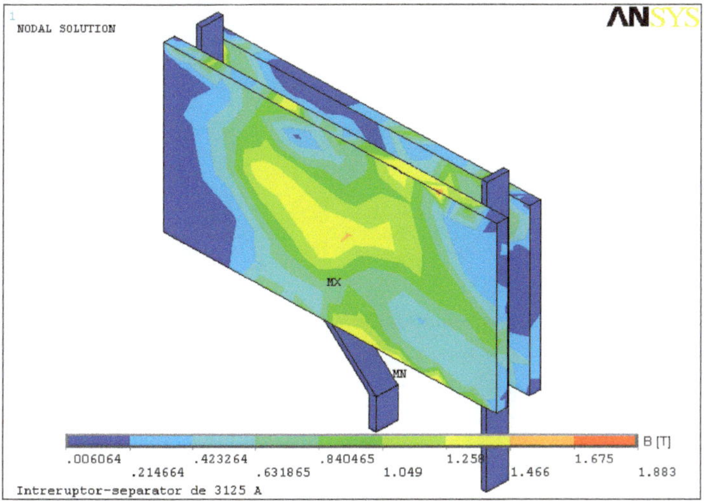


**Figure 21.** Magnetic induction distribution for the ferromagnetic material EmagSilicon

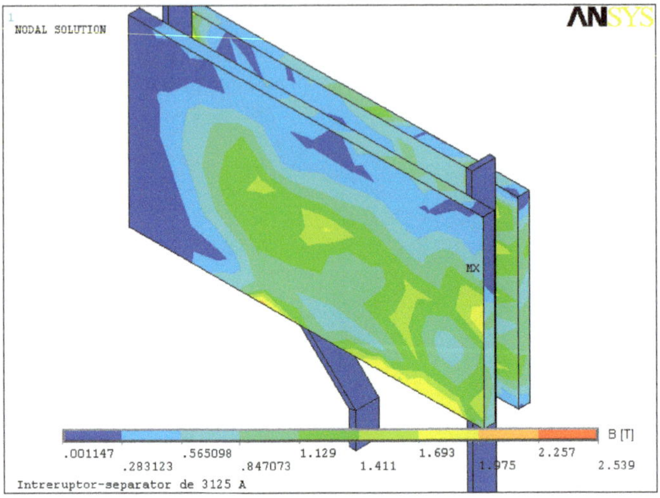


**Figure 22.** Magnetic induction distribution for the ferromagnetic material EmagVanad

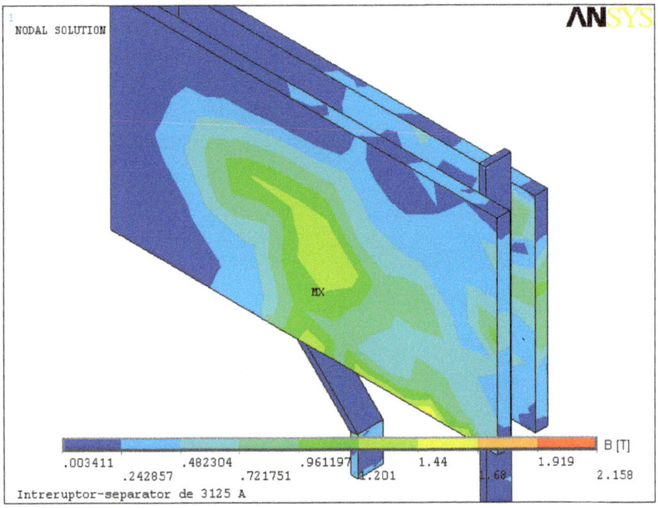


**Figure 23.** Magnetic induction distribution for the ferromagnetic material EmagSa1010

## 5. Conclusion

It is well known that in electromechanical construction of a switching device, the arcing chamber along with current paths and contacts represents the all-important elements concerning switching performances of these in normal operating conditions as well as in operation under faults (Truşcă & Truşcă, 2001).

Comparing the magnetic flux density spectrums in the three cases it can be observed that the maximum arc-quenching effect is obtained by using EmagVanad for the ferromagnetic shapes. For this material an optimal distribution for the magnetic field in the circuit breaker arcing chamber is obtained, which leads to a rapid movement of the electric arc towards the ferromagnetic plates. Arc quenching and arc voltage limiting occur in base of the niche effect principle along with the electrode effect (Hortopan, 1996).

## Author details

Elena Otilia Virjoghe, Diana Enescu, Mihail-Florin Stan and Marcel Ionel
*Valahia University of Targoviste, Romania*

## 6. References

Andrei, H., Flueraşu, C., Flueraşu, C., Enescu, D., Vîrjoghe, E.O., Husu, A.G., Andrei, P.C. & Popovici, D. (2012). Metode numerice, modelări şi simulări cu aplicaţii in ingineria electrică/Numerical Methods, Modelling and Simulation Applied in Electrical Engineering, in *Electra*, Bucureşti, 427-436.

Eckhardt, H. (1978) Numerische Verfahren in der Energietechnik. Stuttgert:Teubner-Verlag.

Flueraşu, C., Flueraşu, C. (2007). Electromagnetism, in *Printech*, Bucureşti.

Gârbea, D. (1990). Analiză cu elemente finite, in Tehnică, Bucureşti, 7-8.

Hănţilă, I.F. (2004). Electrotehnica teoretică, in *Electra*, Bucureşti.

Hortopan, Gh. (1996). Aparate electrice de comutaţie, vol.II, in *Tehnică*, Bucureşti, 58-60.

Ioan D., Ionescu, B. (1993). Numeric methods in electromagnetic field computation, *Politehnica*, Bucureşti.

Mocanu, C.I. (1981). Teoria câmpului electromagnetic, *Didactică şi Pedagogică*, Bucureşti, 456-460.

Moraru, A. (2002). Bazele electrotehnicii - Teoria câmpului electromagnetic, *MatrixRom*, Bucureşti.

Morozionkov, J., Virbalis, J.A. (2008). Investigation of Electric Reactor Magnetic Field using Finite Element Method, *Electronics and Electrical Enginnering*, No.5(85).

Răduleţ, R. (1975). Bazele teoretice ale electrotehnicii, Litografia Învăţământului, Bucureşti.

Silvester P.P., Ferrari, R.L. (1996). Finite Elements for Electrical Engineers, 3rd Edition, *University Press*, Cambridge, 28-29, 118-120, 183-187.

Şora, C., (1982). Bazele electrotehnicii, *Didactică şi Pedagogică*, Bucureşti, 258-260.

Stammberger, H. (1995). Magnetfeld-und Kraftberech-nungen für strombegrenzende Niederspannungs-Schaltgeräte, *VDI Verlag*, Elektrotechnik, Bonn, pp.12-20.

Timotin, A., Hortopan, V., Ifrim, A., Preda, V., (1970). Lecţii de bazele electrotehnicii, *Didactică şi Pedagogică*, Bucureşti.

Truşcă, V., Truşcă, B. (2001). Elektrische anlagen – vorlesungen- , *Printech*, Bukarest.

Vîrjoghe, E.O., (2004). Aparate electrice de comutaţie, *Electra*, Bucureşti, 93-99, 122-129.

Vîrjoghe, E.O., Enescu, D., Ionel, M., Stan, M-F. (2010). 3D Finite Element Analysis for arcing chamber optimization of the current-limiting circuit breaker, *WSEAS TRANSACTIONS on POWER SYSTEMS*, ISSN: 1790-5060, Issue 1, Volume 5.

ANSYS Release 11.0 Documentation.

# Chapter 7

# The IEEE Model for a Ground Rod in a Two Layer Soil – A FEM Approach

António Martins, Sílvio Mariano and Maria do Rosário Calado

Additional information is available at the end of the chapter

## 1. Introduction

The calculation of a ground electrode resistance, using a two layer soil model, has been widely presented in literature. Several methods had been used. Formulas for grid in two layers soil using the synthetic-asymptote approach have been developed in (Salama et al., 1995). Berberovic explored the Method of Moments in the calculation of ground resistance, using higher order polynomials approximation in the unknown current distribution in (Berberovic et al., 2003), and the Galerkin's Moment Method with a variation was used in (Sharma & De Four, 2006). Another theoretical tool commonly used is the Boundary Element Method, as in (Colominas et al., 1998, 2002a, 2002b; Adriano et al., 2003). These authors transformed the differential equation that governs the physical phenomenon into an equivalent boundary integral equation. The Matrix/Integration Method for calculating the mutual resistance segment in one and two layered soil was adopted by (Coa, 2006) and an optimised method of images for multilayer soils was used in (Ma et al., 1996). Even in the study of ionization phenomena, the two layer ground model was used in (Liu et al., 2004). In general these works used the theory of images, which implies infinite series for the expanded Green function as in (Berberovic et al., 2003). Recently, a work presented the effect of low resistance materials filling in a pit surrounding a rod, working with two different soil resistivity's (Al-Arayny et al., 2011). This type of research was also presented in (Zhenghua et al., 2011), that even considered the use of electrolytic materials. A Finite Element Method (FEM) application to grounding can also be found in (Manikandan et al., 2011) to the analysis of wind turbines grounding. In this chapter the FEM is presented in a theoretical basis for cylindrical symmetry problems, using a ground rod resistance calculation as an example. Comparison with experimental result is also made.

This chapter presents the Finite Element Method for the calculation of a rod resistance in a two layer soil. Theoretical basis of the method are presented. The FEM mesh was tested in homogeneous soil, first a cylindrical approach to IEEE model was considered, in order to guarantee that mesh is voltage and energy adapted, and then the whole model was discretized. Resistance was calculated using Joule's law in a FEM energetic approach. The zero volt Dirichlet boundary was located within a distance of 3 cm, 15 cm and 7.6 m from the ground electrode, to analyze if the resistance relations indicated by IEEE in homogeneous soil could be also used in a two layer soil. The simulated results were compared with those obtained from Tagg formula. In homogeneous soil the errors are less than 1%. In two layer soil, resistance error to Tagg formula decreased from values of 28% with zero volt boundary at 3 cm to 22% with zero volt boundary at 15 cm; for the whole model discretized the error is near -18%. The results were unsatisfactory, thus the percentage resistances at these distances cannot be generalized in a two layer soil. However the error between simulated and field measured values is of 4.6%, turning the FEM analysis a valuable simulation tool.

## 2. Finite element method

The first order finite elements method using triangular elements may be regarded as two-dimensional generalizations of piecewise-linear approximation techniques as in reference (Sylvester & Ferrari, 1990), widely used in Electrical Engineering. The method allows several choices for mesh types and an easy treatment for boundary shapes.

Several problems in electrical engineering require the solution of Laplace equation in two or three dimensions with two kinds of boundary conditions, such as prescribed potential values along the referred boundaries, Dirichlet conditions, and vanishing normal derivative along the symmetry planes, Neumann condition. As an example, in power system grounding, where capacitive and inductive effects are not considered, since industrial frequency is too low, the soil potential satisfies Laplace equation. On the other hand grounding systems dimensions are much smaller than power line wavelength, so that propagation phenomenon is not considered. Laplace equation solution is equivalent, according to the minimum potential energy principle, to the following energy ( $W(u)$ ) functional minimization, that stores field energy per unit volume, as in (Sylvester & Ferrari, 1990):

$$W(u) = \frac{1}{2}\iiint_V |\nabla u|^2 \, dV \tag{1}$$

where $u$ is the potential and $V$ the volume.

The integral is evaluated over all the volume defined by the problem boundary.

### 2.1. Discretization

In Fig. 1 the IEEE model for a ground rod is presented. The boundary conditions within a distance of 7.6 m from the electrode have zero potential. At surface, normal derivative potential is also zero, since there is no current flowing in the air.

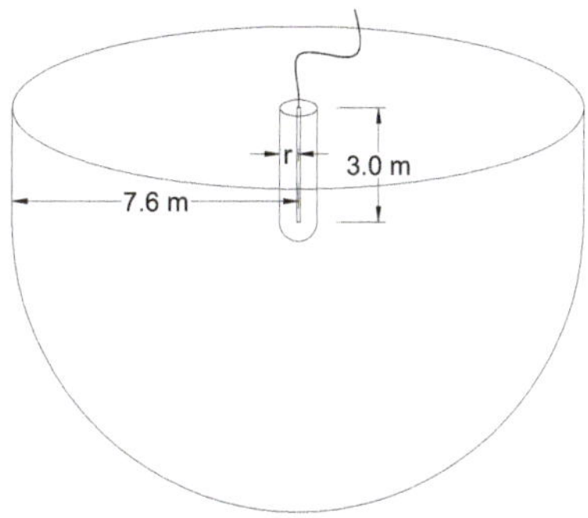


**Figure 1.** IEEE model as in (IEEE Std 142, 2007)

Due to isotropy, the problem was discretized in the '*rz*' plane, since cylindrical coordinates were used. To obtain an approximate solution by FEM the problem region is subdivided into triangular elements. The solution mesh is presented in Fig. 2, near the rod top end.

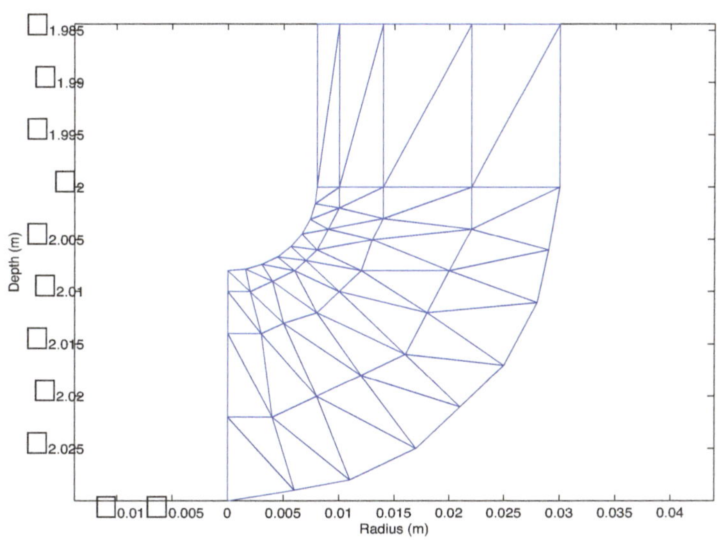


**Figure 2.** Model discretization near rod top end

## 2.2. First order triangular elements

The essence of the method lies in first approximating the potential $u$ within each element in a standardized fashion, and thereafter interrelating the potential distribution in the various elements so as to constrain the potential to be continuous across interelement boundaries, as in reference (Sylvester & Ferrari, 1990). This standardized fashion is done using a first order polynomial for potential within each element, which is the approximate solution to the actual one.

$$u(r,z) = a + br + cz \tag{2}$$

The true potential distribution is thus replaced by a piecewise planar function and within each triangle side, potential is obtained by a linear interpolation of node potentials.

Considering a generic triangular finite element, as shown in Fig. 3, equation (2) must satisfy node potentials. Using the equation in the three nodes a system is obtained, allowing the constants '$a$', '$b$' and '$c$' to be calculated as functions of node potentials.

$$\begin{aligned} U_1 &= a + br_1 + cz_1 \\ U_2 &= a + br_2 + cz_2 \\ U_3 &= a + br_3 + cz_3 \end{aligned} \tag{3}$$

where $U_1$, $U_2$ and $U_3$ are node potentials.

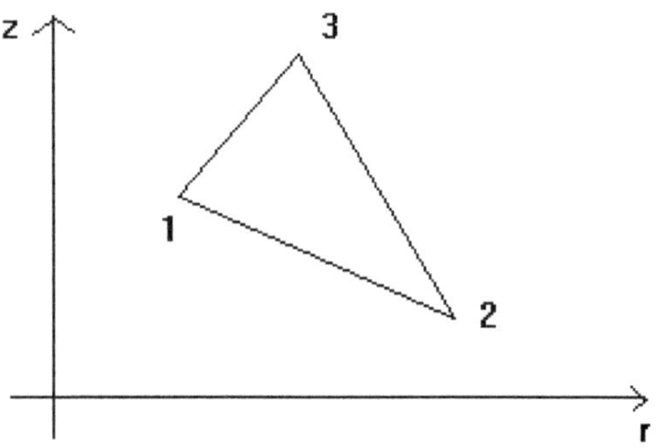


**Figure 3.** Finite element with nodes numbered

Using Cramer's rule the referred constants are easily obtained. The formula for '$a$' is:

$$a = \frac{\begin{vmatrix} U_1 & r_1 & z_1 \\ U_2 & r_2 & z_2 \\ U_3 & r_3 & z_3 \end{vmatrix}}{\begin{vmatrix} 1 & r_1 & z_1 \\ 1 & r_2 & z_2 \\ 1 & r_3 & z_3 \end{vmatrix}} \tag{4}$$

The denominator is recognized as twice the triangle area (2A) as in reference (Sylvester & Ferrari, 1990). It must be pointed that this formula was found by integration so that area become negative if the nodes are clockwise numbered. Equation (2) is useful when node potentials are already known. In order to calculate these potentials is useful to represent potential within the element as a function of nodes potentials, so equation (2) became:

$$u = \left. \begin{Bmatrix} \left[ (r_2 z_3 - r_3 z_2) + (z_2 - z_3) r + (r_3 - r_2) z \right] U_1 + \\ \left[ (r_3 z_1 - r_1 z_3) + (z_3 - z_1) r + (r_1 - r_3) z \right] U_2 + \\ \left[ (r_1 z_2 - r_2 z_1) + (z_1 - z_2) r + (r_2 - r_1) z \right] U_3 \end{Bmatrix} \middle/ 2A \right. \quad (5)$$

The factors multiplying nodes potentials, after being divided by 2A, are known as shape functions, so we can rewrite previous equation using these functions.

$$u(r,z) = \alpha_1(r,z) U_1 + \alpha_2(r,z) U_2 + \alpha_3(r,z) U_3 \quad (6)$$

Or in a more elegant manner:

$$u(r,z) = \sum_{i=1}^{3} U_i \, \alpha_i(r,z) \quad (7)$$

Shape functions have two important properties;

1. Their value is one in the associated node and zero in the two others.
2. In any point inside the finite element their sum is one.

The finite element energy can now be calculated using equation (1). Node potentials, although unknown, are constants so the potential gradient is

$$\vec{\nabla} u = \sum_{i=1}^{3} U_i \vec{\nabla} \alpha_i(r,z) \quad (8)$$

The energy can now be evaluated, using equation (8) and (1). Considering that the integrand has nine sums, due to the square of potential gradient, energy functional for a single element is:

$$W^e(u) = \frac{1}{2} \iiint_V \sum_{i=1}^{3} \sum_{j=1}^{3} U_i U_j \vec{\nabla} \alpha_i . \vec{\nabla} \alpha_j dV \quad (9)$$

Integrating by decomposition:

$$W^e(u) = \frac{1}{2} \sum_{i=1}^{3} \sum_{j=1}^{3} U_i \iiint_V \vec{\nabla} \alpha_i . \vec{\nabla} \alpha_j dV . U_j \quad (10)$$

Defining the $S$ variable as:

$$S_{ij} = \iiint_V \vec{\nabla} \alpha_i . \vec{\nabla} \alpha_j dV \quad (11)$$

As shape functions are first degrees polynomials, gradients are constants, such as their inner product, so the integral is obtained by the calculus of the revolution volume of the finite element, which is $2\pi r_c$ where $r_c$ is the centroid radius. The $S$ variable became:

1. For equal index

$$S_{11}^{e} = \frac{(z_2 - z_3)^2 + (r_2 - r_3)^2}{2A} \pi r_c \tag{12}$$

2. For different index

$$S_{12}^{e} = \frac{(z_2 - z_3)(z_3 - z_1) + (r_3 - r_2)(r_1 - r_3)}{2A} \pi r_c \tag{13}$$

The remaining terms are obtained by index cyclic rotation. The finite element energy is finally represented as:

$$W^{e}(u) = \frac{1}{2} U^{T} S U \tag{14}$$

where $S$ is a 3x3 matrix and $U$ the nodes potential vector.

## 2.3. Elements assembly

The total energy model is the sum of all finite elements energy. Consider the following two elements with disjoint nodal numbering:

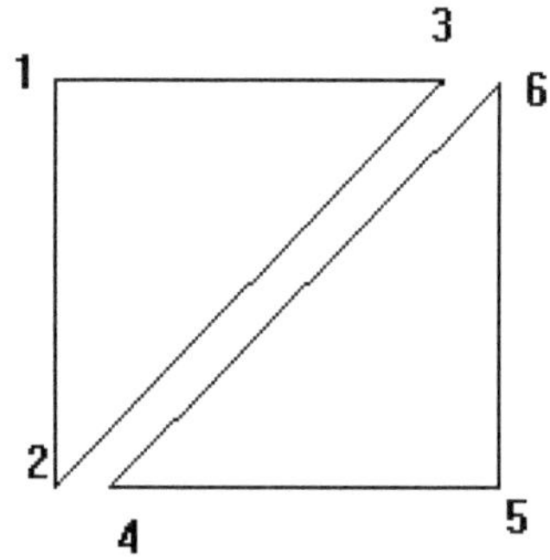


**Figure 4.** Disjoint numbering of finite element nodes

The nodes potential vector is:

$$U_{dis}^{T} = \begin{bmatrix} U_1 & U_2 & U_3 & U_4 & U_5 & U_6 \end{bmatrix} \tag{15}$$

The $S_{dis}$ matrix is formed by the elementary elements matrices.

$$S_{dis} = \begin{bmatrix} \mathbf{S_1} & \mathbf{0} \\ \mathbf{0} & \mathbf{S_2} \end{bmatrix} \tag{16}$$

This matrix is tridiagonal and is called the Dirichlet matrix, as in reference (Sylvester & Ferrari, 1990). The energy of the two elements is:

$$W(u) = \frac{1}{2} U_{dis}^T S_{dis} U_{dis} \tag{17}$$

Potentials continuity implies that the values for the same node are equal. So potential at node three is equal to potential at node six, in Fig. 4, and so on. It can be defined a global nodal numbering for element assembly illustrated in Fig. 5.

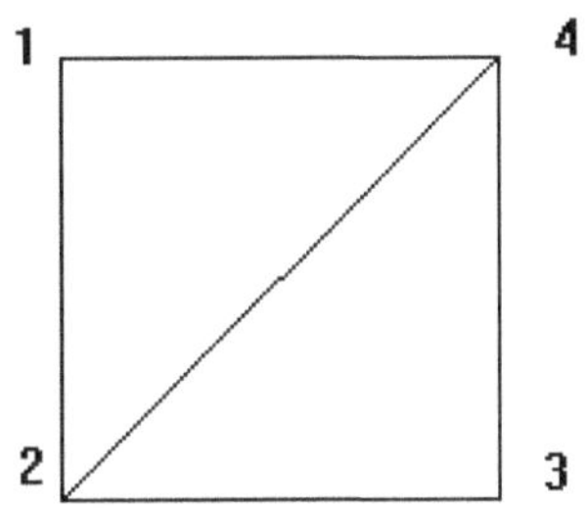


**Figure 5.** Global node numbering

The potential continuity for corresponding nodes will be guaranteed by a linear transformation that relates disjoint nodes numbering with global numbering. In this case it would be:

$$\begin{bmatrix} U_1 \\ U_2 \\ U_3 \\ U_4 \\ U_5 \\ U_6 \end{bmatrix}_{dis} = \begin{bmatrix} 1 & 0 & 0 & 0 \\ 0 & 1 & 0 & 0 \\ 0 & 0 & 0 & 1 \\ 0 & 1 & 0 & 0 \\ 0 & 0 & 1 & 0 \\ 0 & 0 & 0 & 1 \end{bmatrix} \begin{bmatrix} U_1 \\ U_2 \\ U_3 \\ U_4 \end{bmatrix}_{conj} \tag{18}$$

In all cases would be:

$$U_{dis} = C U_{conj} \tag{19}$$

Introducing this relation in equation (15) it can be obtained an energy formulation in function of global numbering nodal potentials.

$$W(u) = \frac{1}{2} U_{conj}^{T} . S U_{conj} \tag{20}$$

with

$$S = C^T S_{dis} C \tag{21}$$

In order to minimize energy expression $W(u)$, it is necessary to calculate the derivative, in equation (21), with respect to potential vector and solve the equation

$$SU = 0 \tag{22}$$

Also, in order to avoid trivial solution $U = 0$ the $S$ matrix should be partitioned in blocks, as well as potential vector:

$$\begin{bmatrix} S_n & S_k \end{bmatrix} \begin{bmatrix} U_n \\ U_k \end{bmatrix} = 0 \tag{23}$$

Where $U_n$ is a vector of unknown node potentials and $U_k$ is the known potential vector. Unknown potential nodes must be the first to be numbered and after the known potentials that satisfy boundary conditions. The $S_n$ and $S_K$ dimensions must allow matrix multiplication. Final solution for the unknown potential is given by:

$$U_n = -S_n^{-1} S_K U_K \tag{24}$$

This FEM solution is called stored energy approach.

### 2.4. Electric field

Knowing the node potentials, the constants in equation (8) are easy to find as well as the electric field strength, which is given by:

$$\vec{E} = -b\hat{i} - c\hat{j} \tag{25}$$

## 3. IEEE model in homogeneous soil

The analysis of a ground rod was carried out with first order triangular finite element only in the 3 cm near the rod, since 25 % of the rod resistance is within this region as in reference (IEEE Std 142, 2007). This avoids the discretization of the IEEE entire model with a zero volt boundary condition at 7.6 m away from the rod, being the rod resistance four times the calculated value.

### 3.1. The IEEE model for cylindrical region

In order to test the FEM mesh, it was considered only the cylindrical region in homogeneous soil. This problem has a theoretical solution, since it is considered as two cylinders centered in the same axis. The inner cylinder has the rod radius and the outer cylinder has a 3 cm radius, which is the zero volt boundary condition. The finite elements are shown in Fig. 6.

The voltage at any point between the cylindrical surfaces is easily obtained by:

$$v(r) = \frac{V_0}{\ln\frac{b}{a}} \ln\frac{b}{r} \tag{26}$$

with 'b' the outer cylinder radius, 'a' the inner cylinder radius, and $V_0$ the potential difference applied between the cylindrical surfaces.

With 'b'=3 cm, 'a'=8 mm and a voltage difference of 220 V between the two cylinders, equation (27) becomes:

$$v(r) = 166.4\ \ln\frac{0.03}{r} \tag{27}$$

where 'r' is the distance to common axis.

The numerical values for voltage were calculated by FEM. For 2 m length cylinders, the potential at 1 cm away from the common axis is presented in Fig. 7.

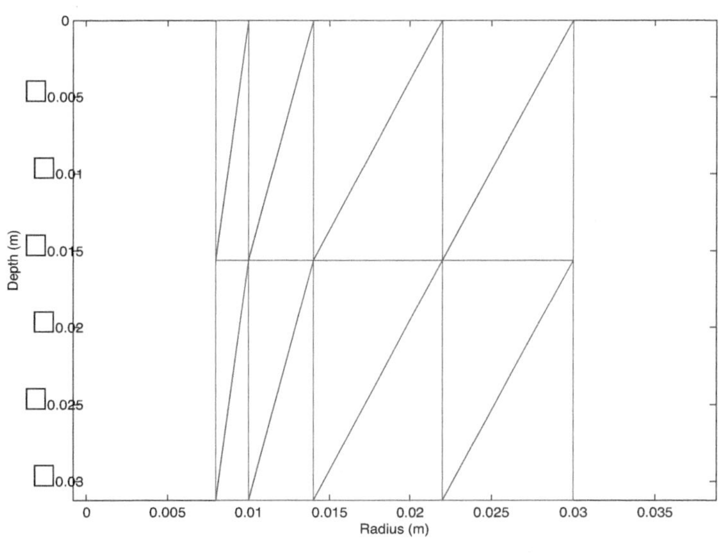


**Figure 6.** Cylindrical region Discretization

The potential is almost constant along a parallel line 1 cm away from rod axis, as expected. In the first point its potential is 183.2 V and the last has the value of 182.1 V. The potential variation with the distance to axis 'r' is presented in Table 1.

| Distance (m) | Theoretical value (Volt) | Mean value (Volt) | Mean error (%) | Maximum error (%) |
|---|---|---|---|---|
| 0.01 | 182.9 | 182.6 | -0.2 | -0.4 |
| 0.014 | 126.9 | 126.6 | -0.2 | -0.6 |
| 0.021 | 59.4 | 59.3 | -0.2 | -0.2 |

**Table 1.** Potential variation between cylindrical surfaces

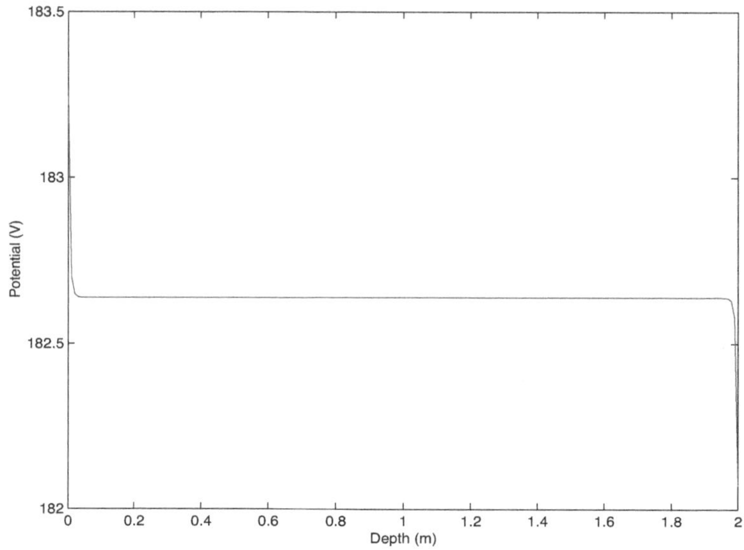


**Figure 7.** Potential along a parallel line 1 cm away from rod axis

The obtained figures for potential at distances of 1.4 cm and 2.1 cm are similar to Fig. 7. It was concluded that this mesh is voltage adapted. For the resistance between the two cylinders was used the FEM energetic approach given by (Martins & Antunes, 1997),

$$R = \frac{v^2}{P_{JOULE}} = \frac{v^2}{\sigma \int_V |E|^2 \, dV} \tag{28}$$

where $v$ is the voltage between the cylinders, $\sigma$ the conductivity of the material between the two cylinders, $E$ the electric field intensity and $V$ the revolution volume due to axial symmetry, generated by each finite element. The obtained value for the electric resistance was 20.8 Ω considering a medium with 200 Ωm for the resistivity. That value should be compared with the theoretical value for the resistance between two cylinders given by (Purcell, 1998):

$$R = \frac{\rho}{2\pi L} \ln \frac{b}{a} \tag{29}$$

with $P$ the electric resistivity, $L$ the length, $b$ the outer surface radius and $a$ the inner cylindrical radius.

The value given by equation (30) is 21.0 Ω. The simulated value of 20.8 Ω is 1% less, which allows the conclusion that the mesh is also adapted in energy.

### 3.2. The complete IEEE model

To achieve the discretization of the entire IEEE model the mesh was altered in the rod bottom as shown in Fig. 8. Triangular finite elements have inner angles greater than 5º avoiding triangle areas close to zero, allowing stiffness matrix to be well defined.

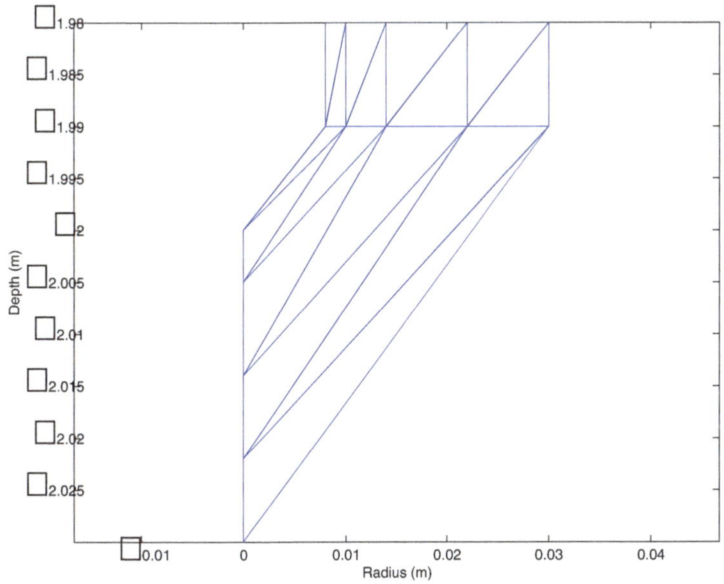


**Figure 8.** Bottom rod FEM discretization

For a rod with these dimensions the Dwight formula (Dwight, 1936) gives a value for resistance of 94.0 Ω. According to the standard (IEEE Std 142, 2007), this formula has 13% excess, so the corrected theoretical value is 83.2 Ω. The FEM simulated value for resistance, with the zero volt boundary at 0.03 m, is 20.8 Ω, representing 25 % of total resistance, so the numerical simulated value is four times 20.8 which results in 83.2 Ω. This value is equal to the theoretical one.

# 4. IEEE model in a two layer soil

## 4.1. Zero volt equipotential at 3 cm

As in previous examples, it was supposed that the Dirichlet border was 3 cm away from the rod axis, accounting for 25% of rod resistance, a supposition that needs validation in a two layer soil. The results, for a 2 m length rod, 8 mm radius, buried at ground level, with an upper soil layer of 100 Ωm resistivity, and a 500 Ωm resistivity in the lower layer, are summarized in Table 2. Theoretical resistance was obtained using Tagg formula, as in reference (Tagg, 1964). The results are unacceptable. The assumption that 25% of resistance in a two layer soil is also in the first 3 cm is probably wrong. Changing the values of the layers resistivity for the same rod, the results are presented in Table 3.

The results are acceptable but would be better if they were greater than the theoretical ones, as a safe margin.

| Upper layer thickness (m) | Theoretical resistance (Ω) | FEM simulated value x4 (Ω) | error (%) |
|---|---|---|---|
| 0.5 | 134 | 104 | -22.4 |
| 1.0 | 93.4 | 69.4 | -25.7 |
| 1.5 | 72.0 | 52.0 | -27.8 |

**Table 2.** Resistance variation for positive voltage reflexion coefficient

| Upper layer thickness (m) | Theoretical resistance (Ω) | FEM simulated valuex4 (Ω) | error (%) |
|---|---|---|---|
| 0.5 | 60 | 51.9 | -13.5 |
| 1.0 | 77 | 69.1 | -10.3 |
| 1.5 | 114 | 103 | -9.6 |

**Table 3.** Resistance variation for negative voltage reflexion coefficient

## 4.2. Zero volt equipotential at 15 cm

The zero volt Dirichlet border was moved to 15 cm, where in homogeneous soil remains 50 % of the rod resistance. The mesh was changed improving the rod bottom discretization as shown in Fig. 2. The results in Table 4 were obtained for a 2 m length rod, 8 mm radius, buried at ground level, with an upper soil layer of 100 Ωm resistivity, and a 500 Ωm resistivity in the lower layer. The analysis of the results shows that the errors are quite high. The assumption that the first 15 cm contain 50 % of rod resistance seems to be incorrect. Interchanging the values of the layers resistivity, new results were obtained, and are shown in Table 5. In this case the errors are acceptable bur not conservative.

| Upper layer thickness (m) | Theoretical resistance (Ω) | FEM simulated value x2 (Ω) | error (%) |
|---|---|---|---|
| 0.6 | 124 | 103 | -17 |
| 1.2 | 83.4 | 67.1 | -20 |
| 1.8 | 63.4 | 49.6 | -22 |

**Table 4.** Resistance variation for positive voltage reflexion coefficient

| Upper layer thickness (m) | Theoretical resistance (Ω) | FEM simulated value x2 (Ω) | error (%) |
|---|---|---|---|
| 0.6 | 62.8 | 59.0 | -6 |
| 1.2 | 89.1 | 85.3 | -4 |
| 1.8 | 162 | 154 | -5 |

**Table 5.** Resistance variation for negative voltage reflexion coefficient

### 4.3. IEEE entire model discretization

In the last simulation the IEEE whole model was discretized. The complete solution mesh is presented in Fig. 9.

In order to validate the mesh, it was used the same conductance for all finite elements. It was chosen the value of 0.005 S/m. With this value the used rod, 2 m length and 8 mm radius, has a resistance, using Dwight formula, of 94 Ω. According to IEEE, this formula has a 13% excess to real value that should be of 83 Ω. The computer program developed returned a value of 87 Ω, which is 4.4 % higher than IEEE value, making the mesh sufficiently accurate.

The next step was to simulate a two layer soil model, with an upper soil layer of 100 Ωm resistivity, and a 500 Ωm resistivity in the lower layer. The results are presented in the Table 6.

| Upper layer thickness (m) | Theoretical resistance (Ω) | FEM simulated value (Ω) | error (%) |
|---|---|---|---|
| 0.5 | 134 | 118 | -11.9 |
| 1.0 | 93.4 | 81.4 | -12.8 |
| 1.5 | 72.0 | 62.0 | -13.9 |

**Table 6.** Resistance variation for positive voltage reflexion coefficient

Interchanging the values of resistivity layers, new results were obtained and are shown in Table 7. The results are acceptable, unfortunately by default. The last one is surprisingly high, but within 20% of theoretical value. It can be concluded than FEM meshes can provide useful results, considering the whole IEEE model in discretizing the problem.

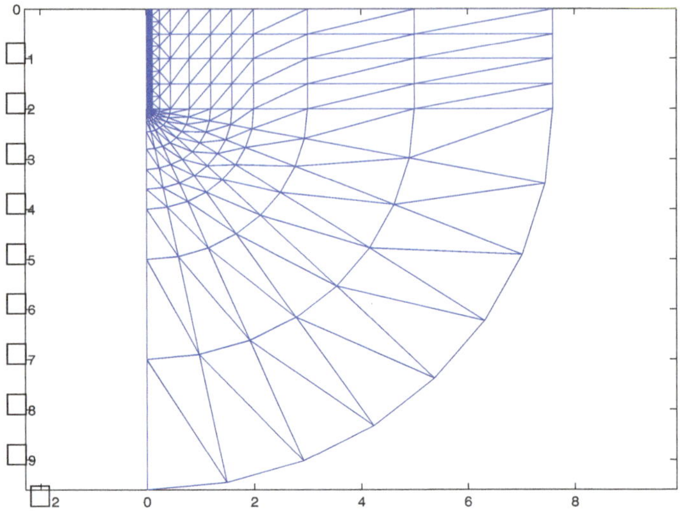


**Figure 9.** Entire IEEE model discretization

| Upper layer thickness (m) | Theoretical resistance (Ω) | FEM simulated value (Ω) | error (%) |
|---|---|---|---|
| 0.5 | 60 | 52.6 | -12.5 |
| 1.0 | 77 | 66.7 | -13.4 |
| 1.5 | 114 | 93.2 | -18.3 |

**Table 7.** Resistance variation for negative voltage reflexion coefficient

## 5. Field measurements

In order to experimentally validate the model, resistivity measurements were done in a sandy soil. The obtained values are presented in Table 8.

| Distance (m) | 0.5 | 1 | 2 | 3 | 4 | 5 | 6 | 8 |
|---|---|---|---|---|---|---|---|---|
| Resistance (Ω) | 552 | 207 | 41 | 7 | 4 | 2 | 3 | 2 |
| Resistivity (Ωm) | 1734 | 1301 | 515 | 132 | 101 | 63 | 113 | 108 |

**Table 8.** Resistivity variation with depth

The distance referred in the first row is the distance between test rods in Wenner method, as in (Telford et al., 1990). The resistivity curve is presented in Fig. 10. To choose a value for top layer resistivity, it was considered only the first resistivity measurement, since it wasn't found an upper asymptote. For bottom layer resistivity it was considered the average resistivity value of the last four points values, which are almost in a horizontal line. The average value is 94.5 Ωm. The upper layer thickness is obtained considering the point where concavity changes. The three first points are almost in a straight line, and concavity is detected only after the third point. It was considered the point where resistivity curve crosses the 400 Ωm ordinate where abscissae seem to be 2.3 m. Using this value in Lancaster-Jones rule one obtains (Lancaster-Jones, 1930):

$$\frac{3h}{2} = 2.3 \tag{30}$$

The upper layer thickness '$h$' is 1.53 m. This value was rounded to 1.5 m.

The ground rod was discretized as indicated in Fig. 9. The measured value was 108 Ω and the simulated value using FEM is 113 Ω, which is 4.6 % higher. The equipotential lines were calculated and presented in Fig. 11.

Equipotential lines in bottom layer are closer to rod, since this layer has a smaller resistivity.

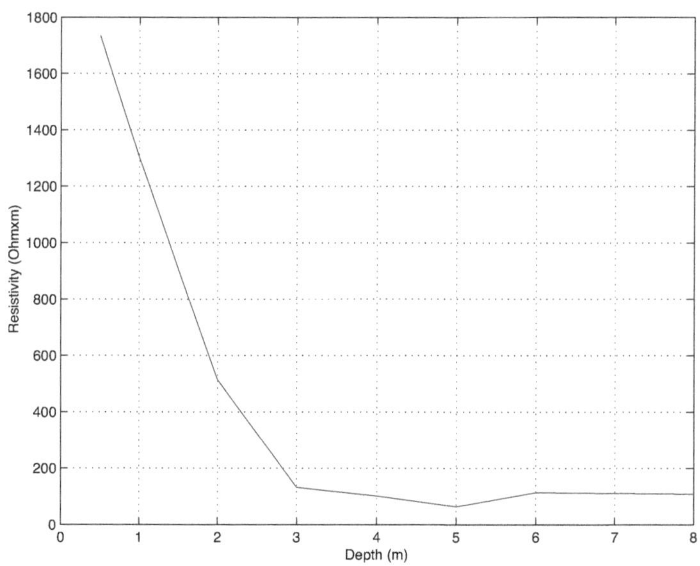


**Figure 10.** Resistivity measurements with depth

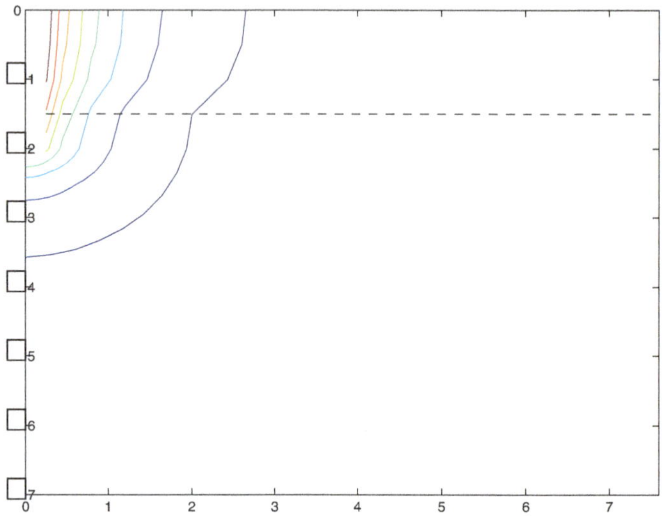


**Figure 11.** Equipotential lines with depth for a 2 m rod

## 6. Conclusions

In this chapter the Finite Element Method for the calculation of a rod resistance in a two layer soil was presented. The developed meshes were tested in homogeneous soil and in a cylindrical problem, which has theoretical solutions, and they are well adapted considering potential distribution or energy dissipation.

In homogeneous soil, 25% of rod resistance is in the 3 cm rod closest soil, according with IEEE and as validated in this work.

For an upper layer soil, with resistivity smaller than the resistivity of the lower layer soil, the assumption that 25% of rod resistance is in the nearest 3 cm and 50% in the nearest 15 cm is wrong and cannot be generalized.

For an upper layer soil, with resistivity bigger than the resistivity of the lower layer soil, the assumption that 25% of rod resistance is in the nearest 3 cm and 50% in the nearest 15 cm is acceptable, but the results are not conservative.

Discretizing the whole IEEE model allowed obtaining results with less than 20% error, but these results are not conservative. The whole mesh was tested considering equal resistivity, obtaining results similar to the ones gotten from the homogeneous soil simulation.

The comparison with the field measurement is good, since simulated value for resistance is only 4.6% higher.

## Author details

António Martins,
*Polytechnic Institute of Guarda, Portugal*

Sílvio Mariano and Maria do Rosário Calado
*University of Beira Interior, Portugal*

## 7. References

Adriano, U.; Bottauscio, O. & Zucca, M. (2003). Boundary Element Approach for the analysis and design of grounding systems in presence of non-homogeneousness, *IEE Proceedings Gener. Transm. Distrb.*, Vol. 150, No. 3, pp. 360-366.

Al-Arayny, A.; Khan, Y.; Qureshi, M. & Malik Puzheri, F. (2011). Optimized Pit Configuration for Efficient Grounding oh the Power System in High Resistivity Soils using Low Resistivity Materials, *4th International Conference on Modeling, Simulation and Applied Optimization*, pp. 1 – 5.

Berberovic, S.; Haznadar, Z. & Stih, Z. (2003). Method of moments in analysis of grounding systems. *Engineering Analysis with Boundary Elements*, Vol. 27, No. 4, (April 2003), pp. 351-360.

Coa, L. (2006). Comparative Study between IEEE Std.80-2000 and Finite Elements Method application for Grounding System Analysis, *Transmission & Distribution Conference and Exposition: Latin America*, pp. 1-5, Caracas, Venezuela, 15-18 Aug. 2006.

Colominas, I.; Aneiros, J.; Navarrina, F. & Casteleiro, M. (1998). A BEM Formulation for Computational Design of Grounding Systems in Stratified Soils, *Proc. Computational Mechanics: New Trends and Applications*, Buenos Aires, Argentine, 1998.

Colominas, I.; Gomez-Calvino, J.; Navarrina, F. & Casteleiro, M. (2002). A general numeric model for grounding analysis in layered soils. *Advances in Engineering Software*, Vol. 33, No. 7–10, (July–October 2002), pp. 641–649.

Colominas, I.; Navarrina, F. & Casteleiro, M. (2002). A Numerical Formulation for Grounding Analysis in Stratified Soils. *IEEE Transactions on Power Delivery*, Vol. 17, No. 2, (April 2002), pp. 587-595.

Dwight, H. (1936). Calculation of Resistances to Ground. *Transactions of the American Institute of Electrical Engineers*, Vol. 55, No. 12, (December 1936), pp. 1319-1328.

IEEE Std 142. (2007). *Grounding of Industrial and Commercial Power Systems*, IEEE.

Lancaster-Jones, E. (1930). The Earth-Resistivity Method of Electrical Prospect. *Mining Magazine*, Vol. 42, pp. 352-355.

Liu, Y.; Theethayi, N.; Thottappillil, R.; Gonzalez, R. & Zitnik, M. (2004). An improved model for soil ionization around grounding systems and its application to stratified soil. *Journal of Electrostatics*, Vol. 60, No. 2-4, (March 2004), pp. 203-209.

Ma, J.; Dawalibi, F. & Southey, R. (1996). On the equivalence of uniform and two-layer soils to multilayer soils in the analysis of grounding systems. *IEE Proceedings Gener. Transm. Distrb.*, Vol. 143, No. 1, (January 1996), pp. 49-55.

Manikandan, P.; Rajamani, M.; Subburaj, P. & Venkatkumar, D. (2011). Design and Analysis of grounding systems for Wind turbines using Finite Element Method, *International Conference on Emerging Trends in Electrical and Computer Technology*, pp. 148-153, Tamil Nadu, India, 23-24 March 2011.

Martins, A. & Antunes, C. (1997). Hemispheric- Cylindrical Numeric Modeling of a Ground Electrode Using Voltage Adapted Finite Elements. *Proccedings of the 5ª Jornadas Hispano-Lusas de Ingenieria Electrica,* pp. 641-648, Salamanca, Spain, July 1997.

Purcell, E. (1988). *Berkeley Physics Course,* Edgar Blucher Ltda.

Salama, M.; Elsherbiny, M. & Chow, Y. (1995). Calculation and interpretation of a grounding grid in two-layer earth with the synthetic-asymptote approach. *Electric Power Systems Research,* Vol. 35, No. 3, (December 1995), pp. 157-165.

Sharma, C. & Four, S. (2006). Parametric Analysis of Grounding Systems in Two-Layer Earth using Galerkin's Moment Method, *Proc. Transmission and Distribution Conference and Exhibition,* pp. 541-547, Dallas, TX, USA, 21-24 May 2006.

Sylvester, P & Ferrari, R. (1990). *Finite Elements for Electrical Engineers*. Cambridge University Press.

Tagg, F. (1964). *Earth Resistances,* George Newues Limited.

Telford, W., Geldart, L., Sheriff, R. (1990). *Applied Geophysics,* Cambridge University Press.

Zhenghua, F.; Ling, L. & Junzhong, F. (2011). Researh on Reducing Grounding Resistance of Transmission Line Tower Grounding Grid, *International Conference on Electrical and Control Engineering,* pp. 1216-1219, Yichang, China, 16-18 September 2011.

# Chapter 8

# FEA in Micro-Electro-Mechanical Systems (MEMS) Applications: A Micromachined Spatial Light Modulator (μSLM)

Hao Ren and Jun Yao

Additional information is available at the end of the chapter

## 1. Introduction

Since the Nobel Prize winner, Richard Feynman gave the presentation "there is plenty of room at the bottom" [1], a variety of micromachined sensors, actuators, and systems have emerged and made encouraging progress in the past 50 years, based on technological innovations and increased market demand [2]. To date, Micro-Electro-Mechanical Systems (MEMS) have been developed into an interdisciplinary subject which involves electrical, mechanical, thermal, optical, and biological knowledge. Due to its significant potential, which has partially been demonstrated by the success of inertial MEMS devices (accelerometers, gyroscopes, *etc* [3, 4]) radio frequency (RF) MEMS devices (switches, filters, resonators, *etc* [5-7]) and optical MEMS devices (Digital Light Processing, DLP [8, 9]), the research in MEMS has attracted worldwide interest. Figure 1 shows a typical process of a MEMS device from design goal to system integration. We can see that the structure and fabrication process of MEMS device are designed according to the design goal. Then before fabrication, we need to perform modeling to the structure. By modeling, we can estimate the performance to see if it satisfies the design goal and then optimize it to achieve the best performance. By performing modeling, substantial time and money can be saved, which increases the throughput and reduces the cost. As a result, modeling is critical for MEMS research.

Modeling applied in MEMS applications can mainly be divided into two categories, theoretical modeling and numerical modeling. The theoretical modeling is to apply exact equations to obtain exact solutions. It is a direct approach which is easy to interpret intuitively [10]. However, it has limitations that solutions can only be obtained for few standard cases, and it is incapable or difficult in the following situations: (1) shape, boundary conditions, and loadings are complex; (2) material properties are anisotropic; (3) structure has more than one material; (4) problems with material and geometric non-linearity; (5) multiphysics situations when more than two physics are coupled together. The theoretical modeling is sometimes applied in MEMS applications when the structure is not complex, and it is also useful to verify the result of FEA.

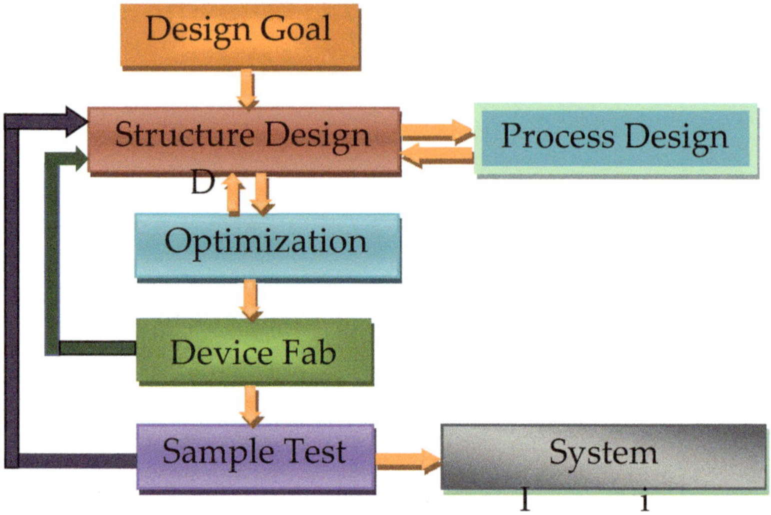


**Figure 1.** A schematic of a typical process of a MEMS device

Numerical modeling is to apply exact equations to obtain approximate solutions only at discrete points called nodes. Contrary to theoretical modeling, numerical modeling can handle situations which theoretical modeling is incapable. Finite Element Method (FEM) and Finite Difference Method (FDM) are two approaches most frequently used in numerical modeling. For both methods, they start from discretization, which derives the solution domain into a number of small elements and nodes. For FDM, differential equation is written for each node, and the derivatives are replaced by difference equations. In contrast, for FEM, it uses integral formulations rather than difference equations to create a system of algebraic equations, and an approximate continuous function is assumed to represent the solution for each element. The complete solution is then generated by connecting or assembling the individual solutions, allowing for continuity at the interelemental boundaries [11]. FEM have quite a few advantages over FDM [10], such as (1) it can give values at any point, while FDM can only give value at discrete node points; (2) FEM can consider the sloping boundaries exactly, while FDM makes stair type approximation to sloping; (3) FEM needs fewer nodes to get good results while FDM needs large number of nodes; (4) FEM can handle almost all complicated problems, while FDM cannot handle complicated problems, such as multiphysics simulation which is the general case in MEMS applications.

Due to the aforementioned advantages, FEA has been widely applied in MEMS applications, including electromagnetic simulation, electrothermal simulation, thermoelectromechanical simulation, piezoelectric/piezoresistive simulation, microfluidics simulation, *etc.* Quite a few commercially available FEA softwares are readily used in MEMS applications, such as Intellisuite, ANSYS, COMSOL, Conventorware. Of them, Intellisuite is specially designated for MEMS simulation with quite a few modules, including Intellimask, Intellifab, MEMeterial, 3D Builder, Thermoelectromechanical modules, *etc.*, which has user friendly interface that users can obtain 3D structure directly from defined masks and fabrication process, and the 3D structure can be directly applied to further simulation [12]. What is more, the simulation result is quite close to experiment. In this chapter, we use Intellisuite for simulation.

In this chapter, we show the importance of FEA in MEMS research through an example of a micromachined spatial light modulator (μSLM). Firstly we will introduce the design and operating principle of the μSLM. Then we will introduce the modeling of the μSLM, including theoretical modeling and FEA modeling. Following is the optimization of the μSLM according to the modeling. Finally we present the fabrication and experiment.

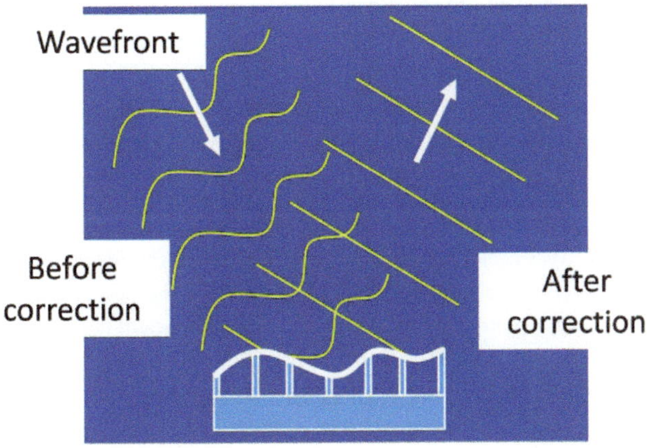


**Figure 2.** A schematic of the operating principle of spatial light modulator (SLM) used in Adaptive Optics systems, SLM is used to corrected the wavefront

Spatial Light modulators (SLMs) play an important role in modern technology, particularly in the field of micro-optical technology. They find applications in optical communication systems, and adaptive optics (AO) systems [13]. AO systems perform closed-loop phase correction of time-varying, aberrated wavefronts using two essential components: a wavefront sensor and a SLM [14]. A schematic of the operating principle of SLM is illustrated in figure 2. When light from stars travel through the atmosphere, aberration is induced by the turbulence in atmosphere, and the wavefront of the light is no longer a plane. If this wavefront is imaged by a telescope, a very blurry image is formed, therefore in Adaptive Optics area, researchers use SLM to correct the wavefront to be plane.

Conventional SLMs based on piezoelectric actuators cost approximately $1000 per actuator and therefore find limited use even at major research centers [15]. In contrast, MEMS technology offers a potentially low cost alternative to existing SLMs: the μSLMs. A large problem for μSLMs is their small stroke (maximum displacement), which greatly undermines the performance of the whole AO system. As a result, researchers have tried quite a few approaches to enlarge the stroke, but these approaches are difficult to implement because of either fabrication difficulty or structural complexity. As a result, we come up with a μSLM based on the leverage principle in this chapter to solve the problem.

## 2. Design and modeling of the μSLM

### 2.1. Design and operating principle of the μSLM

A schematic of the μSLM is shown in figure 3(a). It can be seen that the μSLM is composed of four single out-of-plane actuators and one mirror plate (here only 1/4 mirror plate is shown), which are connected together by a via. From figure 2(b) we can see that one single out-of-plane actuator is composed of two anchors, two microbeams, one lower electrode, one upper electrode (serving as the short arm), and one long arm. The lower electrodes and two anchors are fixed to substrate and the long arm is connected to the two anchors by two microbeams. The size of each actuator is shown in figure 3(b). $W_1$, $L_1$, $a$, and $L$ represent the width and the length of the upper and lower electrodes, and $l_1$, $b_1$, $l_2$ and $b_2$ denote the length and the width of the microbeams and long arms, respectively, while $h$ is the thickness of the structural layer and $d$ is the horizontal distance from the fulcrum to the central line of bottom electrode.

When the lower and upper electrodes are subjected to different potentials, electrostatic attractive force arises. A torque around the microbeams emerges and makes the upper electrode and the long arm rotate around the microbeams, as shown in figure 3(c). As a result, the end of the long arm goes upward, thus forming a lever mechanism, with the microbeams as a fulcrum. If the length of the long arm is much larger than that of the short arm, the downward displacement will be magnified to be a much larger upward displacement. At the same time, microbeams will bend down due to the moment from the electrostatic force. The total displacement of the mirror plate is the vector sum of the upward displacement caused by the rotation and the downward displacement of the microbeams (Here we assume the displacement of the mirror plate is the same as the displacement at the end of the long arm, because in MEMS applications the mass of the mirror plate can be neglected). By a proper design of the structure, the downward displacement of the microbeams will be much smaller than the upward displacement at the end of the long arm, resulting in a larger upward displacement of the mirror plate [16].

### 2.2. Modeling of the μSLM

After presenting the structure and operating principle, we modeled the μSLM both by theoretical modeling and FEA. First we carried out theoretical models. Two approaches were used in the theoretical modeling: the energy method and the superposition method.

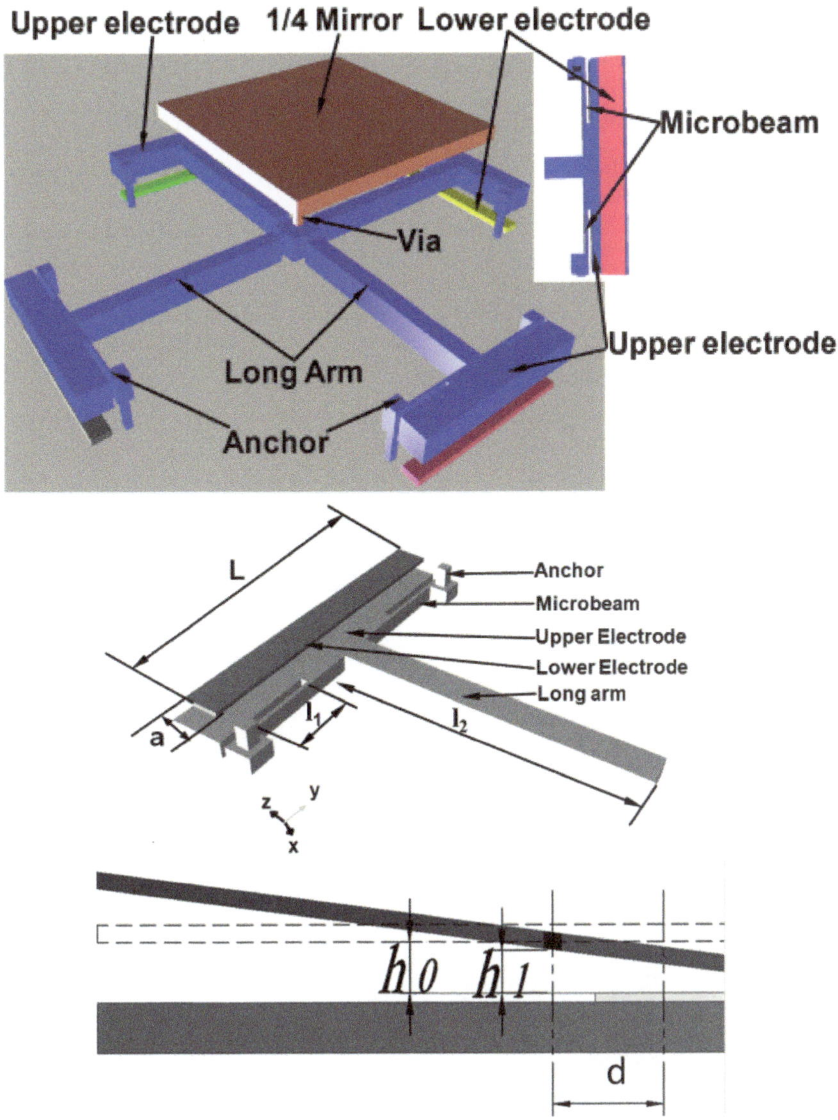


**Figure 3.** (a) A lateral view of a μSLM, it is composed of four single out-of-plane actuators and one mirror plate, each single out-of-plane acutator is composed of two anchors, two microbeams, one lower electrode, one upper electrode (serving as the short arm), and one long arm, (b) A lateral view of a single out-of-plane actuator, (c) cross-section view of the single out-of-plane actuator in figure 2(b) when a voltage is applied between the lower and upper electrode, a small displacement at the end of short arm will be amplified to be a larger displacement at the end of the long arm

In the energy method, the force and moments applied to the structure is shown in figure 4. When different potentials are subjected to the upper and lower electrodes, electrostatic force arises. As mentioned above, the upper electrodes and long arms rotate around microbeams, and at the same time microbeams bend down. The force, moment and torque of anchors tend to resist this movement and the structure will ultimately reach a balance.

When a voltage $V$ is applied to the four upper and the four lower electrodes, the electrostatic attractive force can be calculated by [16]:

$$F_e = \frac{\varepsilon\varepsilon_0 LaV^2}{2h_0^{\,2}}[1 + \frac{2d}{h_0}\theta] \tag{1}$$

In our design, $2d/h_0$=20, and $\theta$<<1. In order to simplify our calculation, the second term can be omitted without bringing much error.

The electrostatic force can be simplified and rewritten as follows

$$F_e = \frac{\varepsilon\varepsilon_0 LaV^2}{2h^2} \tag{2}$$

In order to derive the displacement at the mirror, we used Castigliano's second theorem and set a fictitious load at the central mass, as shown in Figure 4.

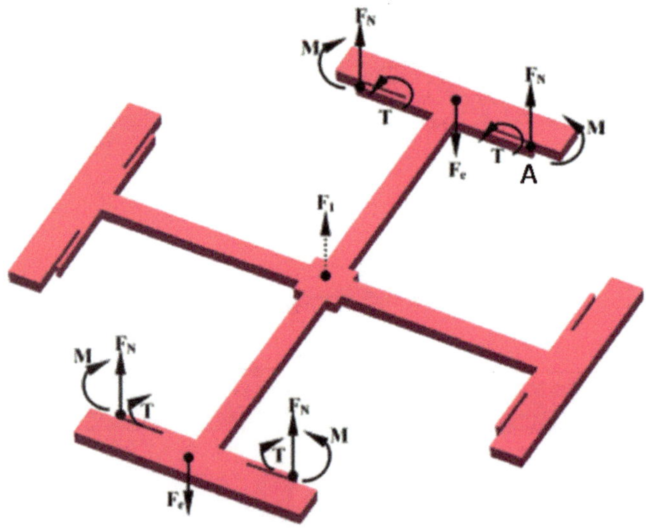


**Figure 4.** Mechanical model of the energy method for the μSLM, forces, moments and torques of one single out-of-plane actuator is shown for simplification.

Considering the symmetry of the structure, the μSLM can be divided into four single out-of-plane actuators and each one can be further subdivided into four parts: two microbeams, one long arm and one upper electrode (short arm). In our case, the width of upper electrodes is much larger than the other three parts and the length is smaller, therefore we assume upper electrodes are rigid.

According to equations of force equilibrium, we get

$$F_N = F_e/2 - F_1/8 \tag{3}$$

As shown in figure 4, the bending moment and the torque applied to two microbeams and the long arm can be given by

$$M_1 = M_2 = F_N x - M,\ T_1 = T_2 = T \tag{4}$$

$$M_3 = F_e(a+x) - 2F_N x - 2T = F_e a + F_1 x/4 - 2T,\ T_3 = 0 \tag{5}$$

According to the virtual work principle, the total strain energy in one single actuator is

$$U_1 = \int_0^{l_1} \frac{M_1^2(x)}{2EI_2} dx + \int_0^{l_1} \frac{M_2^2(x)}{2EI_2} dx + \int_0^{l_2} \frac{M_3^2(x)}{2EI_1} dx + \int_0^{l_1} \frac{T_1^2(x)}{2GJ} dx + \int_0^{l_1} \frac{T_2^2(x)}{2GJ} dx \tag{6}$$

Where $E$, $G$, $J$, $I_1$, and $I_2$ represent the Young's modulus, the shear modulus, the polar moment of microbeams, the inertial moment for the long arm and the microbeams, respectively.

As the total strain energy stored in the lever actuator is four times of that in single actuator, and the bending and torsional angles at point A (see in figure 4) are both zero, according to Castigliano's second theorem

$$\frac{\partial U_{total}}{\partial M} = 0,\ \frac{\partial U_{total}}{\partial T} = 0 \Rightarrow M = \frac{1}{2} F_N l_2,\ T = \frac{1}{8} \frac{(F_1 l_2^2 + 8F_e l_2 d)GJ}{EI_1 l_2 + 2GJl_2} \tag{7}$$

Based on the unit-load method, equation (7) can be rewritten as

$$\begin{aligned} z = &\int_0^{l_1} \frac{M_1(x)}{EI} \frac{\partial M_1(x)}{\partial F_1} dx + \int_0^{l_1} \frac{M_2(x)}{EI} \frac{\partial M_2(x)}{\partial F_1} dx + \int_0^{l_2} \frac{M_3(x)}{EI} \frac{\partial M_3(x)}{\partial F_1} dx \\ &+ \int_0^{l_1} \frac{T_1(x)}{GJ} \frac{\partial T_1(x)}{\partial F_1} dx + \int_0^{l_1} \frac{T_2(x)}{GJ} \frac{\partial T_2(x)}{\partial F_1} dx \end{aligned} \tag{8}$$

By combining the above equations to (8) and letting the fictitious load $F_1$ be zero, we have

$$z = -\frac{F_e l^3}{24EI_2} + \frac{1}{GJ} \frac{F_e l_1^3 l a G^2 J^2}{(EI_1 l + 2GJl_1)^2} + \frac{1}{EI_1}(F_e a - \frac{2F_e l_1 a GJ}{EI_1 l + 2GJl_1})(\frac{l_1^2}{2} - \frac{l_1^3 GJ}{EI_1 l + 2GJl_1}) \tag{9}$$

In the superposition method, the displacement of the mirror plate is the sum of the upward displacement of the mirror plate and the downward displacement of the microbeams.

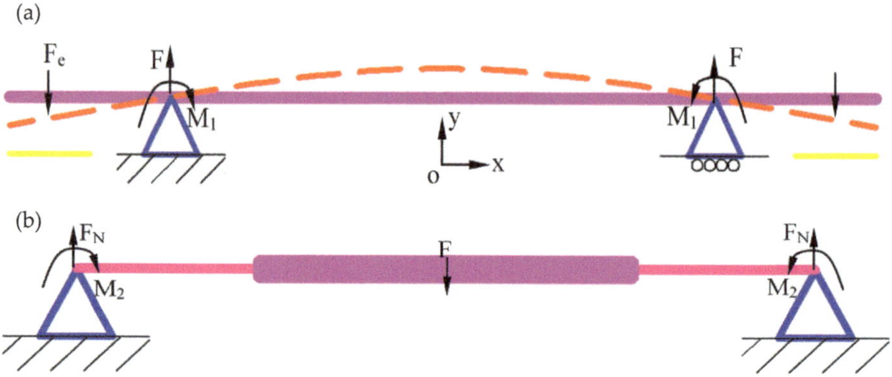


**Figure 5.** Mechanical model of the superposition method: (a) shows the forces and moments of the long and short arm, (b) shows the forces and moments of the 2 microbeams.

First we will analyze the force and moment applied on the long arm, as shown in figure 5. According to the boundary condition

$$y_1(0)=0, y_1'(0)=0, y_1'(l_2)=\theta_1, \theta_1 = M_1 l_1 / 2GI_p \tag{10}$$

We can obtain the displacement of the center of the two long arms:

$$y = \frac{1}{2}\frac{\varepsilon w(b-a)(b+a)V^2 l_2{}^2}{4h^2 EI}\left(1-\frac{1}{1+EIl_1 / 2l_2 GI_p}\right) \tag{11}$$

The mechanical model of 2 microbeams is illustrated in Figure 5(b). It is an indeterminate beam with variable sections, therefore we can derive the displacement of center of the two microbeams using transfer matrix method [17]. First we divide it into 4 sections. The transfer matrix from the left end of section 1 to the right end of section 4 can be obtained

$$C = A_1 \times A_2 \times B \times A_2 \times A_1 \tag{12}$$

Where $A_1$, $A_2$, and $B$ are transfer matrix between left end and right of section 1, left end and right of section 2, right end of section 2 and left end of section 3, which can all be calculated by law of transfer with cross section state vector.

After applying boundary condition, $y_{1L}$=0, $\theta_{1L}$=0, $y_{4R}$=0, $\theta_{4R}$=0 we can calculate force and moment applied on the left end of section 1 and right end of section 4. Then substituting the force and moment to

$$C_1 = A_2 \times A_1 \tag{13}$$

We can calculate the transfer matrix from the left end of section 1 to the right end of section 2. Therefore according to

$$Z_{2R} = C_1 \times Z_{1L} \tag{14}$$

Where $Z_{2R}$, $Z_{1L}$ are the section vectors of the left end of section 1 and the right end of section 2, respectively. We can calculate the displacement of the center of various section beams. By substituting parameters in to these equations, we calculated that the downward displacement is 32.4 nm when applied a voltage of 20 V.

Finally the displacement calculated by the superposition method is derived as

$$z = \frac{1}{2}\frac{\varepsilon w a W_1 V^2 l_1^{\,2}}{2h^2 EI_1}\left(1 - \frac{1}{1 + EI_1 l / 2l_1 GJ}\right) - y_2 \tag{15}$$

Here $y_2$ is calculated by the transfer matrix method.

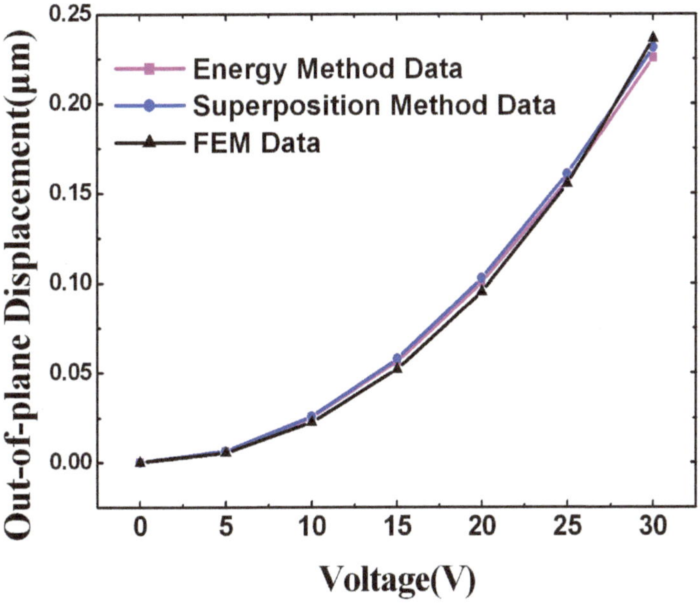


**Figure 6.** Results of the two theoretical models and FEA results, the discrepancy between theoretical modeling and FEA is small

After deriving the theoretical models, we carried FEA modeling using Intellisuite. Then the FEA result is validated the theoretical modeling by comparing them with FEA. As shown in figure 5, both the two theoretical models are in good agreement with FEA, and this validates the results of the theoretical modeling and FEA. Then we combined the theoretical modeling and FEA to optimize the structure.

## 3. Structure optimization

In this section, we discuss the optimization of the structure, including the optimization of the microbeams, the long arm, the end of the long arm, the connection between the four single out-of-plane actuators, and the upper electrode.

### 3.1. The long arm

#### *3.1.1. Width of the long arm*

When studying the influence of the width of the long arm, we calculated the differential of equation (9) to $b_2$

$$\frac{\partial z}{\partial b_2} = \frac{\partial z}{\partial I_1}\frac{\partial I_1}{\partial b_2} = -\frac{1}{2}\frac{EF_e l_1^2 l_2^2 a}{(EI_1 l_1 + 2GJl_2)^2}\frac{h^3}{12} \tag{16}$$

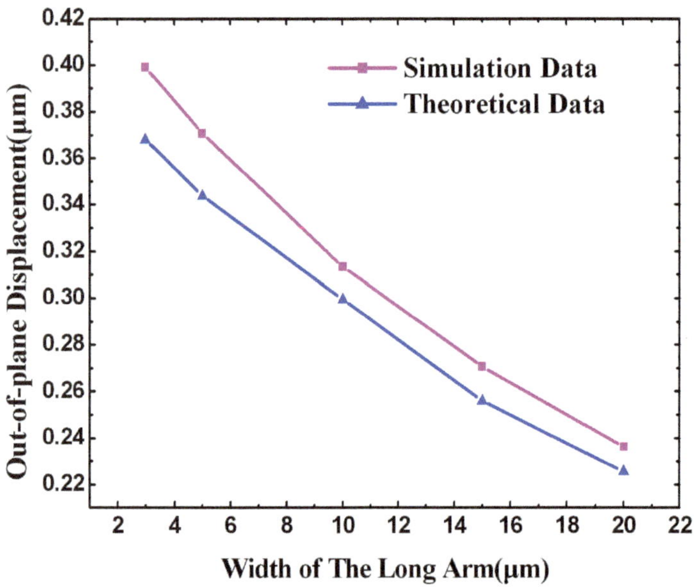


**Figure 7.** Theoretical and simulation data of width of long arm versus displacement, as the with of the long arm increases, the displacement decreases

It is obvious that the differential is constantly negative. Therefore the larger the width of the long arm, the smaller the displacement is when subjected to the same voltage. The simulation and theoretical data of the mirror plate displacement versus the width of long arm is shown in figure 7. Therefore as the long arm becomes wider, the out-of-plane displacement decreases. However, if the width of the long arm is very small, it tends to be more fragile and more likely to break during fabrication and test.

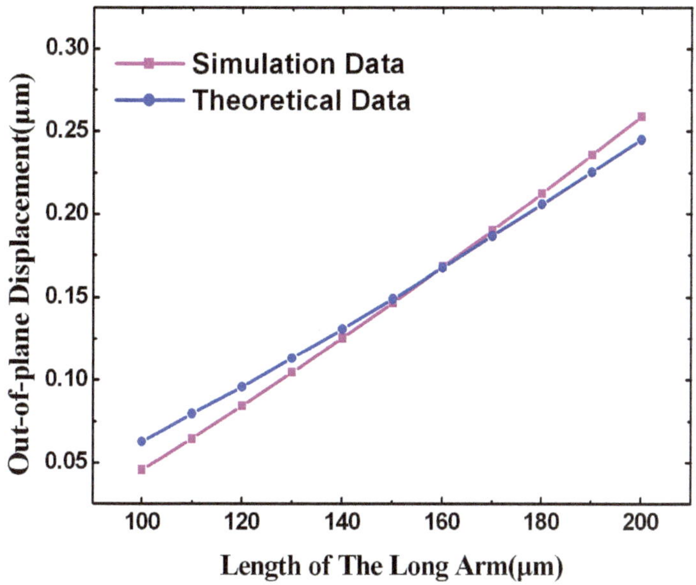


**Figure 8.** Theoretical and simulation data of length of long arm versus displacement, as the length of the long arm increases, the out-of-plane displacement increases

### *3.1.2. Length of the long arm*

Then the influence of the length of the long arm is discussed. As it is well known, the longer the long arm is, the larger out-of-plane displacement will be. The simulation and theoretical displacement are in good agreement with what is expected, as shown in figure 8. However, as mentioned above, a too long arm is very fragile and is more likely to be broken.

## 3.2 .The microbeams

After deriving characteristics of long arm, we went on with the microbeams. In this section, the influence of the width and length of microbeams is discussed. First, we come to the width of microbeams. It's easy to see that as the width of microbeam becomes larger, so does the torsional stiffness, which will thwart the rotation of the microbeams and make the out-of-plane displacement smaller. This is verified by FEA.

Second, the influence of the length of microbeams is discussed. Figure 9 is the simulation data of out-of-plane displacement versus microbeam length. As the microbeam length increases, so does the out-of-plane displacement, however, the increase rate slows down. This is because when the length of microbeam is small, the torsional stiffness is relatively larger, as mentioned above, thwarting the microbeams from rotating. In contrast, as the length increases, the downward bending displacement of microbeam increases, as shown in equation (9). As a result, the upward displacement is partly offset by downward bending of microbeams.

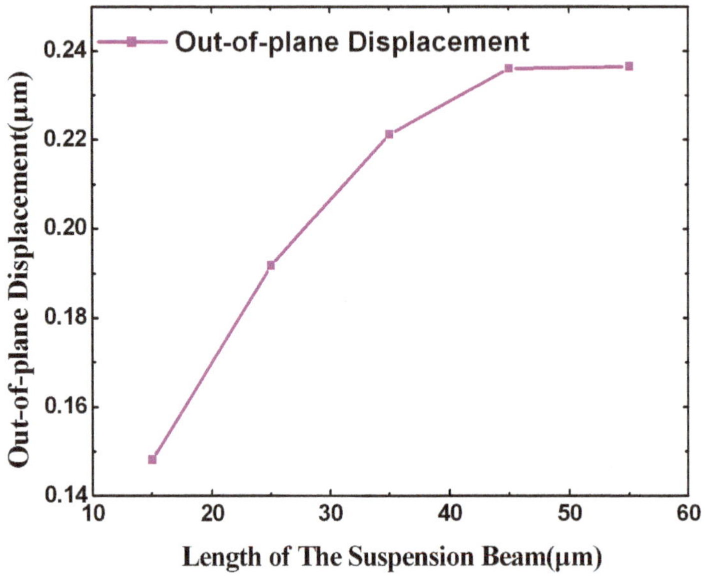


**Figure 9.** Simulation data of length of microbeams versus displacement, displacement increases with the increase of the length of microbeams, and the rate of increment slows down gradually.

There is one thing to notice when the microbeams are too short. When applied a relatively high voltage, the microbeams may fracture due to torsional stress. According to mechanics of materials, the maximum torsional stress can be calculated as follows

$$\tau_{\max} = \frac{T}{\alpha b_1 h^2} \tag{17}$$

where $\tau_{max}$, $\alpha$, $T$, $b_1$ and $h$ represent the maximum torsional stress, coefficient related to $b_1/h$, torque of the microbeams, width and height of microbeams. Also we know that

$$\theta = \frac{Tl_1}{GJ} \tag{18}$$

where $\theta$ and $T$ are rotational angle and torque, respectively.

Assuming the end of short beam has a displacement of 0.5 μm, the rotation angle is about 0.0125 radian, then it can be calculated that $\tau_{max}$=45.97 MPa. Here the parameters we use are

$\alpha$=0.231, G=65 GPa. According to ref. [18], using these two equations above, we calculate that $\tau_{max}$=60.2 MPa and the portion of stress-strain curve is about 0.00076. We set G=79 GPa [19] and $\alpha$=000.299, other parameters are based on reference [19]. Then according to reference [19], the fracture stress of polysilicon is two to ten times smaller than single crystal silicon, therefore at this time, when microbeam is shorter than 22.9 μm, it will fracture when applied a high voltage.

### 3.3. The end of the long arm

According to theoretical modeling and FEA, we found that the end of the long arm greatly confines the displacement. As a result, we made three optimizations to the structure.

First, we presented a long arm with variable sections. The structure of the long arm is shown in figure 10, which consists of two different sections, one has a width of 3 μm and the other 20 μm. The total length of the two sections is set to be 190 μm. When the length of the thin 3 μm width section changes, the theoretical calculation and simulation of the thin long arm length versus the displacement is plotted in figure 11, the theoretical data is calculated using equation (A. 4) (appendix) in reference [16]. We can see that the displacement of the mirror plate increases remarkably when the length of thin long arm varies from 0 μm to 40 μm, this is attributed to the fact that the implementation of a thin long arm at the end of long arm makes the confinement of opposite levers to decrease, therefore, it makes the long arm easier to rotate. In contrast, the displacement changes little, just from 0.5 μm to 0.54 μm, when the length of thin long arm varies from 40 μm to 120 μm. This is because when the length of thin long arm increases, the confinement of the end of long arms decreases, it makes the out-of-plane displacement to increase. However, when the thin long arm become longer, the bending of the long arm increases and this makes the out-of-plane displacement decrease. The increased displacement, which is caused by a decreased confinement, is pulled back by the decrease displacement caused by the increased bending for longer thin arms. When the thin long arm is longer than 120 μm, the out-of-plane displacement demonstrates a remarkable decrease, as this thin long arm makes too much bending. Since it doesn't change much from 40 μm to 120μm, we set the thin long arm 44 μm, for the reason that according to design rules, the shorter the thin long arm, the more robust it is in fabrication. This is the first structure after optimization (Structure 1).

The second optimization of the structure was to add a crab-leg beam, as illustrated in figure 12. By this means, the structure is more compact while at the same time achieving a larger displacement. We can see that both the displacement and amplification factor increase as the length of the crab-leg beam becomes longer. As a result, we set the length of the crab-leg beam to be 66 μm according to design rules, and we obtained the second optimized structure (Structure 2). The third optimization was to add a gimbal-like serpentine beam, as demonstrated in figure 13 [11]. This further reduces the confinement of the end of the long arm, thus achieving a larger displacement.

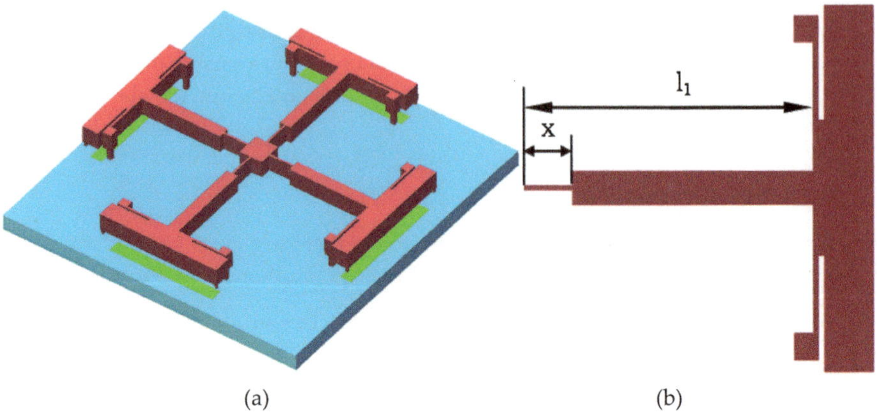


**Figure 10.** Schematic of the first optimized structure: (a) Lateral view of the structure in optimization, (b) Top view of a single actuator in optimization

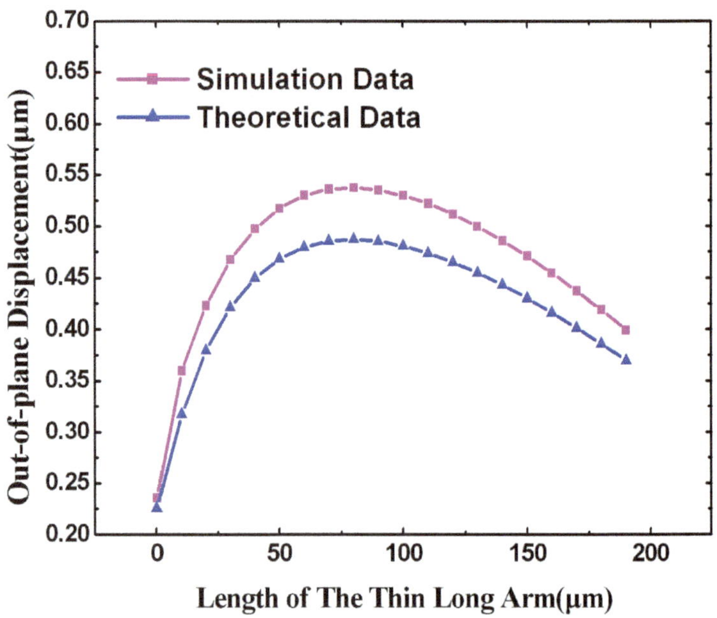


**Figure 11.** Theoretical and simulation data of length of thin long arm versus displacement, as the length of the thin long arm increases, the displacement firstly increase, then after hitting a maximum, it decreases as length of thin long arm increase

### 3.4. The connection between the four single out-of-plane actuators

After optimizing the end of the long arm, we investigated the connection of four actuators. It is found that the first mode of the natural frequency of the structure 3 was the rotation along the dotted line $l_1$ and $l_2$ in figure 13, other than the out-of-plane movement. This was because there was only one connection in the structure, and the restriction to this mode was smaller than the mode of out-of-plane movement. This was not desirable for inducing mechanical instability when working at a high frequency. We made an optimization to the structure, by making four separate connections to connect the four actuators to the mirror, as depicted in figure 13. Through this method, we can enlarge the restriction of the first mode, thus making out-of-plane mode to be the first mode. After this optimization, the piston mode became the first mode, which had a resonant frequency of 4.8 kHz.

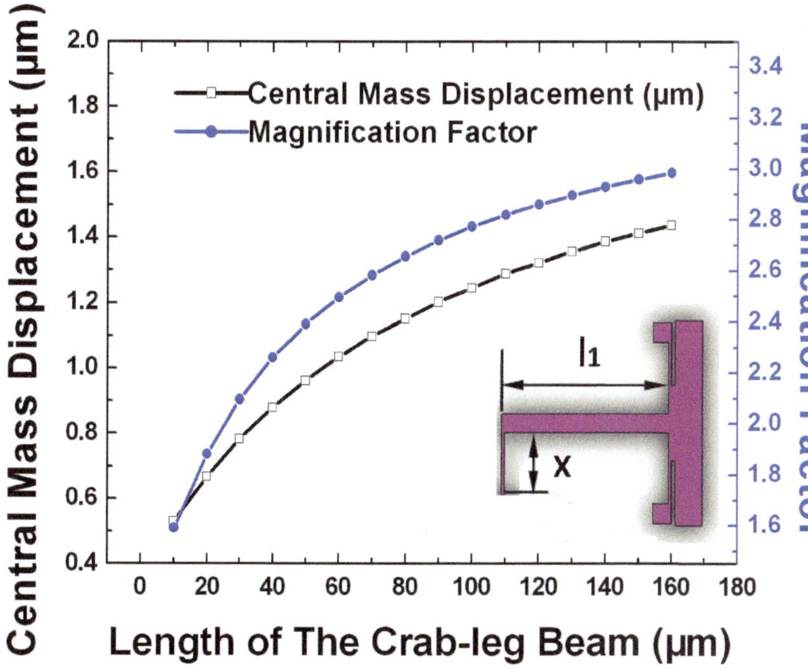


**Figure 12.** Displacement and Magnification factor of the second optimized structure

### 3.5. The upper electrode

Then we made an optimization to the upper electrode by introducing the third layer polysilicon to the upper electrode, thus enlarging the gap from 2 μm to 2.75 μm, as illustrated in figure 13, by this approach, we obtained the third optimized structure (Structure 3).

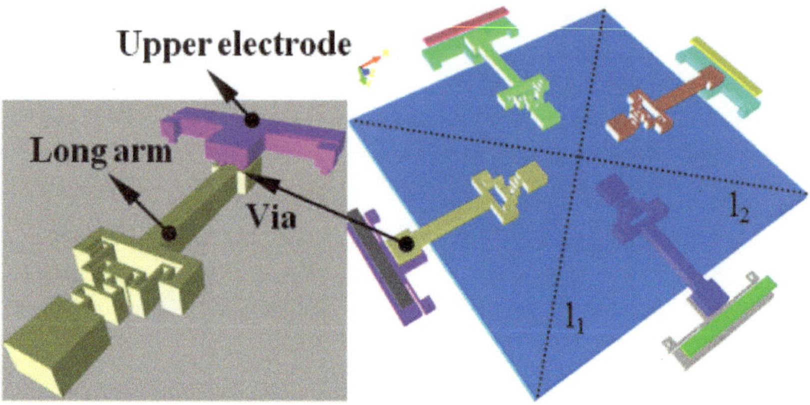


**Figure 13.** Lateral view of the third optimized μSLM

### 3.6. Results after optimization

The results after optimization is shown in figure 14. From this figure we can see that after optimization, the maximum displacement are 1.58 μm , 1.87 μm, and 4.5 μm, which are 3.04, 3.6, 8.65 times, respectively, higher than the structure before optimization. After optimization, we did experiment on the fabrication and test of the structures.

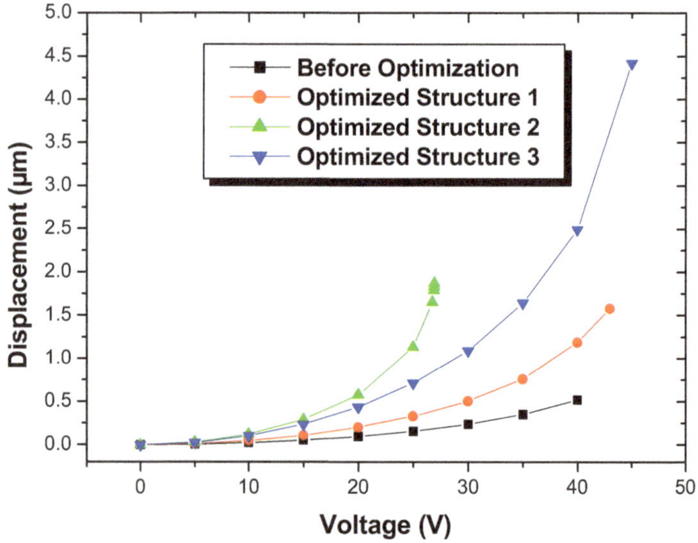


**Figure 14.** Simulation results before and after optimization, after optimization, the maximum displacement are 1.58 μm , 1.87 μm, and 4.5 μm, which are 3.04, 3.6, 8.65 times higher than the structure before optimization

## 4. Fabrication and test

The three optimized structures were fabricated by a three-layer polysilicon surface microfabrication process. First of all, 600 nm low-stress silicon nitride is deposited on an n-type (100) wafer with a diameter of 150 mm to form electrical isolation layer. Then 500nm polysilicon film is deposited as the first polysilicon layer (Poly0). Afterward, Poly0 is patterned by photolithography and etched. Then 2 μm of phosphosilicate glass (PSG) is deposited as sacrificial layer. Then the first silicon dioxide layer (Oxide1) is patterned by lithography and etched to form dimples. The following step is to deposit a 2 μm polysilicon layer (Poly1) as the second polysilicon layer, which is etched afterwards to form the leverage mechanism. At last the Oxide1 layer is sacrificed in a bath of 49% HF to release the structural layer and the structure is dried by supercritical $CO_2$ drying technique. Figure 16 shows the SEM photograph of the structures. We can see the shapes of the structures are good with little curvatures, indicating that the stress gradient and stress variation along the beams is negligible.

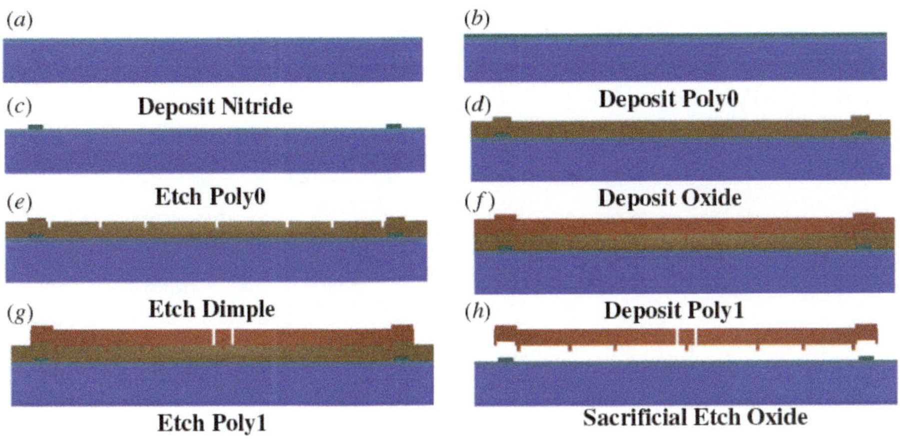


**Figure 15.** Fabrication process of the μSLM, this is a three-layer surface microfabrication process, and this schematic is based on the fabrication of the first optimized structure

Then we use an optical interferometer to measure the displacement versus voltage for the three optimized structures. The test is performed using the Zygo Newview 7300 (Zygo Inc., CT, USA) A light source, in this case an incoherent broadband LED light source is split at the objective so that some of the light passes to a reference mirror and some is focused onto the surface of the sample under measurement. Light from the mirror (embedded into the interference lens) and the sample surface is reflected back into the instrument and imaged onto a camera. If the distances from the light splitter to the mirror and from the splitter to the surface are equal so that there is no optical path difference (OPD) then the camera will observe an interference pattern. This occurs when the objective is held so that the focal plane of the objective lies in the same plane as the surface. In order to perform a measurement of the surface observed by the field of view of the objective, the objective lens is translated vertically and linearly so that the focal plane moves through the entire height range of the surface being measured. As it does so, the interference fringes will move and follow the height profile of the surface and this information is processed by the instrument to calculate the height profile to a very high precision (0.1 nm).

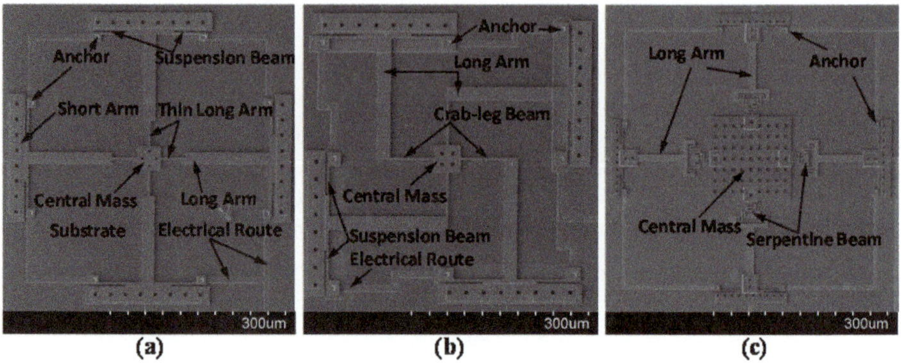


**Figure 16.** SEM photographs of the three optimized structures

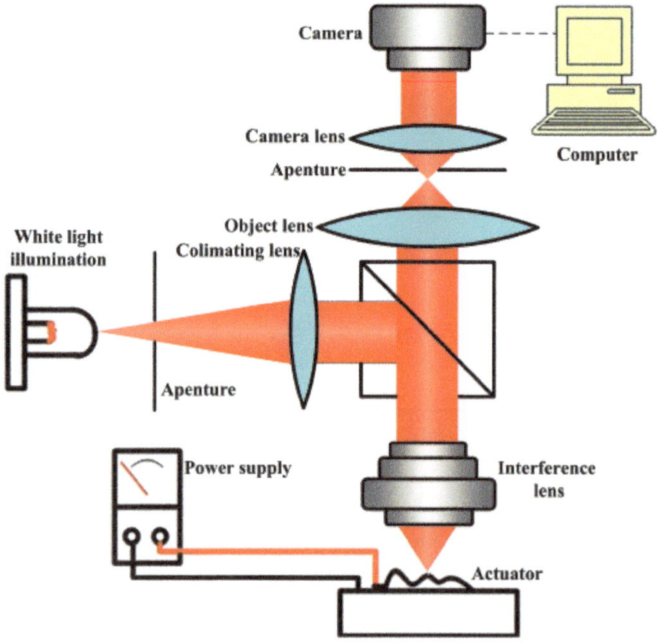


**Figure 17.** Optical configuration of the white light interferometer

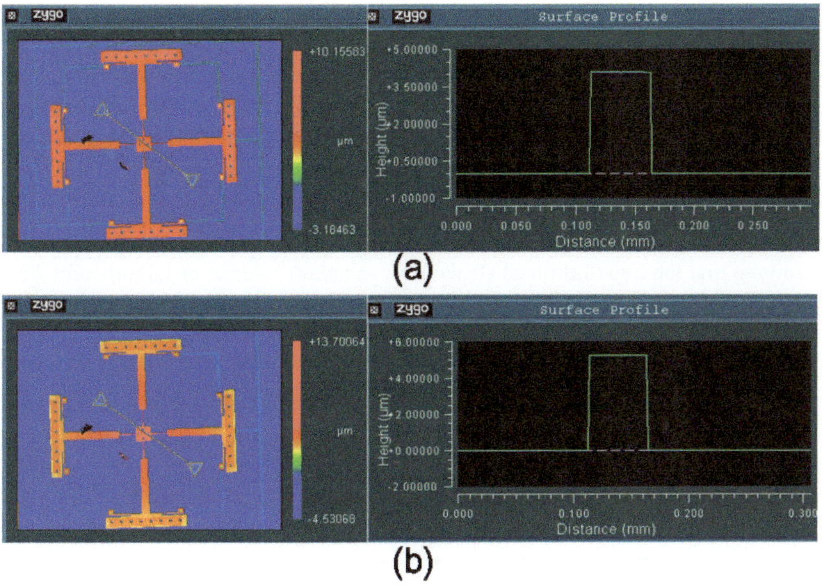


**Figure 18.** Optical profiles for Structure 1 before and after deformation: (a) Optical profile before applying voltage, we can see the profile is approximately a plane (b) Optical profile when applied a voltage of 47 V, we can see the short arm goes down and the long arm and central mass goes up.

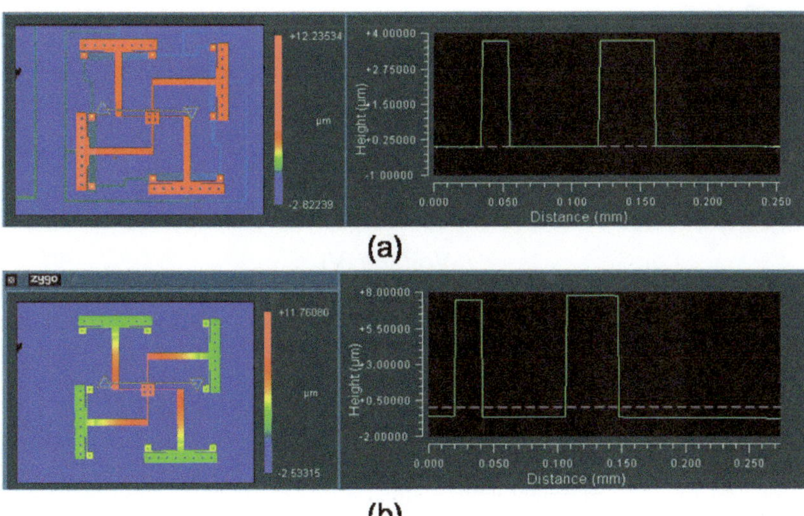


**Figure 19.** Optical profiles for Structure 2 before and after deformation: (a) Optical profile before applying voltage, we can see the profile is approximately a plane (b) Optical profile when applied a voltage of 24.75 V, we can see the short arm goes down and the long arm and central mass goes up.

The optical profile of Structure 1 before and after applying voltage is shown in figure 18. Before applying voltage, the structure profile is approximately a plane. Then after applying a voltage of 47 V, due to the electrostatic force, the upper electrode (short arm) goes down and the long arm and central mass goes up. Similar result is observed for Structure 2, as shown in figure 19. For the Structure 3, after applying voltage, the mirror plate did not go up, instead, it went down. After analysis, we believe there may be problem for the isolation of the mirror plate and the substrate. We will fix this problem in the future.

The displacements versus voltage for the first two optimized structures are shown in figure 20. We can see that the two optimized structures can obtain a stroke of 1.45μm, and 2.21μm, which are more than two times, and three times larger, respectively, than the stroke before optimization. Through this example, we can clearly see the importance of FEA in MEMS research: it saves time and money, while at the same time can handle complex/nonlinear structures.

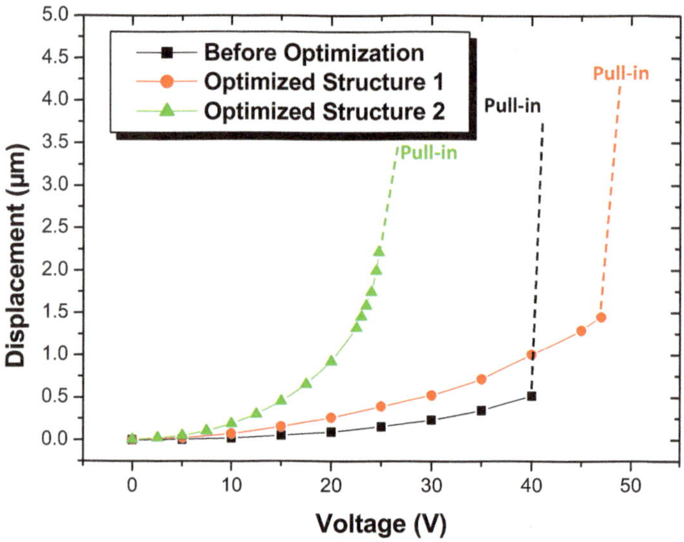


**Figure 20.** Experimental results of the first two structures after optimization in comparison with the structure before optimization, the maximum displacement is more than two and three times larger than the maximum displacement before optimization.

## 5. Conclusion

In this chapter we mainly discussed the significance of FEA in MEMS research through an example of a micromachined spatial light modulator (μSLM). We have used FEA to model the μSLM structure, verify theoretical models, and perform optimizations. After fabrication, we found that the stroke after optimization was more than 3 times larger than the stroke before optimization. As is demonstrated, FEA makes MEMS research to be time and cost efficient and thus has been widely applied in MEMS research.

## Author details

Hao Ren
*State Key Lab of Optical Technologies for Microfabrication, Institute of Optics and Electronics, Chinese Academy of Sciences, Chengdu, China*

*School of Electrical, Computer, and Energy Engineering, Arizona State University, Tempe, AZ, USA*

Jun Yao
*State Key Lab of Optical Technologies for Microfabrication, Institute of Optics and Electronics, Chinese Academy of Sciences, Chengdu, China*

## 6. References

[1] R.P. Feynman, There is plenty of room at the bottom. *J. Microelectromech. Syst.* 1 60-66 1992
[2] D.J. Nagel and M. E. Zaghloul, MEMS: micro technology, mega impact. *IEEE Circuits Devices* 17 14-25 2001
[3] N. Yazdi, F. Ayazi and K. Najiafi Micromachined Inertial Sensors. *Proceedings of IEEE* 86(8): 1640-1659 1998,
[4] J. Chae, H. Kulah and K. Najafi, A monolithic three-axis micro-g micromachined silicon capacitive accelerometer. *J. Microelectromech. Syst.* 14 (2): 235-242 2005
[5] B. Piekarski, D. Devoe, M. Dubey R. Kaul and J. Conrad, Surface micromachined piezoelectric resonant beams filters, *Sensors and Actuators A: Physical*, 91:313-320 2001
[6] J.Y. Park, G.H. Kim, K.W. Chung and J.U. Bu, Monolithically integrated micromachined RF MEMS capacitive switches, *Sensors and Actuators A: Physical*, 89:88-94 2001
[7] M. B. Kassem and R.R. Mansour, Two movable-plate nitride-loaded MEMS variable capacitor, *IEEE Transactions on Microwave Theory and Techniques*, 52(3):831-837 2004
[8] L.J. Hornbeck, W.E. Nelson, Bistable eeformable mirror device. *OSA Technological Digest Series*, 86:1687-1704 1988
[9] J.M. Younse, Mirrors on a chip, *IEEE Spectrum*, 30:27-31 1993
[10] S.S. Bhavikatti. *Finite Element Analysis*. New age international publishers, New Delhi 2005
[11] S. Moaveni. *Finite element analysis, Theory and Application with ANSYS*. Prentice Hall, Upper Saddle River, New Jersey 1999
[12] http://www.intellisense.com/
[13] H. Ren, Z.G. Ni, J.M. Chen, A.L. Gong and J. Yao, A micro spatial light modulator based on leverage principle, *Key Engineering Materials* 483:137-142 2011
[14] R.K. Tyson, *Principles of Adaptive Optics* (2nd ed.), Academic Press, New York 1998
[15] N. Doble, M. Helmbrecht, M. Hart and T. Juneau, Advanced wavefront correction technology for the next generation of adaptive optics equipped ophthalmic instrumentation. *Proc. SPIE* 5688 125-132 2005

[16] H. Ren, F.G. Tao, W.M. Wang and J. Yao, An out-of-plane electrostatic actuator based on the lever principle, *J. Micromech. Microeng.* 21 045019 2011
[17] S. Timoshenko and J. N. Goodier, *Theory of Elasticity*, McGraw-Hill, New York, 1951.
[18] J. M. Gere and S. P. Timoshenko *Mechanics of materials* 2nd edition (Brooks/Cole Engineering Division) 1984
[19] W. Alexander and S. Harald. Torsional stress, fatigue and fracture strength in silicon hinges of a micro scanning mirror *Proc. of SPIE* 5342:176-185 2004

Chapter 9

# Finite-Element Modelling and Analysis of Hall Effect and Extraordinary Magnetoresistance Effect

Jian Sun and Jürgen Kosel

Additional information is available at the end of the chapter

## 1. Introduction

The Hall effect was discovered in 1879 by the American physicist Edwin Herbert Hall. It is a result of the Lorentz force, which a magnetic field exerts on moving charge carriers that constitute the electric current [1, 2]. Whether the current is a movement of holes, or electrons in the opposite direction, or a mixture of the two, the Lorentz force pushes the moving electric charge carriers in the same direction sideways at right angles to both the magnetic field and the direction of current flow. As a consequence, it produces a charge accumulation at the edges of the conductor orthogonal to the current flow, which, in turn, causes a differential voltage (the Hall voltage). This effect can be modeled by an anisotropic term added to the conductivity tensor of a nominally homogeneous and isotropic conductor.

The Hall effect is widely used in magnetic field measurements due to its simplicity and sensitivity [2]. Hall sensors are readily available from a number of different manufacturers and are used in various applications as, for example, rotating speed sensors (bicycle wheels, gear-teeth, automotive speedometers, and electronic ignition systems), fluid flow sensors, current sensors or pressure sensors.

Recently, a large dependence of the resistance on magnetic fields, the so-called extraordinary magnetoresistance (EMR), was found at room temperature in a certain kind of semiconductor/metal hybrid structure [3]. Sharing a similar origin with the Hall effect, the EMR effect is mainly based on the Lorentz force generated by a perpendicularly applied magnetic field, which causes a current deflection. This results in a redistribution of the current from the metal shunt into the semiconductor causing a resistance increase. It is important to note that the fundamental principle of EMR is the change of the current path in the hybrid structure upon application of a magnetic field rather than the change of magnetoconductivity $\sigma$ of either the semiconductor or the metal [4-5]. This effect has drawn much attention due to its potential advantages over other solid-state magnetic field sensors [6-9]. Noise is rather low in EMR devices, since they are made of nonmagnetic materials, and there is no contribution from magnetic noise as it is in contemporary tunnel magnetoresistance or giant magnetoresistance devices [10], and there is less thermal noise than in Hall sensors due to the lower resistance provided by the conducting shunt. The saturation field exceeds 1 T, resulting in a large working range and, compared to Hall sensors, EMR sensors provide a higher sensitivity.

EMR devices have been fabricated using high-mobility and narrow-gap semiconductors shunted by a highly conductive metal bulk [7]. These properties of the semiconductor guarantee that the material shows a large Lorentz force along with a comparatively good conductivity. Experiments on the EMR effect were initially performed in a macroscopic composite Van der Pauw disk made of a semiconductor disk with a concentric metallic circular inhomogeneity embedded (Figure 1(a)), and four electrodes were used to apply current and measure voltage. Though this structure provided good results, its realization in microscopic and nanoscopic length scales is unreasonable. Using bilinear transformation, a bar-type geometry, which is a semiconductor bar shunted by a metal stack on one side (Figure 1(b) and (c)), has been derived from the Van der Pauw disk showing a similar EMR effect and being simpler in terms of fabrication [11].

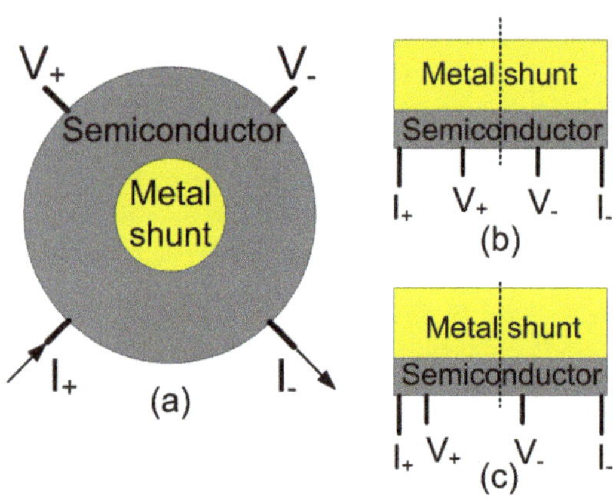


**Figure 1.** 4-contact EMR devices with (a) Van der Pauw disk geometry, (b) symmetric bar geometry and (c) asymmetric bar geometry. The dark lines labeled with $I_+$, $I_-$ $V_+$ and $V_-$ represent the two current electrodes and two voltage probes, respectively. The dashed lines represent the central axes of the devices.

The EMR effect is an extrinsic property that strongly depends on the geometry of the device and the placements of the electric contacts. Several studies have been carried out in order to investigate the influence of contact configurations and the geometry of the metallic region on the performance of the EMR device [12-14]. The typical 4-contact device consists of a high mobility semiconductor bar with a metal shunt attached to one side. With respect to the placement of the two current electrodes (I) and the two voltage electrodes (V), two major kinds of four-contact configurations can be distinguished: IVVI (Figure (b) and (c)) and VIIV, where the voltage electrodes are between the current electrodes or vice versa, respectively.

Both the Hall and EMR effects are based on the Lorentz force, which can be considered by introducing an anisotropic conductive behavior in the material models. Depending on the characteristics of the materials being used and the complexity of the device geometry, the analytical calculation of the Hall and EMR effect can be very complicated. Several studies have shown that the finite element method (FEM) provides a proper tool to model and simulate the Hall and EMR effects.

In this chapter, we will introduce the 2-D and 3-D FEM models, which describe the anisotropic behavior of the conductors under an external magnetic field, and which can be applied to both Hall and EMR effects. Especially in case of the EMR effect, the 3-D simulations provide a considerably higher accuracy than the 2-D ones due to the inhomogeneous and unsymmetrical structures used. The developed 3-D model is also verified by experimental results. The models will be applied to study the effects of geometry on the performance of Hall and EMR devices. Thereby, the focus will be put on the EMR effect, since it has been much less investigated so far.

## 2. Theories of Hall and EMR effect

Both the Hall and EMR effects result in an anisotropic conductivity, which is caused by the magnetic field through the Lorentz force. The force acting on a single electric carrier can be expressed as

$$\boldsymbol{F} = \mathrm{q}(\boldsymbol{E} + (\boldsymbol{v} \times \boldsymbol{B})) \tag{1}$$

where $\boldsymbol{F}$ is the force vector acting on the charged carrier, $q$ is the charge of the carrier, $\boldsymbol{E}$ is the vector of the applied electric field, $\boldsymbol{v}$ is the instantaneous drift velocity vector of the moving carrier and $\boldsymbol{B}$ is the magnetic field vector. The term $q\boldsymbol{E}$ is called the electric force, while the term $q\boldsymbol{v} \times \boldsymbol{B}$ is called the magnetic force $\boldsymbol{F}_m$.

In a solid conductor as shown in Figure 2, the current expressed in terms of the drift velocity is

$$I = j \times A = nqvwd \tag{2}$$

where $j$ is the current density and $n$ is the number of charge carriers. The cross-section area of the conductor is $A = wd$, whereby $w$ and $d$ are the width and thickness of the conductor, respectively. Then the drift velocity is found as

$$v = \frac{I}{nqwd} \tag{3}$$

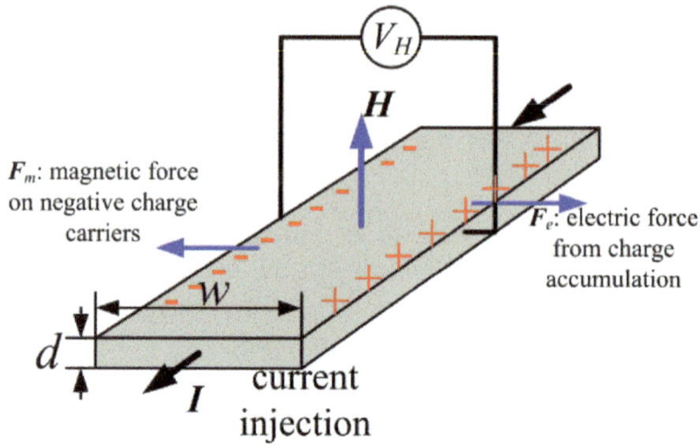


**Figure 2.** Schematic view of the bar (n-type) Hall device. A constant current, *I*, is applied as well as a constant magnetic field, *H*, perpendicular to the surface. This will cause an accumulation of charge carriers transverse to the current direction and a Hall voltage, $V_H$. $F_m$ and $F_e$ indicate magnetic force and electric force, respectively. Note that the direction of the current *I* in the diagram is that of conventional current, so that the motion of electrons is in the opposite direction.

When a magnetic field is applied in the direction perpendicular to the current flow, the Lorentz force causes deflection of the current. As a consequence, charges of opposite sign accumulate at two surfaces or edges of the conductor orthogonal to the current flow creating an electric field, the Hall field $\boldsymbol{E_H}$. When the magnetic force $\boldsymbol{F_m}$ is equivalent to the electric force $\boldsymbol{F_e}$ generated by the Hall electric field, an equilibrium state is achieved. In this case

$$\boldsymbol{F}_{\mathrm{m}} = \boldsymbol{F}_{\mathrm{e}} \Leftrightarrow q\boldsymbol{v} \times \boldsymbol{B} = \frac{V_H q}{w} \tag{4}$$

$V_H$ is the output voltage caused by the electric field of the accumulated charges. Using equation (3) and (4), $V_H$ can be expressed in terms of the applied current and the magnetic field

$$V_H = \frac{IB}{nqd} \tag{5}$$

For n-type semiconductors the charge carrier is negative (electron) and equation (5) can be substituted as

$$V_H = \frac{-IB}{ned} \tag{6}$$

where *e* is the electron charge.

In case of p-type semiconductors, $V_H$ would, then, be positive for positive values of *I* and *B*. Another way of describing this is by means of the Hall angle, $\theta_H$ (Figure 3). The direction of the current is collinear with the applied electric field ***E***, but not collinear with the total electric field $\boldsymbol{E_t}$, because of the contribution from the Hall electric field $\boldsymbol{E_H}$. The angle between the total electric field and the applied electric field is called the Hall angle, $\theta_H$,

$$\theta_{\mathrm{H}} = \arctan\left|\frac{E_{\mathrm{H}}}{j}\right| \tag{7}$$

and for small values it is approximated by

$$\theta_{\mathrm{H}} \approx \left|\frac{E_{\mathrm{H}}}{j}\right| \tag{8}$$

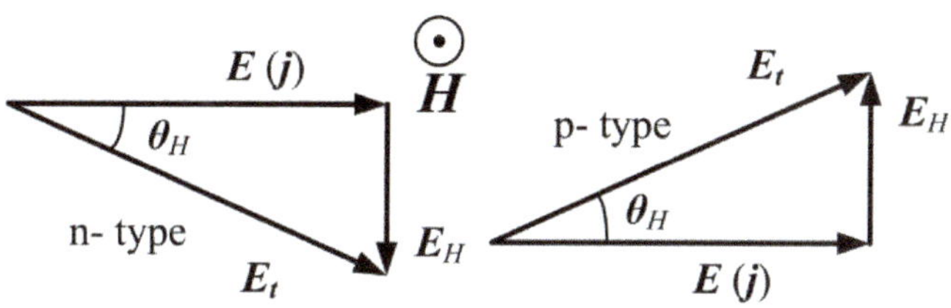


**Figure 3.** The Hall electric field $E_H$ generated for n-type and p-type conductor based Hall devices. $j$ is the current density, $\theta_H$ is the Hall angle indicating the difference between the applied electric field and the total electric field.

Let us suppose we have a semiconductor with a metal stack as an inhomogeneity embedded in it, as shown in Figure 4(a). The conductivities of the semiconductor and the metal are denoted as $\sigma_s$ and $\sigma_m$, respectively, and $\sigma_m >> \sigma_s$. In low magnetic fields, the current flowing through the conductor is concentrated into the metallic region with the metal acting as a short circuit. The current density $j$ is parallel to the total electric field $\boldsymbol{E}$ as indicated in Figure 4(a). The metal inhomogeneity is essentially an equipotential body due to its high conductivity. Thus, the direction of $\boldsymbol{E}$ at the semiconductor/metal interface is normal to the interface. At high magnetic fields, the current is deflected by the Lorentz force, which results in a directional difference between $j$ and $\boldsymbol{E}$ by the Hall angle. For sufficiently high fields this angle approaches 90° in which case $j$ is parallel to the semiconductor/metal interface and the current is deflected around the metal inhomogeneity which acts like an open circuit (Figure 4(b)). The transition of the metal from a short circuit at low fields to an open circuit at high fields gives rise to the very large magnetoresistance or the so-called extraordinary magnetoresistance effect.

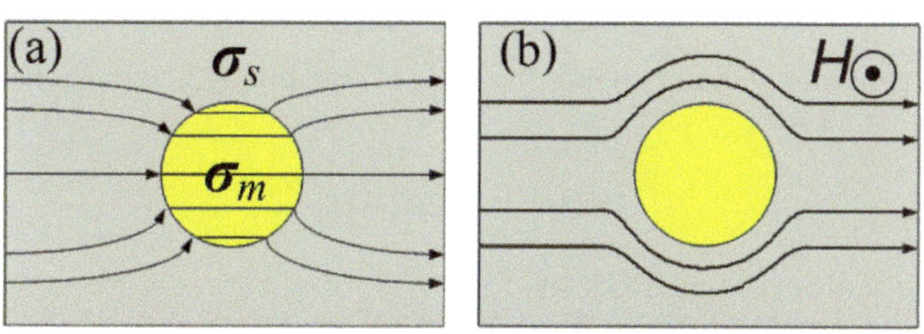


**Figure 4.** The current flow distribution in a semiconductor/metal hybrid. The gray and yellow areas express the semiconductor and metal, respectively. The dark lines show the paths of current. (a) At low magnetic fields, the current is parallel to the electric field E and the metal acts as a short circuit. (b) At high field, the current is mainly flowing in the semiconductor, the hybrid acts as an open circuit.

In the ideal case, the Hall effect is independent of geometry (compare equation (5) and (6)). However, the EMR effect is strongly geometry dependent, and analytical models for the EMR effect are highly complicated, especially when a complex geometry is considered. Therefore, the Finite Element Method (FEM) has been employed before to study the EMR effect [4, 14-16].

## 3. FEM models for Hall and EMR effect

The conductivity $\sigma$ of a homogenous, isotropic solid conductor is given by

$$\sigma \equiv \frac{1}{\rho} = (n_e\mu_e + n_h\mu_h)q \tag{9}$$

where $\rho$ is the resistivity, $n_e$ and $n_h$ are the densities of electrons and holes and $\mu_e$ and $\mu_h$ are the mobility of electrons and holes, respectively.

High sensitivity Hall and EMR devices are commonly made from n-type semiconductors since the mobility of electrons is typically much larger than that of holes yielding larger effects [17]. Therefore, only the dominant carrier-electron needs to be taken into consideration in the FEM model.

By direct integration of the Boltzmann equation [18], the current density due to carriers at a single quadratic energy extremum is given by

$$j = \boldsymbol{\sigma}_0(\boldsymbol{E} + \frac{j \times \boldsymbol{H}}{n_e e}) \tag{10}$$

where $\boldsymbol{\sigma}_0 = n_e e \boldsymbol{\mu}$ is the conductivity without magnetic field, and $\boldsymbol{\mu}$ is the mobility tensor. In an isotropic conductor the mobility components are identical for all spatial axes

$$\boldsymbol{\mu} = \mu_e \begin{bmatrix} 1 & 0 & 0 \\ 0 & 1 & 0 \\ 0 & 0 & 1 \end{bmatrix} \tag{11}$$

According to Ohm's law, the vector of the current density is expressed as

$$j = \boldsymbol{\sigma}(\boldsymbol{H}) \cdot \boldsymbol{E} \tag{12}$$

By re-arrangment of equation (10) into the form of equation (12), the magnetoconductivity tensor $\sigma(\boldsymbol{H})$ is found to be

$$\boldsymbol{\sigma}(\boldsymbol{H}) = n_e e\left(\boldsymbol{\mu}^{-1} + \boldsymbol{H}\right)^{-1} \tag{13}$$

where $\boldsymbol{H}$ is a magnetic field matrix defined as

$$\boldsymbol{H} = \begin{bmatrix} 0 & -H_z & H_y \\ H_z & 0 & -H_x \\ -H_y & H_x & 0 \end{bmatrix} \tag{14}$$

$H_x$, $H_y$, and $H_z$ are the components of the applied field in x-, y-, and z-directions, respectively. By substitution of equations (11) and (14) into equation (13), the components of the corresponding magnetoconductivity tensors are found to be

$$\begin{aligned}
\sigma_{11} &= (\mu_e^3 H_x^2 + \mu_e)p \\
\sigma_{12} &= \mu_e^2 (H_z + \mu_e H_x H_y)p \\
\sigma_{13} &= -\mu_e^2 (H_y - \mu_e H_x H_z)p \\
\sigma_{21} &= -\mu_e^2 (H_z - \mu_e H_x H_y)p \\
\sigma_{22} &= (\mu_e^3 H_y^2 + \mu_e)p \\
\sigma_{23} &= \mu_e^2 (H_x + \mu_e H_y H_z)p \\
\sigma_{31} &= \mu_e^2 (H_y + \mu_e H_x H_z)p \\
\sigma_{32} &= -\mu_e^2 (H_x - \mu_e H_y H_z)p \\
\sigma_{33} &= (\mu_e^3 H_z^2 + \mu_e)p \\
p &= n_e e(\mu_e^2 H_x^2 + \mu_e^2 H_y^2 + \mu_e^2 H_z^2 + 1)^{-1}
\end{aligned} \tag{15}$$

Since Hall and EMR devices are made of thin film structures, the thickness (z-direction) is typically neglected, and the complex 3-dimensional magnetoconductivity model is reduced to a concise 2-D model, where only the carrier transport in the x-y plane and a perpendicular magnetic field $H_z$ are considered. By substituting x- and y-components of the magnetic field in equation (14) with zero, the magnetoconductivity of a 2-D model is found to be

$$\sigma(H) = \frac{\sigma_0}{1+\beta^2} \begin{bmatrix} 1 & -\beta & 0 \\ \beta & 1 & 0 \\ 0 & 0 & 1 \end{bmatrix} \tag{16}$$

where $\beta = \mu_e \cdot H_z$.

In a steady state condition, the problem of determining the electrostatic potential $\varphi(x,y,z)$ in the conductor reduces to the solution of a Laplace's equation:

$$\nabla \left[ \boldsymbol{\sigma} \cdot \nabla \varphi(x,y,z) \right] = 0 \tag{17}$$

which can be solved by means of the finite-element method under specific boundary conditions and initial conditions [19].

## 4. Finite-element simulation of Hall effect

In this section, a cross-shaped hall device is investigated by means of the FEM using the 2-D model. The model of the Hall device consists of a semiconductor cross with the geometry shown in Figure 5. The dimensions of the 2 arms are 30×10 $\mu m^2$. The tips of the arms are connected with narrow metal blocks acting as electric contact pads. The sensing area is defined by the intercross region of the two arms, which has an area of 10×10 μm². The material parameters are: $\mu_s$ = 4.55 $m^2V^{-1}s^{-1}$ and $n_s$ = 2.55×$10^{22}$ $m^{-3}$ for the semiconductor; and $\mu_m$ = 5.3×$10^{-3}$ $m^2V^{-1}s^{-1}$ and $n_m$ = 5.9×$10^{28}$ $m^{-3}$ for the metal (gold).

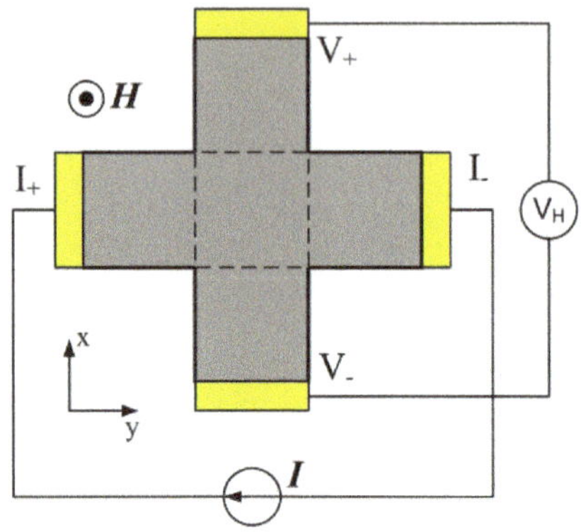


**Figure 5.** Schematic view of a cross-shaped Hall device. $I_+$, $I_-$, $V_+$, and $V_-$ indicate the two current leads and two voltage probes, respectively.

A current of 100 μA is applied as boundary conditions at $I_+$ and $I_-$ current pads, which is a feasible assumption based on the electromigration limit of Au as well as heating constraints. The $I_-$ current pad is grounded. All other outer boundaries were set to electrical insulation ($\sigma$ = 0, Neumann condition). The models were meshed with free triangular elements, which conform well to a large range of model geometries, and consisted of approximately 2.6×$10^4$ elements and 5.3×$10^4$ degrees of freedom. The electric potential along the device was obtained by solving equation (17). Figure 6 shows the potential distribution inside the device with and without application of a magnetic field. The Hall voltage is calculated as the

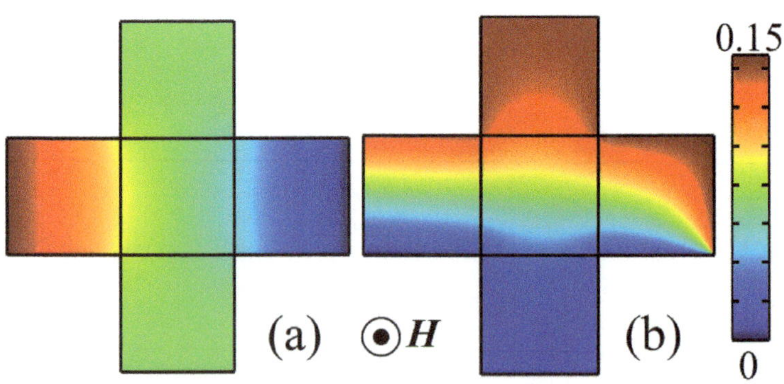


**Figure 6.** Potential distribution inside the Hall device (a) without and (b) with application of a magnetic field. The color bar indicates the strength of the voltage potential (V). The current flow is from the left to right side.

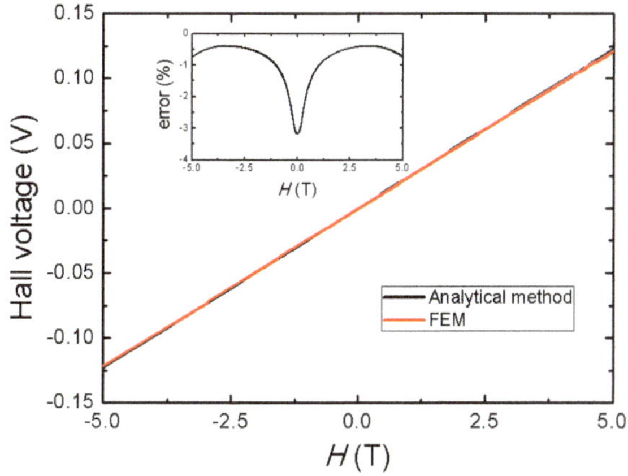


**Figure 7.** The Hall voltage as a function of the magnetic field calculated by an analytical and by FEM. The inset shows the error between the results from these two methods.

difference of the two potential values at voltage pads, i.e. $V_H = \varphi(V_+) - \varphi(V_-)$. Figure 7 shows the Hall voltage as a function of the magnetic field. The result from an analytical calculation is also provided for comparison. Within the magnetic field range of ±5 T, the analytical result is consistent with the one of the FEM showing a linear relation with the field change. The inset of Figure 7 shows the error between the results obtained with these two methods. The error is found to be only 1% to 3%, which shows the good agreement of the FEM with the analytical method.

## 5. Finite-element simulation of EMR effect

### 5.1. FEM simulation with 2-D model

In this section, a semiconductor/metal hybrid EMR device made of a semiconductor bar shunted by a concentric metallic inclusion (see Figure 4) is simulated with the 2-D FEM model in order to investigate the principle of the effect. The material parameters used in the simulation were the same as in section 3, i.e. $\mu_s = 4.55\ m^2V^{-1}s^{-1}$, $n_s = 2.55\times10^{22}\ m^{-3}$, $\mu_m = 5.3\times10^{-3}\ m^2V^{-1}s^{-1}$ and $n_m = 5.9\times10^{28}\ m^{-3}$.

The device model consists of a semiconductor with an area of 100×100 μm² and a concentric metallic disk with a radius of 30 μm. A current density of $10^9\ A/m^2$ is applied to the left edge of the semiconductor as the boundary condition, and the right edge is grounded. All other outer boundaries were set to electrical insulation.

The current distributions in the semiconductor/metal hybrid device at various external magnetic fields are shown in Figure 8(a)-(d). At zero magnetic field (Figure 8(a)), the current flow is predominantly concentrated in the metal shunt, which can be seen clearly from the high current density in the metallic region and the low one above and below the disk in the semiconducting region. The electric field lines are perpendicular to the semiconductor/metal interface and the current flow is parallel to the electric field lines. As the result, owing to the high conductivity of the metal, the current distribution at zero field constitutes the low resistance state of this structure.

As the external field rises to 0.1 T (Figure 8(b)), the current is still predominantly flowing through the metal shunt; however the magnetic field deflects the current path to some extent. Now, a larger proportion of the current is being forced to flow around the conductor through the semiconductor region of high resistance. Due to the current deflection, the direction of the electric field, which is still perpendicular to the surface of the metal shunt, and the one of the current are not parallel anymore and deviate from each other by the Hall angle.

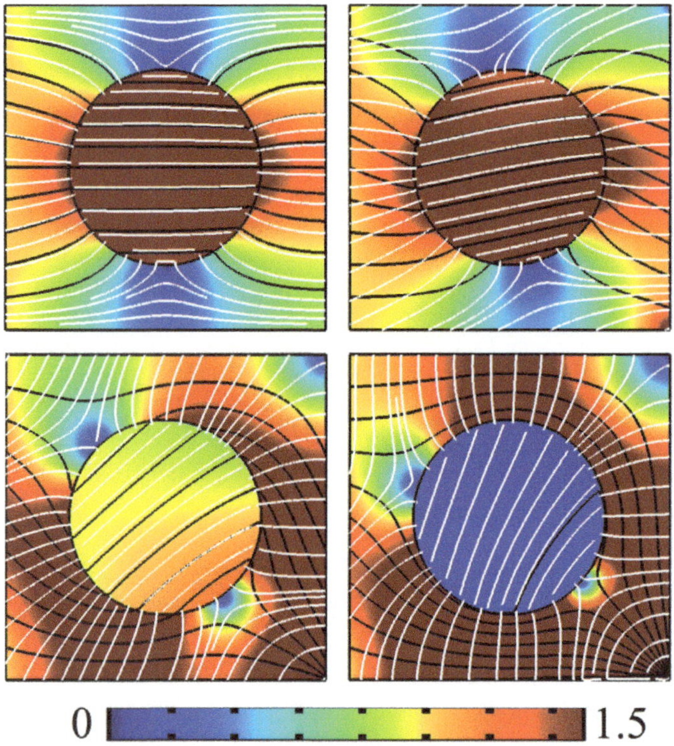


**Figure 8.** The current distributions in the semiconductor/metal hybrid device at external magnetic fields of (a) 0 T, (b) 0.1 T, (c) 0.5 T, and (d) 5 T. The dark solid lines indicate the paths of current. The white lines are the electric field lines. The color bar indicates the strength of the current density from 0 (blue) to $1.5 \times 10^6$ A/m$^2$ (red).

The Hall angle increases as the magnetic field becomes stronger. At 0.5 T (Figure 8(c)) highest current density is no longer observed in the metal region since the current is forced to flow through the semiconductor despite the higher resistance.

Finally, at a very strong applied field of 5 T (Figure 8(d)), he Hall angle approaches 90° forcing the current to flow almost entirely around the metal inhomogeneity. This results in a substantially higher resistance of the device compared to the case of no magnetic field applied.

### 5.2. FEM simulation with 3-D model

In this section, the EMR effect in a bar-type device is simulated under spatial magnetic fields with a 3-D FEM model.

In order to validate the simulation results, they are compared with experimental ones obtained from an EMR device specifically fabricated for this purpose. The EMR device consists of a semiconductor bar of 1.5 μm in thickness, 300 μm in length, and 23 μm in width (Figure 9). The metal shunt is a Ti (50 nm)/Au (250 nm) stack with a width of 80 μm. An overlap, which typically arises due to the fabrication process, of 3 μm exists between the semiconductor and metal shunt. Two metal leads with a length of 5 μm are located at the two corners of the semiconductor bar to apply the current and measure the voltage $V$. At the metal/semiconductor interface, the device showed a contact resistivity of $10^{-7}$ Ωcm², and this value was also implemented in the model.

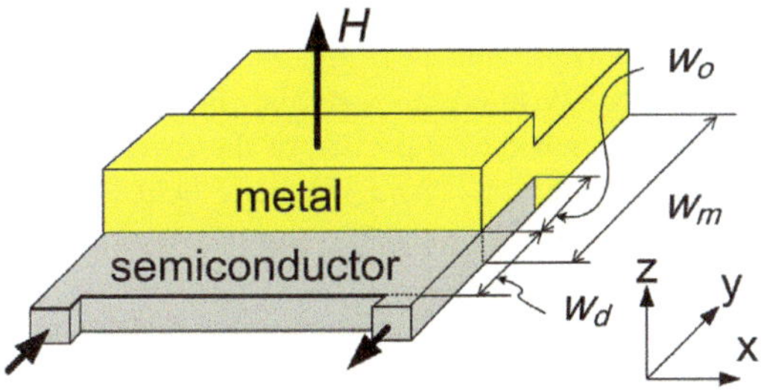


**Figure 9.** A bar-type EMR device with overlap between metal and semiconductor (typical for practical devices). The current injection direction is indicated by the solid arrows. The thickness of metal and semiconductor are denoted as $t_m$ and $t_s$, respectively, and the width of metal and semiconductor are denoted as $w_m$ and $w_s$, respectively. $w_o$ represents the overlap between metal and semiconductor, note $w_d + w_o = w_s$.

A constant current with the density of $6.67\times10^7$ A/m² (corresponding to 100 μA) was applied as the boundary conditions at the faces of the current leads. All other outer boundaries were set to be electrically insulating ($\sigma$ = 0). The model was meshed with tetrahedral elements, whereby the mesh density was varied adaptively in order to account for the large differences of the dimensions of the structural components. The material parameters used in the simulation were the same as the ones of the experimental sample: $\mu_s$ = 0.82 m²/V·s and $n_s = 5.6\times10^{22}$ m$^{-3}$ for the semiconductor, $\mu = 5.30\times10^{-3}$ m²/V·s and $n = 5.90\times10^{28}$ m$^{-3}$ for gold, and $\mu = 2.90\times10^{-3}$ m²/V·s and $n = 5.12\times10^{27}$ m$^{-3}$ for titanium. Homogenous magnetic fields in x, y, and z direction changing from -1 to 1 T with steps of 0.1 T were applied. The result of a 2-D FEM simulation with a magnetic field applied in z-direction is also provided for comparison, in which the width of the semiconductor is 20 μm and, due to the limitations of the 2-D model, only a gold shunt without a Ti adhesion layer is considered.

The EMR ratio, in general, is expressed as

$$MR(\boldsymbol{H}) = \frac{R(\boldsymbol{H}) - \mathrm{R}(0)}{R(0)}, \quad R(\boldsymbol{H}) = \frac{V(\boldsymbol{H})}{I} \tag{18}$$

where $R(\boldsymbol{H})$ is the resistance at spatial magnetic field $\boldsymbol{H}$ and $R(0)$ the resistance at zero magnetic field ($R(0) = R(\boldsymbol{H} = 0)$). $V(\boldsymbol{H})$ is the voltage at magnetic field $\boldsymbol{H}$ calculated from the solution of equation (17), and $I$ is the value of current injected into the device. The output sensitivity $\delta$ is defined as the change of the output voltage with respect to a small variation $\Delta\boldsymbol{H}$ of the field

$$\delta(\boldsymbol{H}) = \frac{V(\boldsymbol{H} + \Delta\boldsymbol{H}) - V(\boldsymbol{H})}{\Delta\boldsymbol{H}} \tag{19}$$

In the case of the 2-D simulation, the magnetic field vector needs to be replaced by the perpendicular field component $H_z$ in z-direction.

Figure 10 shows the simulated current paths in the EMR device as a function of external fields in different directions. (Note, different dimensions than the ones mentioned above are used for the device in order to provide clear illustrations). The dark streamlines show the current flow in the device. Without an external field, the current flows from the semiconductor into the metal, whereby a portion of the current flows through the overlap between the semiconductor and metal. As a magnetic field is applied, deflections of the current paths occur due to the Lorentz force. It is important to note that in case of magnetic fields applied in the x-direction (Figure 10 (a)), the current redistributions for positive and negative fields are asymmetric with respect to the current distribution at zero field, which is a result of the asymmetric geometry of the structure in x-view. However, in case of magnetic fields applied in z-direction or y-direction, current redistribution is symmetric.

Figure 11(a) shows the EMR ratio as a function of the perpendicular magnetic field $H_z$. Both the 2-D and 3-D FEM models provide accurate results for smaller field values. At higher fields the 3-D model is considerably more accurate than the 2-D model. For example, the EMR ratios at $H_z = 1$ T are 59.1%, 61.8%, and 76.2% from experimental measurement, 3-D FEM and 2-D FEM calculation, respectively. The larger EMR effect found with the 2-D simulation is mainly due to neglecting the lower conductivity of the titanium layer compared to gold and the semiconductor/metal overlap, which results in a wrong estimation of the device's resistance at zero field. The dependences of the EMR ratio to magnetic fields applied in planar directions are shown in Figure 11(b). The ratios are 11.5% and 8.2% at $H_x = 1$ T and 9.0% and 7.4 % at $H_y = 1$ T for experimental measurements and 3-D FEM calculations, respectively. Their values are about 15% to 20% of the ones obtained for a perpendicular field. A stronger EMR effect is observed at $H_x$ than at $H_y$. This can be attributed to the difference in the strength of the Lorentz force acting on the carriers near the interface between the semiconductor and metal shunt, which is the region where the

Lorentz force has the largest impact on the current redistribution. The current density at the interface is highest near the source and drain contacts, where the current flow is mainly in y-direction. Therefore, the Lorentz force on the carriers at the interface is larger when the magnetic field is in x-direction rather than in y-direction. Also, the magnetoresistance curve for the fields in x-direction is asymmetric due to the asymmetric geometry.

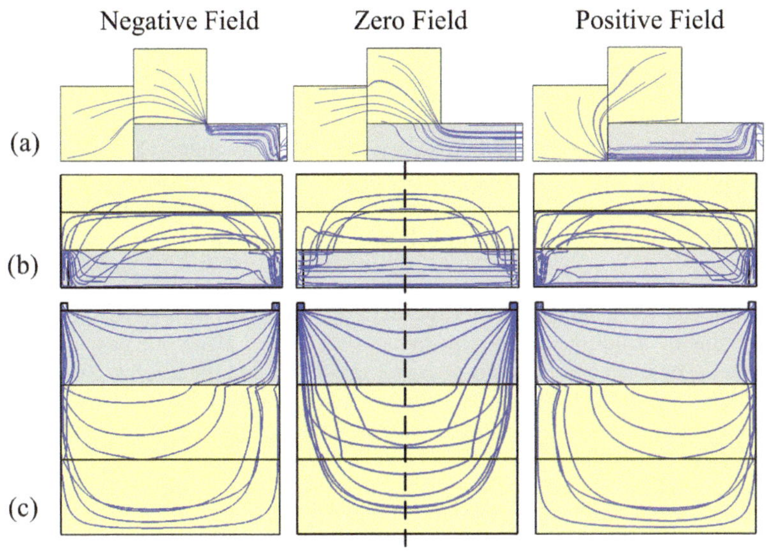


**Figure 10.** Current redistributions at magnetic field in (a) x-direction (left side view of Figure 9), (b) y-direction (front view of Figure 9), (c) z-direction (bottom view of Figure 9). The solid lines show the path of current flow. The yellow block represents the metal shunt and the grey block the semiconductor. ($t_m = 2t_s$, $w_m = w_s$, $w_o/ w_s = 50\%$).

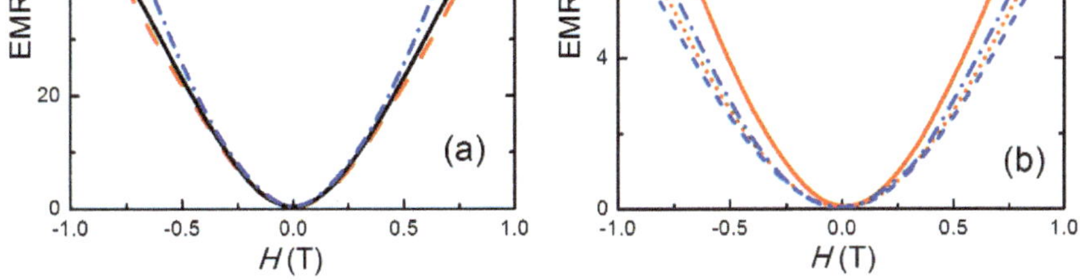


**Figure 11.** EMR ratios as functions of (a) perpendicular field $H_z$ and (b) planar fields $H_x$ and $H_y$.

The simulated results for planar fields are slightly smaller than the experimental ones. We assume this can be attributed to the defects in the semiconductor, which are more abundant in the vicinity of the interface. Since the current flow is not uniform in regions with defects, the current component in z-direction is larger as compared to the ideal case causing additional current redistribution due to Lorentz forces in case of planar fields.

## 6. FEM analyses on geometric dependence of EMR effect

As mentioned before, the EMR effect strongly depends on the geometry of the device and the locations of the electrodes. In this section, the performance of a bar-type EMR device is simulated for different device geometries and electrode locations. The performance is evaluated with regard to the output sensitivity (equation (19)) of the device, rather than the often-analyzed EMR ratio (equation (18)), since it is more relevant than the EMR effect for potential applications ranging from reading heads to smart biomedical sensors.

The model of the bar-type EMR device consists of a semiconductor bar shunted by a metal stack. The structure and geometric parameters of the device model are shown in Figure 12. The device is symmetric about the y-axis, and the x-axis is placed along one of the edges of the semiconductor. The widths of the semiconductor and metal and the lengths of the device are denoted $W_s$, $W_m$, and $L$, respectively. The current leads I and voltage probes V were placed along the edge of the semiconductor bar. Depending on their arrangement, the contact configuration can be classified into two types, namely, IVVI (Figure 12(a)) and VIIV (Figure 12(b)). While the two outer contacts are placed at the edges of the semiconductor, the locations of the inner ones are varied, whereby $\alpha_{VL}$, $\alpha_{VR}$, $\alpha_{IL}$ and $\alpha_{IR}$ are the distances of the left and right voltage probes and current leads in IVVI and VIIV configuration, respectively, from the y-axis.

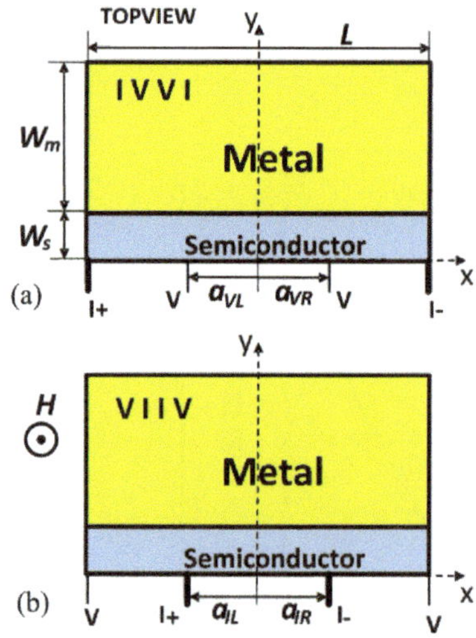


**Figure 12.** Geometry of the semiconductor/metal (a) IVVI and (b) VIIV configurations, where I and V represent current lead and voltage probe, respectively.

The material parameters are: $\mu_s$ = 4.55 $m^2V^{-1}s^{-1}$ and $n_s$ = 2.55×$10^{22}$ $m^{-3}$ for the semiconductor, and $\mu_m$ = 5.3×$10^{-3}$ $m^2V^{-1}s^{-1}$ and $n_m$ = 5.9×$10^{28}$ $m^{-3}$ for the metal (300 K). The width $W_m$ is 15 μm while $W_s$ and $L$ are varied during the simulation. A width of 0.1 μm is assigned to the current leads, while the voltage probes were considered as a point-like contact with zero width.

A current of 10 μA is applied as the boundary conditions at $I_+$ and $I_-$, which is a feasible assumption based on the electromigration limit of the metal as well as heating constraints. The current lead $I_-$ is grounded. All other outer boundaries are electrically insulating ($\sigma$ = 0, Neumann condition), while the interface between the semiconductor and metal is modeled with a contact resistivity of $10^{-8}$ $\Omega{\cdot}cm^2$, which is a prerequisite for showing a strong EMR effect [15]. The models are meshed with free triangular elements, which conform well to a large range of model geometries, and consist of approximately 1.5×$10^5$ elements and 3×$10^5$ degrees of freedom. The mesh density is varied adaptively.

As pointed out before, the EMR effect and the output sensitivity of the device are strongly geometry dependent. Hence, the shape of the device and the placements of current leads and voltage probes are critical. We investigate the relationships between the output sensitivity and different semiconductor length/width ratios $\alpha = L/W_s$, the placements of the voltage probes in an IVVI configuration, and the placements of the current leads in a VIIV configuration as a function of the magnetic field. The output sensitivity is calculated using equation (19) with $\Delta H$ = 1×$10^{-4}$ T.

## 6.1. Width of the metal shunt

In this section, the influence of the width of the metal shunt is studied. An IVVI configuration is employed with a semiconductor of width $W_s$ = 5 μm, length $L$ = 75 μm, and symmetrically placed voltage probes with $\alpha_{VR} = -\alpha_{VL}$ = 12.5 μm. The width of the metal shunt $W_m$ is varied from 0.01×$W_s$ (50 nm) to 50×$W_s$ (250 μm). Figure 13 shows the sensitivity as a function of different $W_m/W_s$ ratios at 0.05T and 1T. As can be seen, the sensitivity increases as the width of the shunt increases up to a certain value before it saturates. However, when $W_m$ is thicker than 0.1×$W_s$, the influence is rather small with an increase of less than 1% for strong fields and around 3% for weak fields, indicating that the current is mainly confined to a thin layer in the metal shunt close to the semiconductor. Similar results were found for a model with a different semiconductor bar of $W_s$ = 3 μm using the same values for all other parameters. Hence, it can be concluded that the dimension of the metal shunt should be larger than 0.1$W_s$ in order to maintain a good performance of the EMR device. A thickness larger than 5$W_s$ is not necessary due to its negligible influence on the device sensitivity. In the following simulations, a value of $W_m$ = 5$W_s$ is being used.

## 6.2. Length/width ratio of the device

In order to investigate the influence of the length/width ratio $\alpha$, a device with symmetric IVVI configuration is studied. The length $L$ is varied from 30 μm to 105 μm with a step size of 3 μm in order to obtain different values of $\alpha$, while $W_s$ and $\alpha_{VR}$ are kept at 3μm and $L$/6, respectively. The width of the metal is kept at $W_m = 5W_s = 15$ μm. Figure 14 shows the sensitivity as a function of $\alpha$ at magnetic fields of 0.05 T, 0.5 T, and 1 T. The optimum value of $\alpha$ with respect to sensitivity shows some dependence on the magnetic field. Values of 5, 10 and 20 provide maximum sensitivity for magnetic fields of 0.05, 0.5 and 1 T, respectively. The optimal value increases for stronger fields. This can be understood by the help of the insets in Figure 14, which indicate the relation between the length/width ratio and the current path in the EMR device. In very simplified terms, if no magnetic field is applied, the current flows straight through the semiconductor into the conductor shunt via the shortest possible way. Due to the Hall angle θ resulting from a magnetic field, the current path will be deflected into the semiconductor, which causes an increased resistance. It's important to keep in mind that a given external field will result in a certain Hall angle θ, independent of the geometry. In case of a low value $\alpha$, the deflected current will be confined to the semiconductor bar (Figure 14 inset (a)). As $\alpha$ increases, the current path through the semiconductor becomes longer and, eventually, reaches the interface between the semiconductor and metal, leading to the optimal value of $\alpha$, at which small changes of the magnetic field cause the largest changes in resistance (Figure 14 inset (b)). It can be seen from Figure 14 inset (c) that a further increase in length only increases the path of the current flow through the conductor, which doesn't contribute to the resistance. The sensitivity decreases as $\alpha$ increases. Since the Hall angle is larger at stronger fields, the optimal value of $\alpha$ is larger at stronger fields.

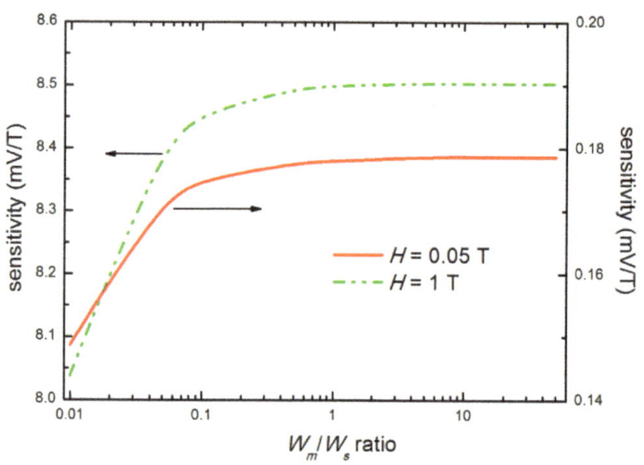


**Figure 13.** Output sensitivity of the EMR device as a function of the ratio $W_m/W_s$ ($W_s = 5$ μm)at $H = 0.05$ and 1 T. The arrows indicate the corresponding axis.

The investigation of the length/width ratio is also carried out with a model of varied width $W_s$ and fixed length $L = 75$μm. This simulation provided exactly the same results as the previous one indicating that the scale of a device has no influence on the performance of an EMR device. Therefore, EMR devices can be fabricated according to the spatial resolution requirements of the specific applications or restrictions due to the fabrication technology.

This not only gives flexibility with respect to design but also higher accuracy for specific applications.

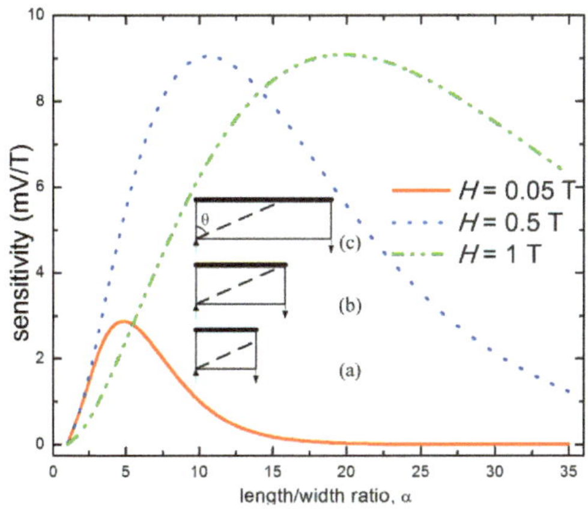


**Figure 14.** Sensitivity as a function of the length/width ratio $\alpha$ at various external fields. The insets show the current deflection in the semiconductor bar of the EMR device with (a) low length/width $\alpha$, (b) optimal $\alpha$, (c) large $\alpha$ under a constant Hall angle $\theta$. The dashes indicate the deflected current path injected from the inlet current lead. The dark thick lines represent the interfaces between semiconductor and metal (not shown here).

## 6.3. Placement of voltage probes in IVVI configuration

An IVVI device with length $L$ = 45 μm and width $W$ = 3 μm ($\alpha$ = 15) is simulated to investigate the influence of the placement of the voltage probes. Here, $\alpha_V$ is used to denote the distance of the voltage probes from the y-axis (Figure 12 (a)). Firstly, a symmetric configuration is used, i.e. $\alpha_V = \alpha_{VL} = \alpha_{VR}$. $\alpha_V$ is changed from 0.5 to 22.5 μm in steps of 1 μm, and its influence on the sensitivity is shown in Figure 15. The sensitivity increases as the probes move further away from the y-axis at both weak and strong fields, and it can be concluded that the larger the separation of the probes the higher the performance. It is interesting to note that there is a considerable increase in sensitivity as the two probes approach their respective corners. In order to get a better understanding of this effect, Figure 16 shows the electric potential distribution along the edge of the semiconductor bar for different magnetic fields. Most of the potential change occurs in the semiconductor close to the current leads. This is due to the smaller cross-section available for the current to pass through, resulting in higher current densities and, consequently, larger electric fields. As the magnetic field increases, the current density along the right edge of the semiconductor increases even further, while on the left edge of the semiconductor the opposite happens (compare Figure 10(c) for positive fields). In between, the two edges, at higher fields, the amount of current flowing through the semiconductor increases thereby facing an increased resistance and, hence, larger potential differences are observed. Since the largest changes of the potential arise around the corners, higher sensitivity can be obtained by placing the voltage probes closer to the corners (Figure 15). Hence, a symmetric EMR sensor with high sensitivity can be reduced to a two-contact device where the contacts are utilized for current injection as well as voltage measurement.

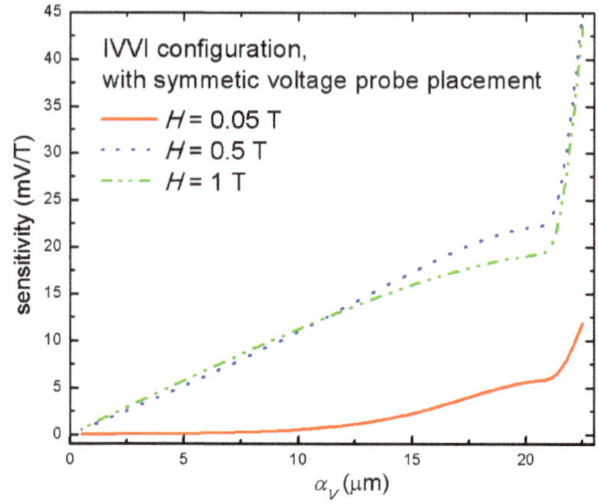


**Figure 15.** Sensitivity as function of the voltage probe placement at various external fields for the EMR device with symmetric IVVI configuration.

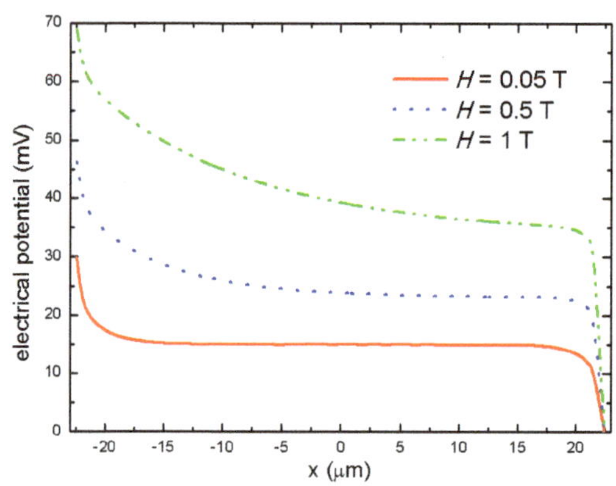


**Figure 16.** The potential distribution along the edge of the semiconductor bar in the EMR device at different magnetic fields.

It has previously been reported that an asymmetric contact configuration increases the EMR effect [12]. In order to study the influence of an asymmetric arrangement of the voltage probes, an IVVI configuration is simulated with $\alpha_{VR}$ fixed at 7.5 μm while $\alpha_{VL}$ is varied from -22.5 μm to 22.5 μm. The current is injected at the current lead with varying position. Figure 17 shows the sensitivity as a function of $\alpha_{VL}$. Compared to the results found for the symmetric arrangement (Figure 15), the maximum sensitivity at high field (1 T) has reduced. In the case of low fields, the sensitivity has slightly increased due to the asymmetric arrangement.

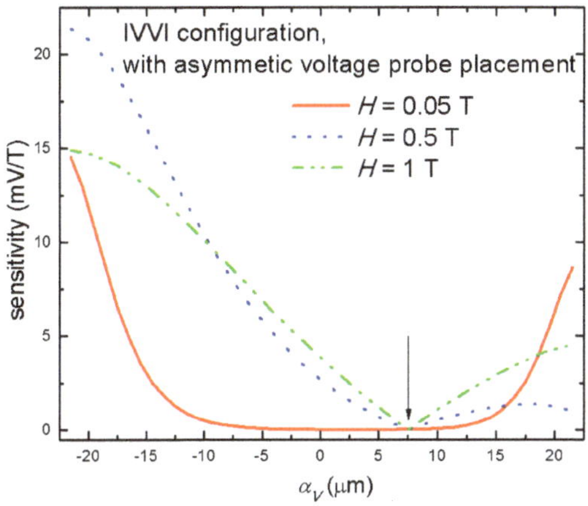


**Figure 17.** Sensitivity as a function of the placements of voltage probes at various external fields for the EMR device with asymmetric IVVI configuration. The arrow indicates the location of the fixed voltage probe.

### 6.4. Placement of current leads in VIIV configuration

The effect of the placements of the current leads $\alpha_I$ in a symmetric VIIV device is studied using the same geometric parameters for the model as before in section 5.3. $\alpha_I$ denotes the distance of the current leads from the device's center (Figure 12(b)). The sensitivity depends on $\alpha_I$ in a very similar way as it did on $\alpha_V$ in the case of the IVVI configuration and the same trends can be observed (Figure 18). Again, this result shows the high sensitivity that can be obtained by using a simple two-contact electrode arrangement. A study of asymmetric current lead arrangement shows that the sensitivity can only be improved for low magnetic fields (Figure 19), which is also similar to what is found for the asymmetric IVVI configuration. The large change in sensitivity as the leads are placed closer to the corners of the device can be explained in the same way as in case of the IVVI configuration. It can be concluded that VIIV and IVVI configurations are very similar in general and optimally utilized with two electrodes at the corners of the semiconductor.

## 7. Finite-element analysis on the effect of the semiconductor/metal interface

In this section, an IVVI bar-type EMR device is investigated. The geometry and dimensions of the device model are: $W_m$ = 15 μm, $W_s$ = 3 μm, $L$ = 75 μm and $\alpha$ = 25 μm. The current leads ($I_+$, $I_-$) have a width of 0.1 μm. The voltage probes ($V_1$, $V_2$) are considered as a point-like contact. A current $I$ of 100 μA is applied through setting the boundary condition of the current leads. At the semiconductor/metal interface, different values of the contact resistivity $\rho$ between $10^{-11}$ Ω·cm$^2$ and $10^{-5}$ Ω·cm$^2$ are applied. In practice, $10^{-8}$ Ω·cm$^2$ is considered a very low ohmic contact resistivity.

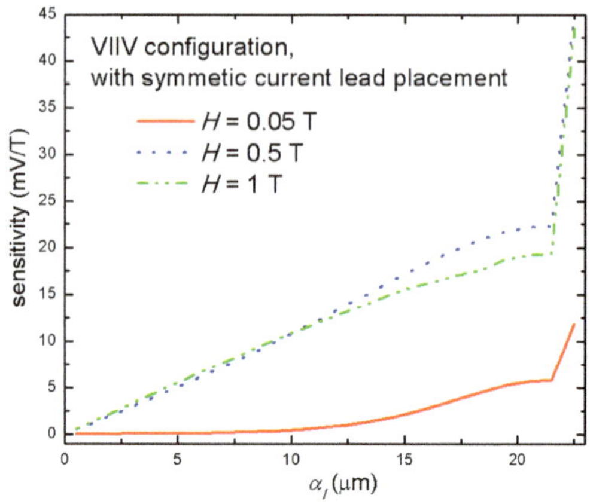


**Figure 18.** Sensitivity as a function of the placement of the current leads at various external fields for the EMR device with symmetric VIIV configuration.

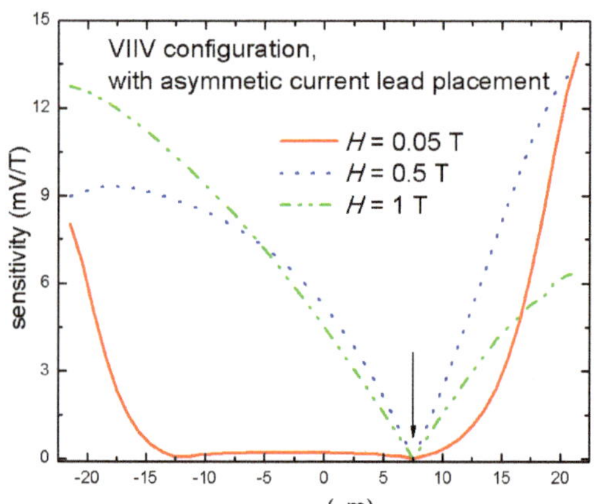


**Figure 19.** Sensitivity as a function of the placement of current leads at various external fields for the EMR device with symmetric VIIV configuration. The arrow indicates where the location of the fixed current lead.

The model was meshed with free triangular elements and consisted of approximately $6.8\times10^5$ elements and $1.4\times10^6$ degrees of freedom. The mesh density was varied adaptively.

The external magnetic field $H$ of 1 T is applied in the positive z-direction to the model. The voltage output value of $V(H)$ between the probes $V_1$ and $V_2$ was evaluated and the EMR effect and sensitivity were calculated by the help of equation (18) and (19).

Figure 20 shows the distribution of the current density in devices with different contact resistivities at $H = 0$ T and $H = 1$ T. The current density is symmetric in case of $H = 0$ T whereas it is deflected by the Lorentz force at $H = 1$ T yielding increased density in the semiconductor and decreased density in the metal. With a low contact resistivity ($10^{-8}$ $\Omega \cdot cm^2$), the current distribution in the device at zero and high fields are almost the same as in the ideal case with no contact resistivity. As the resistivity increases at the semiconductor/metal interface, it acts like a barrier and the current is increasingly inhibited to enter the metal shunt. Since the current density becomes larger in the semiconductor region, the value of $R(0)$ increases, thereby reducing the EMR effect. These dependencies can also be seen in Figure 21, which shows the current density distribution for the y/z-cross-section along the y-axis. It is worth to note that the area under each curve is identical to $I$. As the contact resistivity increases, the current density in the semiconductor increases, and a similar effect is obtained by increasing the magnetic field.

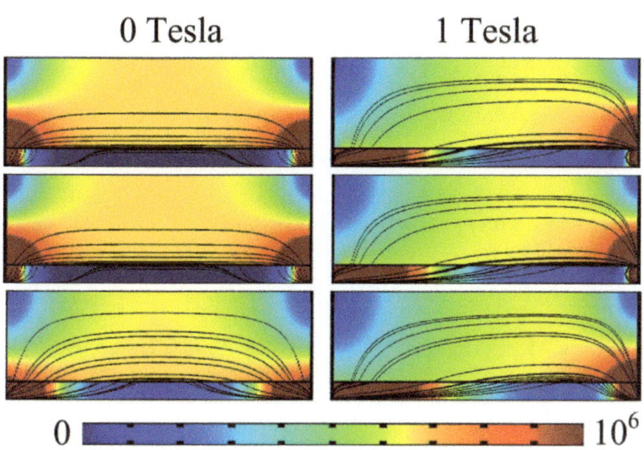


**Figure 20.** Current density distributions in different devices with various contact resistivities. (a) Ideal contact (no resistivity), (b) $10^{-8}$ $\Omega \cdot cm^2$, (c) $10^{-5}$ $\Omega \cdot cm^2$. The color bar represents the strength of the current density ($A/m^2$). Dark streamlines show the path of current. Left column: zero external field. Right column: external field is 1 T.

Figure 22 shows the EMR effect and the sensitivity as functions of the contact resistivities at $H = 1$ T. Nonlinear curve fitting is applied to find the exponential functions that approximate both the EMR effect and the sensitivity. For contact resistivities between $10^{-11}$ $\Omega \cdot cm^2$ and $10^{-7}$ $\Omega \cdot cm^2$, the EMR effect is almost constant. In case of the sensitivity, this range is from $10^{-11}$ $\Omega \cdot cm^2$ to $10^{-8}$ $\Omega \cdot cm^2$. As the contact resistivity increases beyond $10^{-7}$ $\Omega \cdot cm^2$ and $10^{-8}$ $\Omega \cdot cm^2$, the EMR effect and the sensitivity, respectively, decrease exponentially. In particular, a device with contact resistivity of $10^{-11}$ $\Omega \cdot cm^2$ shows an EMR effect of $1.1 \times 10^5$ %. The EMR effect is $1.1 \times 10^6$ % for the device with a contact resistivity of $10^{-8}$ $\Omega \cdot cm^2$ and it decreases to $6.0 \times 10^5$ % and 423 % for the devices with $10^{-6}$ $\Omega \cdot cm^2$ and $10^{-5}$ $\Omega \cdot cm^2$, respectively. The EMR effect is almost reduced by 95% compared to the ideal device ($10^{-11}$ $\Omega \cdot cm^2$) when the contact resistivity increases to $10^{-6}$ Ω·cm². Even in this case, the value is comparably high with an order of magnitude larger than values reported for contemporary GMR sensors. However, the EMR effect in the device with $10^{-5}$ Ω·cm² contact resistivity is only 0.4% of the one of an ideal device. For a contact resistivity lower than $10^{-8}$ Ω·cm², the sensitivity shows a value of app. 8.6 mV/T. It drops rapidly as the contact resistivity increases and has a value of 7.4 mV/T at $10^{-5}$ Ω·cm².

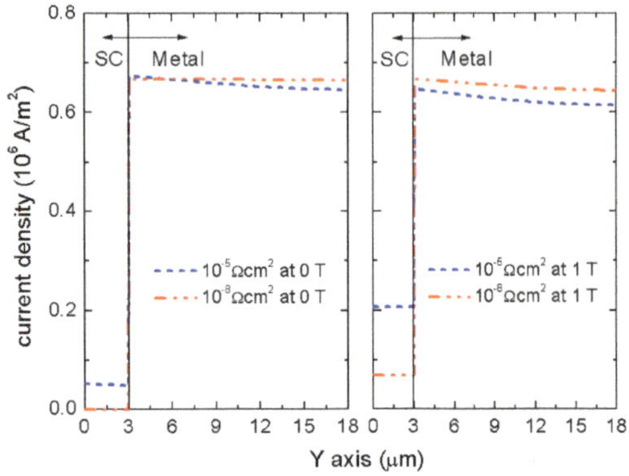


**Figure 21.** Current density distribution along the symmetry axis (y-axis) of the EMR devices with different values of the contact resistivity of the interface between semiconductor (SC) and metal at magnetic fields of (a) 0 T and (b) 1 T.

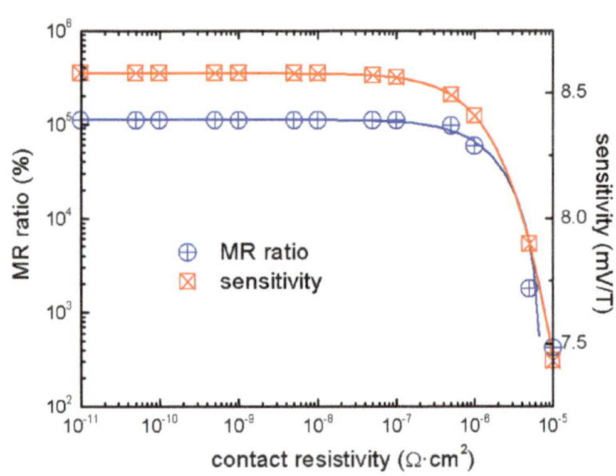


**Figure 22.** EMR effect and sensitivity as a function of the contact resistivity at 1 Tesla (MR ratio and contact resistivity axes are logarithmic).

## 8. Conclusion

In this chapter, we discussed the basic concepts of the Hall effect and EMR effect. We introduced a mathematical model to describe the inhomogeneous conductivity of a conductor under the application of a magnetic field, which can be used to simulate both effects. The model is applicable for 2-D simulations as well as 3-D simulation. Especially for complex designs, the FEM is a convenient technique to carry out the simulation. While in case of Hall devices a 2-D simulation is normally sufficient, the structure of EMR devices requires a 3-D simulation in order to obtain accurate results. Specifically, for the EMR ratio we found an error of almost 30 % at 1T for the 2-D model and about 5 % for the 3-D model compared to experimental results. The model has been employed to simulate Hall and EMR devices. The focus of the chapter has been put on the investigation of EMR devices, since, compared to Hall devices, they have not been studied yet.

The modeled EMR device is of bar-type geometry since this is the most promising one for future applications taking into account device miniaturization and fabrication. The sensitivity of the device is investigated thoroughly, since it is the most significant parameter for many sensor applications and has not been addressed previously. Four different geometric parameters are investigated and the results show that an optimal performance requires certain aspects of the design to be customized for specific applications. The length/width ratio $\alpha$ of the semiconductor layer is found to have an optimal value between 10 and 20 for strong magnetic fields and 5 for weak ones. The placement of the voltage probes and current leads crucially influences the output sensitivity. In general, IVVI and VIIV devices are affected in a similar way. The two inner probes need to be placed as close as possible to the corners of the device in order to provide a high sensitivity. This leads to the surprising result that an EMR sensor with high sensitivity can be reduced to a two-contact device with the two contacts located at the corners of the semiconductor layer. An asymmetric arrangement of the contacts does not yield an improved performance at high fields, but has a positive effect on the low field sensitivity.

The simulations also show that the EMR effect and output sensitivity won't drop until the contact resistivity between the semiconductor and metal exceeds $10^{-8}$ $\Omega\cdot cm^2$. This result is consistent with experimental results reported previously. Interestingly, a device with contact resistivities up to $10^{-6}$ $\Omega\cdot cm^2$ still could show a considerable large sensitivity which is 95% of that in the device with contact resistivity of $10^{-8}$ $\Omega\cdot cm^2$. Attempts to reduce the contact resistivity below this value will not yield any further improvement of the performance of the device, which is often governed by the sensitivity. This is an important finding since, in practice; a good ohmic contact with low resistivity requires a costly device fabrication with carefully controlled material growth and a rapid thermal annealing process after the deposition of the material.

## Author details

Jian Sun* and Jürgen Kosel
*Physical Sciences and Engineering Division, King Abdullah University of Science and Technology, Thuwal, Kingdom of Saudi Arabia*

* Corresponding Author

## 9. References

[1] Edwin H (1879) On a New Action of the Magnet on Electric Currents. American Journal of Mathematics. 2 (3): 287–292.
[2] Ramsden E (2006) Hall-effect Sensors: Theory and Applications. Oxford: Newnes.
[3] Solin S, Thio T, Hines D, Hermans J (2000) Enhanced Room-Temperature Geometric Magnetoresistance in Inhomogeneous Narrow-Gap Semiconductors. Science. 289: 1530-1532.
[4] Moussa J, Ram-Mohan L, Sullivan J, Zhou T, Hines D, Solin S (2001) Finite-Element Modeling of Extraordinary Magnetoresistance in Thin Film Semiconductors with Metallic Inclusions. Phys. Rev. B 64: 184410.
[5] Rowe A, Solin S (2005) Importance of Interface Sampling for Extraordinary Resistance Effects in Metal Semiconductor Hybrids. Phys. Rev. B. 71: 235323.
[6] Moussa J, Ram-Mohan L, Rowe A, Solin S (2003) Response of an Extraordinary Magnetoresistance Read Head to a Magnetic Bit. J. of App. Phys. 94(10): 1110-1114.
[7] Solin S, Hines D, Rowe A, Tsai J, Pashkin Y (2003) Nanoscopic Magnetic Field Sensor Based on Extraordinary Magnetoresistance. J.Vac. Sci. Technol. B, 21(6): 3002-3006.
[8] Moller C, Kronenwerth O, Heyn C, Grundler D (2004) Low-Noise Magnetic-Flux Sensors Based on the Extraordinary Magnetoresistance Effect. Appl. Phys. Lett. 84: 3343-3346.
[9] Solin S, Hines D, Rowe A, Tsai J, Pashkin Y, Chung S, Goel N, Santos M (2002) Nonmagnetic Semiconductors as Read-Head Sensors for Ultra-High-Density Magnetic Recording. Appl. Phys. Lett. 80(21): 4012-4014.
[10] Smith N, Arnett P (2001) White-Noise Magnetization Fluctuations in Magnetoresistive Head. Appl. Phys. Lett. 78(10): 1448-1450.
[11] Zhou T, Hines D, Solin S (2001) Extraordinary Magnetoresistance in Externally Shunted Van Der Pauw Plates. App. Phys. Lett. 78(5): 667-669.
[12] Holz M, Keronenwerth O, Grundler D (2003) Optimization of Semiconductor Metal Hybrid Structures for Application in Magnetic-Field Sensors and Read Heads. Appl. Phys. Lett. 83: 3344-3346.
[13] Hoener M, Kronenwerth O, Heyn C, Grundler D, Holz M (2006) Geometry-Enhanced Magnetoresistance of Narrow Au/InAs Hybrid Structures Incorporating a Two-Dimensional Electron System. J. Appl. Phys. 99: 036102.
[14] Hewett T, Kusmartsev F (2010) Geometrically Enhanced Extraordinary Magnetoresistance in Semiconductor-Metal Hybrids. Phys. Rev. B. 82: 212404.
[15] Sun J, Kosel J (2011) Finite Element Analysis on the Influence of Contact Resistivity in an Extraordinary Magnetoresistance Magnetic Field Micro Sensor. J. Supercond. Nov. Magn. DOI:10.1007/s10948-011-1256-8.
[16] Sun J, Gooneratne C, Kosel J (2011) Design Study of a Bar-Type EMR Device. IEEE Sensors Journal.12(5): 1356-1360.
[17] Zawadzki W (1974) Electron transport phenomena in small-gap semiconductor. Advances in Physics. 23: 435-522.
[18] Abeles B, Meiboom S (1956) Galvanomagnetic Effects in Bismuth. Phys. Rev. 101: 544–550.
[19] Zienkiewicz O, Taylor R, Zhu J (2005) The Finite Element Method: Its Basis and Fundamentals. Oxford: Elsevier Butterworth-Heinemann

Chapter 10

# Steady-State and Transient Performance Analysis of Permanent-Magnet Machines Using Time-Stepping Finite Element Technique

Kazumi Kurihara

Additional information is available at the end of the chapter

## 1. Introduction

Recently, global warming has become an important problem. High-efficiency machines have been needed in a large variety of industrial products in order to save electrical energy. For many applications, permanent-magnet (PM) synchronous machines can be designed which is smaller in size but more efficient as compared to induction machines [1-3]. Besides, PMs have been employed as an alternative to current carrying coils for magnetic field excitation in synchronous machines for over 50 years. The lack of slip rings, brushes and field winding losses have always been viewed as distinct advantages over that of conventional wound field machines. However, when the machine size becomes small, the efficiency becomes low. This is mainly due to the reason that the iron loss and the copper loss are large, because the iron core of the stator in the small machine generally does not have annealing and the resistance of the stator windings is comparatively large.

This chapter presents a successful design of the high-efficiency small but novel Interior permanent-magnet (IPM) machines using Neodymium-Boron-Iron (NdBFe) magnets. It is designed to operate with both high-efficiency line-start IPM motors [3] and generators with damper bars [4]. Time-stepping finite element analysis has been used to successfully predict the dynamic and transient performances of the prototype machines. Time-stepping finite element analysis [3-6] has been used to successfully predict the dynamic and transient performances of the prototype IPM machines. The computed performance has been validated by tests in the prototype machine.

## 2. IPM machine configuration

The photograph of an IPM rotor, the cross section of a quarter of the high-efficiency motor and the demagnetization curve of the NdBFe magnet used for finite-element analysis are shown respectively in Figures 1-3, respectively [3].

A frame size of a 600 W, 3-phase, 4-pole, Y-connected, 50 Hz, 200 V squirrel- cage induction machine was used for testing the IPM rotor shown in Figure 1. The four-pole magnets arrangement in the rotor is oriented for a high-field type IPM synchronous machine. The experimentally developed rotor has the following distinctive design features [3]:

1. The fluxes from both sides of the magnet are concentrated effectively in the middle of the magnetic poles of the rotor.
2. The reluctance of the d axis is larger than that of the q axis, because the d- axis flux passes across the magnet with high reluctance. Large reluctance torque can be obtained.
3. The conducting material between the magnet and the rotor core is made from aluminum and has both functions of the flux barrier and cage bar.

**Figure 1.** IPM rotor

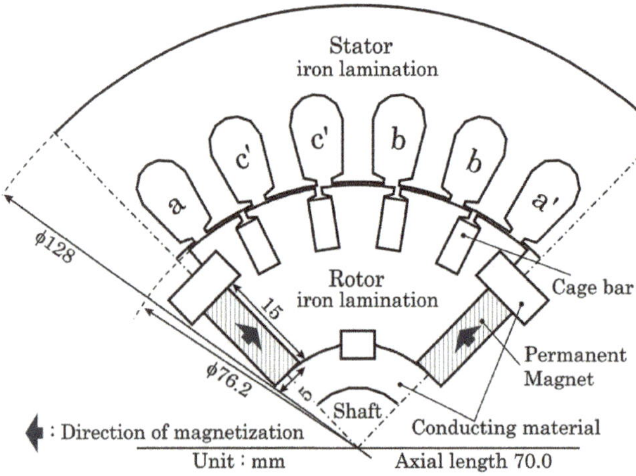


**Figure 2.** Configuration of high-efficiency IPM machine

Furthermore, the number and configuration of rotor slots have been successfully designed by using the finite-element method so that the waveform of the electromotive force (EMF) due to the PMs was close to the sine waveform and the cogging torque was low.

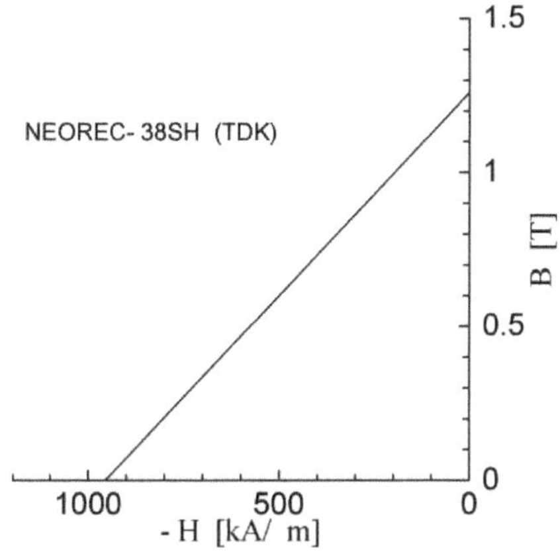


**Figure 3.** Demagnetization curve of NdBFe magnet

## 3. Method for analysis

The analysis for taking the eddy currents into account, in general becomes essential to solve the three-dimensional problem. In this paper, it is assumed that the eddy currents flow approximately in the axial direction, because the rotor shown in Figure 1 is equipped with end rings. This reduces the analysis to a two-dimensional problem. The fundamental equations for the magnetic field are represented in the two-dimensional rectangular co-ordinates as

$$\frac{\partial}{\partial x}\left(\nu\frac{\partial A}{\partial x}\right)+\frac{\partial}{\partial y}\left(\nu\frac{\partial A}{\partial y}\right)=-J_0-J_e-J_m \tag{1}$$

$$J_e=-\sigma\frac{\partial A}{\partial t} \tag{2}$$

$$J_m=\nu_0\left(\frac{\partial M_y}{\partial x}-\frac{\partial M_x}{\partial y}\right) \tag{3}$$

where $A$ is the $z$ component of magnetic vector potential $\boldsymbol{A}$, $J_0$ is the stator-winding current density, $J_e$ is the eddy current density, $J_m$ is the equivalent magnetizing current density, $M_x$, $M_y$ are $x$ and $y$ components of the magnetization $\boldsymbol{M}$, respectively. $\sigma$ is the conductivity, and $\nu$ is the reluctivity. The value of $\nu$ in the PM is assumed the same as the reluctivity of free space $\nu_0$. $J_m$ is assumed zero, outside the PM.

The effect of the eddy current for the rotor ends is taken into account by multiplying by the coefficient $k_c$ as described below. It is done to reduce the analysis to two-dimensional. The equivalent resistance $R_2$ for rotor bars including the rotor end rings can be given below if the bars are distributed at equal intervals in the rotor [7].

$$R_2 = R_b + R_e \frac{Z_2}{(2p\pi)^2} \tag{4}$$

where $R_b$ is the resistance of a bar, $R_e$ is the resistance of the end rings, $Z_2$ is the number of rotor slots and $p$ is the pole pair number.

Therefore, $k_c$ is given by

$$k_c = \frac{R_b}{R_2} \tag{5}$$

This coefficient $k_c$ is found effective to take into account the rotor-bar current for the fundamental space harmonic. Moreover, it has been found that the agreement between computed and measured results of the starting performance characteristics in the IPM motor is good [3]. Therefore, it is considered that design use of the $k_c$ is acceptable, even if the higher space harmonics exists [5]. The value of coefficient $k_c$ is 0.55 in this paper.

### 3.1. Voltage, current and dynamic equations and calculation steps for IPM synchronous motor

Figure 4 shows the circuit of the three-phase line-start IPM synchronous motor. It has three stator phase windings, which are star connected with neutral. The voltage and current equations of the IPM motor are given as

$$e_a + r_1 i_a + L_1 \frac{\partial i_a}{\partial t} = v_a - v_n \tag{6}$$

$$e_b + r_1 i_b + L_1 \frac{\partial i_b}{\partial t} = v_b - v_n \tag{7}$$

$$e_c + r_1 i_c + L_1 \frac{\partial i_c}{\partial t} = v_c - v_n \tag{8}$$

$$i_a + i_b + i_c = 0 \tag{9}$$

where $v_a$, $v_b$, and $v_c$ are the phase voltages, subscripts $a$, $b$, and $c$ represent stator quantities in lines $a$, $b$, and $c$, respectively. $v_n$ is the potential of the neutral $n$, when the potential of the neutral of the supply source is zero, $i_a$, $i_b$, and $i_c$ are the line currents, $r_1$ and $L_1$ are the resistance and end-winding leakage inductance of the stator winding per phase, respectively. $e_a$, $e_b$, and $e_c$ are the induced phase voltages; and $e_a$ is given by the line integral of the vector potential round $c_a$ which is along the stator windings of phase $a$ [5]

$$e_a = \oint_{c_a} \frac{\partial A^t}{\partial t} ds = \oint_{c_a} \frac{A^t - A^{t-\Delta t}}{\Delta t} ds \tag{10}$$

where $A^t$ is $A$ at time $t$. $\Delta t$ is the time step. $e_b$, and $e_c$ can be obtained similarly, as in [5].

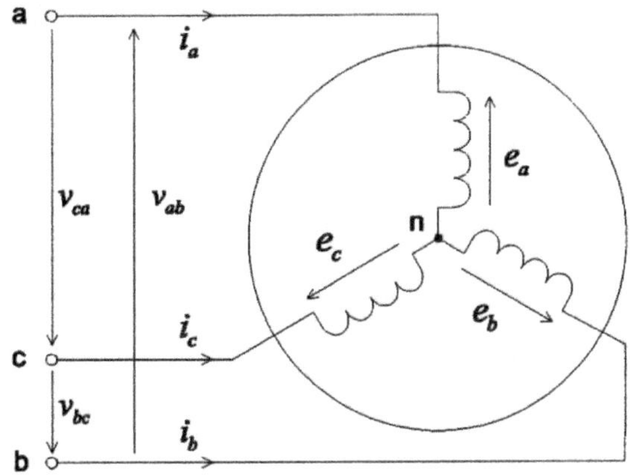


**Figure 4.** Circuit of three-phase line-start IPM synchronous motor

For operation from a balanced three-phase system,

$$v_a + v_b + v_c = 0. \tag{11}$$

$v_n$ can be obtained by adding each side of (6)-(8) and then applying (9) and (11)

$$v_n = \frac{e_a + e_b + e_c}{3}. \tag{12}$$

One obtains the following equation by substituting (12) in (6)-(8) [5]:

$$\frac{2}{3}e_a - \frac{e_b + e_c}{3} + r_1 i_a + L_1 \frac{\partial i_a}{\partial t} - v_a = 0 \tag{13}$$

$$\frac{2}{3}e_b - \frac{e_c + e_a}{3} + r_1 i_b + L_1 \frac{\partial i_b}{\partial t} - v_b = 0 \tag{14}$$

$$\frac{2}{3}e_c - \frac{e_a + e_b}{3} + r_1 i_c + L_1 \frac{\partial i_c}{\partial t} - v_c = 0. \tag{15}$$

The dynamic equation is given as [3]

$$T = J\frac{d\omega_r}{dt} + B_0\omega_r + T_l \tag{16}$$

where $T$ is the instantaneous electromagnetic torque, $J$ is the rotational inertia, $\omega_r$ is the rotor angular speed, $B_0$ is the friction coefficient, and $T_l$ is the load torque. The torque $T$ is calculated by using the *Bil* rule [8]. The angular speed, $\omega_r$ is given by

$$\omega_r = \frac{d\theta}{dt} \tag{17}$$

where $\theta$ is the rotational angle of the rotor.

One obtains the following equation by substituting (17) in (16):

$$T = J\frac{d^2\theta}{dt^2} + B_0\frac{d\theta}{dt} + T_l. \tag{18}$$

In this paper, the forward difference method is used to obtain the rotational angle at time $t$ because the vector potential, currents and rotational angle at time $t$- $\Delta t$ are all known

$$\frac{d\theta^{t-\Delta t}}{dt} = \frac{\theta^t - \theta^{t-\Delta t}}{\Delta t} \tag{19}$$

$$\frac{d^2\theta^{t-\Delta t}}{dt^2} = \frac{\theta^t - 2\theta^{t-\Delta t} + \theta^{t-2\Delta t}}{(\Delta t)^2}. \tag{20}$$

One obtains the following equation by substituting (19) and (20) in (18) [6]:

$$\theta^t = \frac{1}{J + B_0\Delta t}[(T^{t-\Delta t} - T_l^{t-\Delta t})(\Delta t)^2 + (2J + B_0\Delta t)\theta^{t-\Delta t} - J\theta^{t-2\Delta t}]. \tag{21}$$

In the case when the effect of the friction is negligibly small, the above equation can be represented simply as follows:

$$\theta^t = \frac{(\Delta t)^2}{J}\left(T^{t-\Delta t} - T_l^{t-\Delta t}\right) + 2\theta^{t-\Delta t} - \theta^{t-2\Delta t}. \tag{22}$$

One can obtain the vector potential, currents and rotational angle by solving (1), (13)-(15), and (18) using the time-stepping finite element technique [3].

Next, the calculation steps for this analysis are shown in Figure 5.

1. First, the terminal voltage $V_l$, its initial phase angle $\phi_0$, $T_l$, and $\Delta t$ are set, respectively. Each voltage for the three stator phase windings can be represented by

$$v_a^{\ t} = \sqrt{\frac{2}{3}}V_l\cos\left(\omega t + \phi_0\right) \tag{23}$$

$$v_b^{\ t} = \sqrt{\frac{2}{3}}V_l\cos\left(\omega t + \phi_0 - \frac{2}{3}\pi\right) \tag{24}$$

$$v_c^{\ t} = \sqrt{\frac{2}{3}}V_l\cos\left(\omega t + \phi_0 - \frac{4}{3}\pi\right). \tag{25}$$

2. The vector potential $A$ at $t = 0$ is set, where the static field caused by only PMs is given as the initial value.
3. At $t = \Delta t + t$, the value of $\theta^t$ at new $t$ is set.
4. At $t = \Delta t + t$, each voltage at new $t$ is set.
5. The initial values for $A^t$, $i_a^t$, $i_b^t$, and $i_c^t$ are set.
6. The matrix equation constructed by the time-stepping finite element technique is solved [5].
7. The convergence of $A^t$ is tested. Unless $A^t$ converges, the process returns to step 6).
8. After the convergence of $A^t$, $i_a^t$, $i_b^t$, and $i_c^t$, $T^t$ can be calculated. Then, the $\theta^t$ is determined from (22).
9. The calculation process from step 3) to step7) continues till the steady-state currents are obtained.

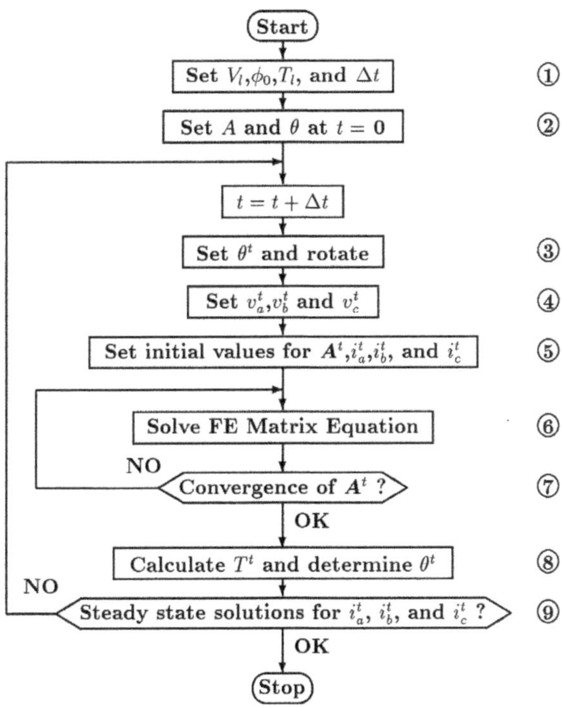


**Figure 5.** Flowchart of three-phase line-start IPM synchronous motor

## 3.2. Voltage and current equations and calculation steps for IPM synchronous generator

Figure 6 shows the circuit of the three-phase IPM synchronous generator. The voltage and current equations of the IPM generator are given as

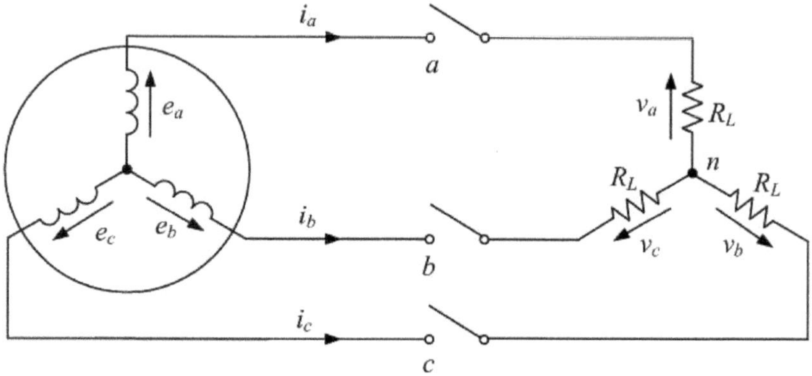


**Figure 6.** Circuit of three-phase IPM synchronous generator

$$e_a = r_1 i_a + L_1 \frac{\partial i_a}{\partial t} + v_a + v_n \tag{26}$$

$$e_b = r_1 i_b + L_1 \frac{\partial i_b}{\partial t} + v_b + v_n \tag{27}$$

$$e_c = r_1 i_c + L_1 \frac{\partial i_c}{\partial t} + v_c + v_n \tag{28}$$

For a balanced three-phase resistance load,

$$v_a = R_L i_a,\ v_b = R_L i_b,\ v_c = R_L i_c \tag{29}$$

where $R_L$ is a load resistance per phase.

$v_n$ can be obtained by substituting (29) in (26)-(28), adding each side of (26)-(28) and then applying (9)

$$v_n = \frac{e_a + e_b + e_c}{3} \tag{30}$$

One obtains the following equation by substituting (30) in (26)-(28).

$$\frac{2}{3}e_a - \frac{e_b + e_c}{3} - (r_1 + R_L) i_a - L_1 \frac{\partial i_a}{\partial t} = 0 \tag{31}$$

$$\frac{2}{3}e_b - \frac{e_c + e_a}{3} - (r_1 + R_L) i_b - L_1 \frac{\partial i_b}{\partial t} = 0 \tag{32}$$

$$\frac{2}{3}e_c - \frac{e_a + e_b}{3} - (r_1 + R_L) i_c - L_1 \frac{\partial i_c}{\partial t} = 0 \tag{33}$$

One can obtain the vector potential, currents by assuming a constant speed and then solving (1), (31)-(33) using the time-stepping finite-element technique [5].

Next, the calculation steps for this analysis are shown in Figure 7.

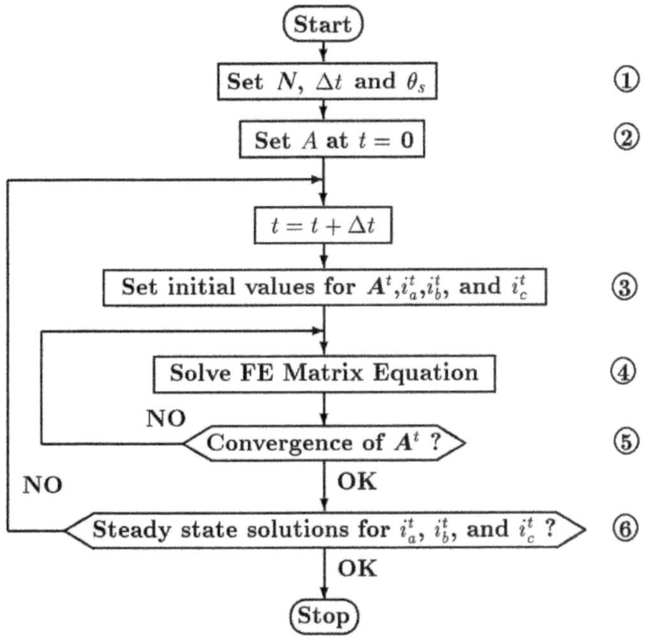


**Figure 7.** Flowchart of three-phase IPM synchronous generator

1. First, $N$, $\Delta t$ and the corresponding rotational step $\theta_s$ are set, respectively.
2. The vector potential $A$ at time $t = 0$ is set, where the static field caused by only PMs is given as the initial value.
3. At $t = t + \Delta t$, the initial values for $A^t$, $i_a^t$, $i_b^t$ and $i_c^t$ at new $t$ are set.
4. The matrix equation constructed by the time-stepping finite-element technique is solved [5].
5. The convergence of $A^t$ is tested. Unless $A^t$ converges, the process returns to step 4).
6. After the convergence of $A^t$, $i_a^t$, $i_b^t$ and $i_c^t$ are obtained. The calculation process from step 3) to 5) continues till the steady-state currents are obtained.

## 4. Steady-state synchronous and transient performance

This paper contains the steady-state synchronous and transient performance characteristics of the IPM synchronous machine shown in Figure 2. The good agreement between computed and measured results validates the proposed method for the finite-element analysis to predict the machine performance exactly.

### 4.1. EMF due to PMs

Figure 8 shows the terminal voltage waveform generated by PMs in driving the IPM synchronous machine at 1500 r/min by the external motor. It is shown that the agreement between the computed and measured values of the generated voltage is excellent.

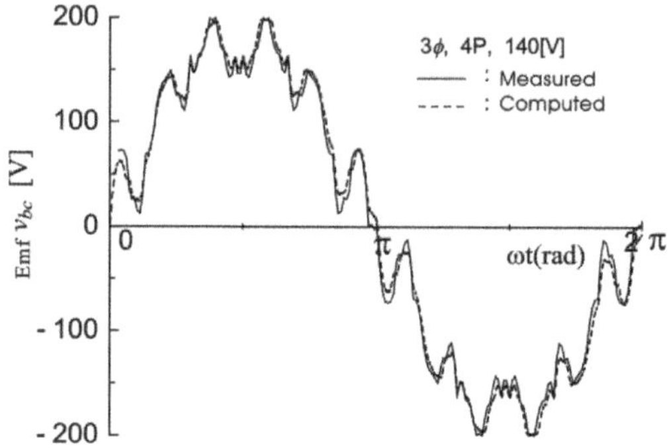


**Figure 8.** EMF generated by PMs

### 4.2. Steady-state synchronous and transient performance of Line-start IPM synchronous motor

Figure 9 shows the load performance characteristics at 140V. It is clear from Figure 9 that the power factor is almost unity at all loads. The efficiency and power factor of the IPM motor were 86.2% and 0.986, respectively for the output of 600 W. The efficiency-power-factor product is 85.0%. It is about 35% higher than that for the induction motor. These values of the IPM motor are very high when compared to those of the induction motor for the same 600 W nameplate rating [3]. Figure 10 shows the computed and measured results of the input current versus the output power at 140V. It is shown that the agreement between the measured and computed results is excellent.

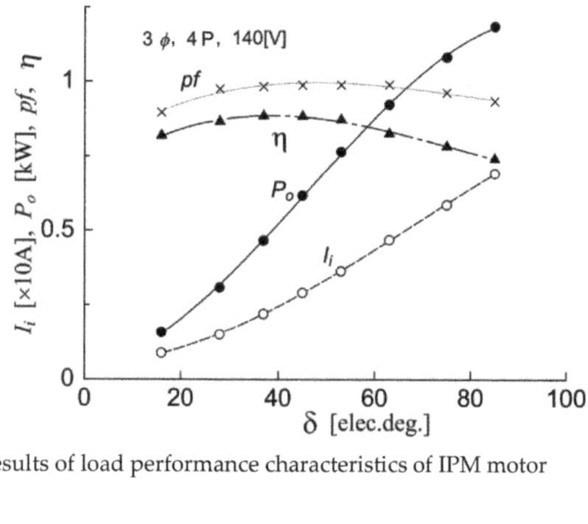


**Figure 9.** Measured results of load performance characteristics of IPM motor

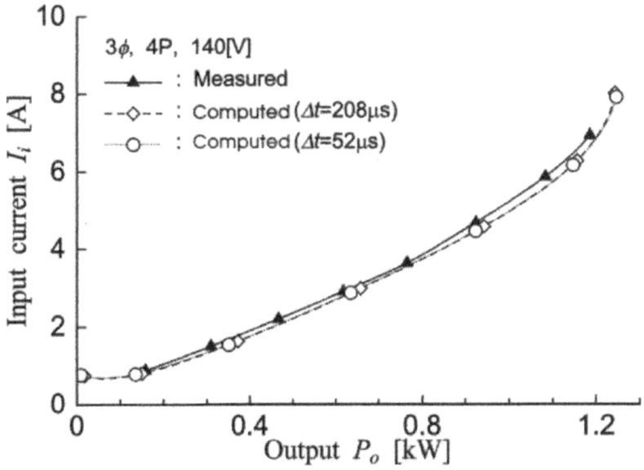


**Figure 10.** Computed and measured results of current versus output power of IPM motor

In the figure 10, two kinds of computed curves are given, and the agreement is also good. It is used to determine the suitable value of the $\Delta t$. This value must be determined by taking into account the effects due to the space harmonics [5]. The space harmonics effect is also the source of the cogging and ripple torques in the IPM motor. It can be compensated by skewing the stator by one slot pitch. Therefore, $\Delta t$. should be smaller than to $\Delta t_s$ to include the influence of the ripple harmonics on the starting with $\Delta t_s$, which is to move by one stator slot pitch at synchronous speed of the motor

$$\Delta t_s = \frac{(1/f)}{(N_s/p)} \tag{34}$$

where $f$ is the line frequency, $N_s$ is the number of stator slots. Herein, the following four values for the $\Delta t$ are chosen: 208, 104, 52 and 26 $\mu$s are an eighth, a sixteenth, a thirty-second, and a sixty-fourth of $\Delta t_s$, respectively. It is evident from Figure 10 that the choice of 208 $\mu$s is suitable at synchronous speed. However, this value is not sufficient in starting the IPM motor with large load inertia.

Figure 11 shows the computed speed-time responses at no load condition with the eddy-current brake disc coupled to the shaft, when the stator of the motor was supplied with balanced three-phase voltages at rated frequency of 50 Hz and rated voltage 140 V. The inertia of the disc is about 18 times the experimental rotor inertia, and the initial phase angle $\phi_0$ of (23-25) is $\pi/2$ in the figure. It is seen that the agreement between the curves of 52 $\mu$s and 26 $\mu$s is good and that those are superposed. The choice of a time step of 52 $\mu$s is suitable when the starting of the IPM motor.

Figure 12 shows the computed and measured speed-time responses with time during run-up and synchronizing period when $\Delta t$ and $\phi_0$ are 52 $\mu$s and $\pi/2$, respectively. It can be seen that the good agreement between the measured and computed results exits.

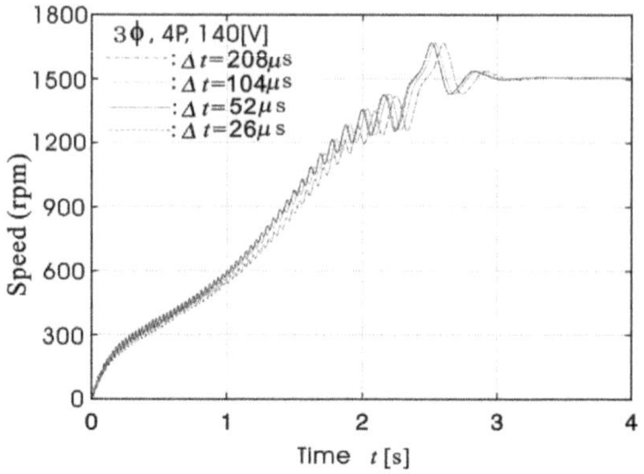


**Figure 11.** Computed speed-time response of IPM motor

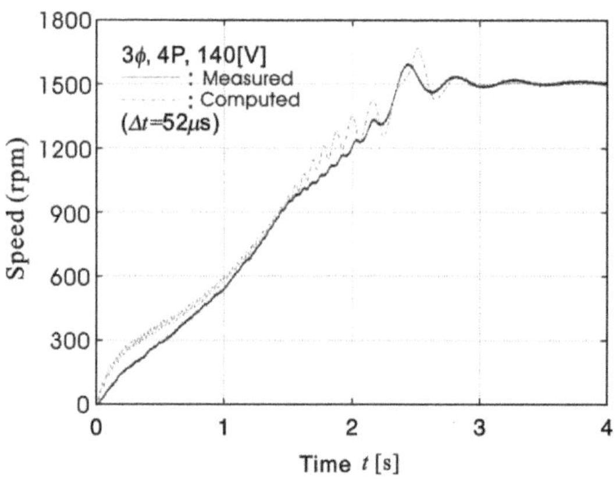


**Figure 12.** Computed and measured speed-time response of IPM motor

### 4.3. Steady-state synchronous and transient performance of IPM synchronous generator

Figure 13 shows the experimental setup for measuring the steady-state load performance characteristics of the IPM generator shown in Figure 2. A 2.2 kW three-phase two-pole 50 Hz 200 V squirrel-cage induction motor and a torque detector were used. The IPM generator has been driven at 1500 r/min by the PWM inverter-driven induction motor.

Figures 14-17 show the terminal voltage and line current, respectively, when the IPM generator with the cage-bars was changed from no-load to resistance load of 15 Ω per phase in Figure 6 at $t$ = 0s. The values of the resistance per phase for the maximum load was 15Ω. A synchronous motor has been used as the prime mover in the experiment.

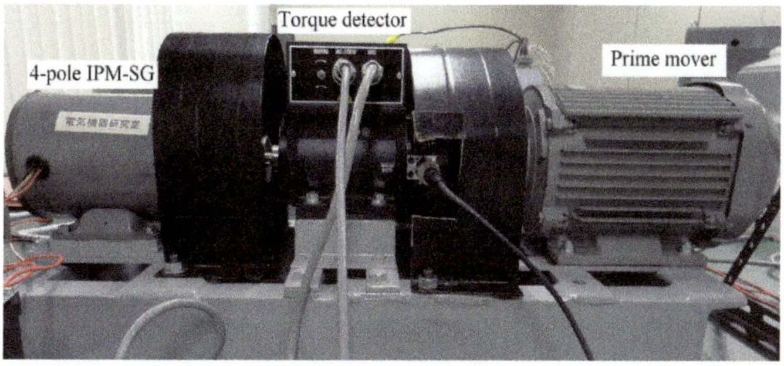


**Figure 13.** Experimental setup for IPM synchronous generator

Figures 14 and 15 show the measured and computed results of the terminal voltage, respectively. The phase angle of the terminal voltage in computing the terminal voltage and current is fitted to the experimental one. It is seen that the good agreement exists between the measured and computed results of the terminal voltage except the difference of the phase. This is the reason why the rotor speed lags synchronous speed in the experiment when load changes rapidly.

Figures 16 and 17 show the measured and computed results of the line current, respectively. The line current is zero before $t$ = 0s because of no-load. It is seen that the amplitude of the measured current was slightly pulsating because of the mechanical dynamic transient. On the other hand, a constant speed has been assumed in simulation. It is, however, seen that the good agreement exists between the measured and computed results of the current except the difference of the phase.

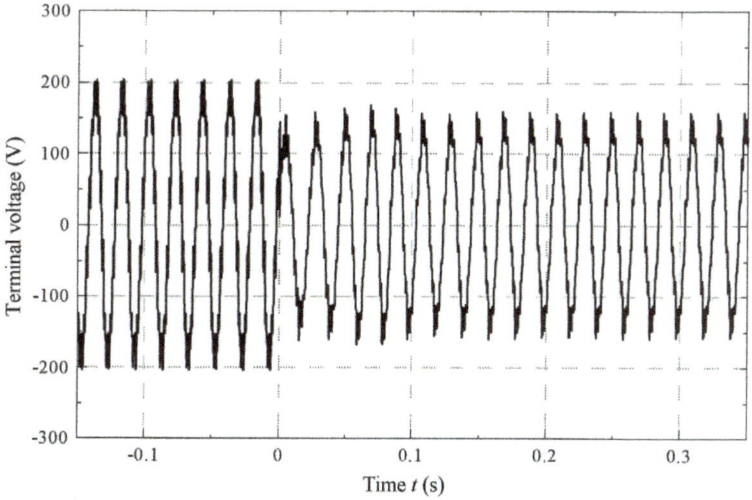


**Figure 14.** Measured results of terminal voltage versus time in IPM generator

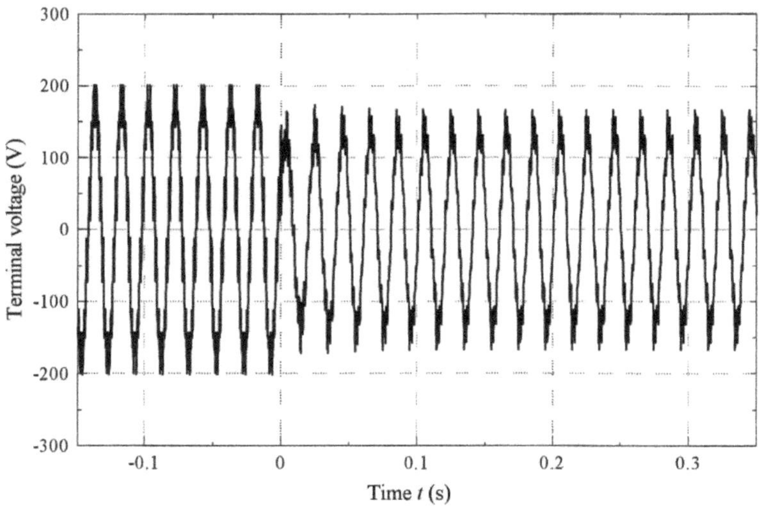


**Figure 15.** Computed results of terminal voltage versus time in IPM generator

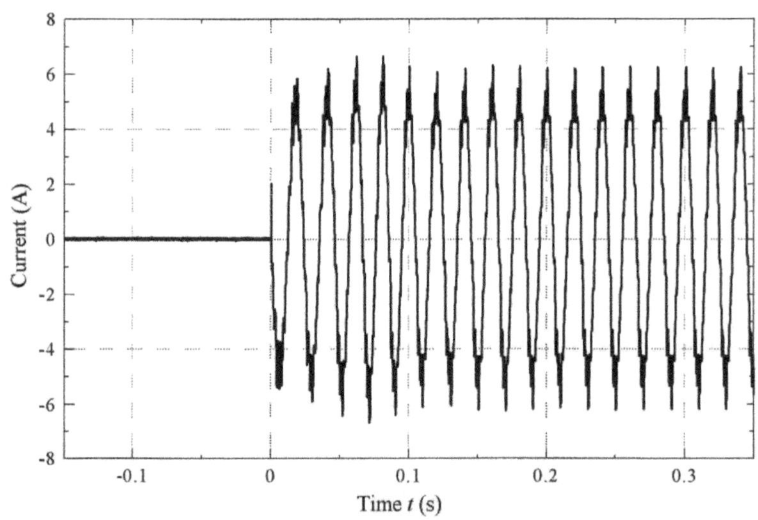


**Figure 16.** Measured results of line current versus time in IPM generator

Figures 18 and 19 show the measured and computed results of the steady-state terminal voltage and line current respectively. It is seen that the good agreement exists between the measured and computed results of the terminal voltage and line current. It is shown that the higher harmonic components by the higher space harmonics [5] were included.

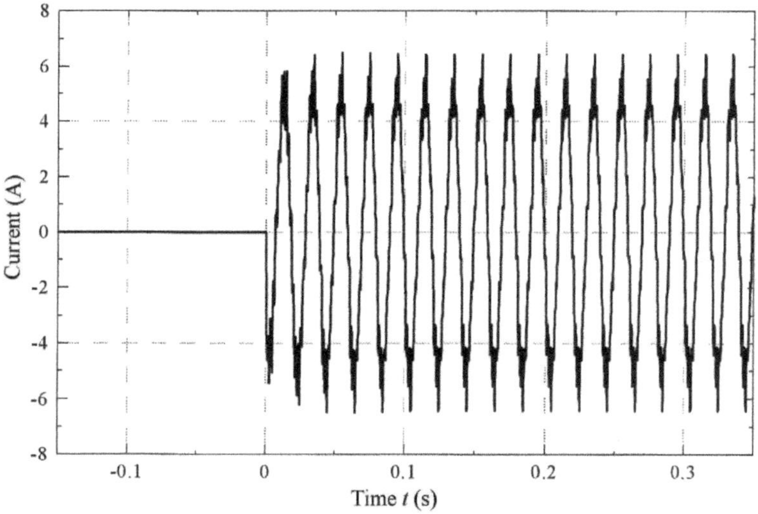


**Figure 17.** Computed results of line current versus time in IPM generator

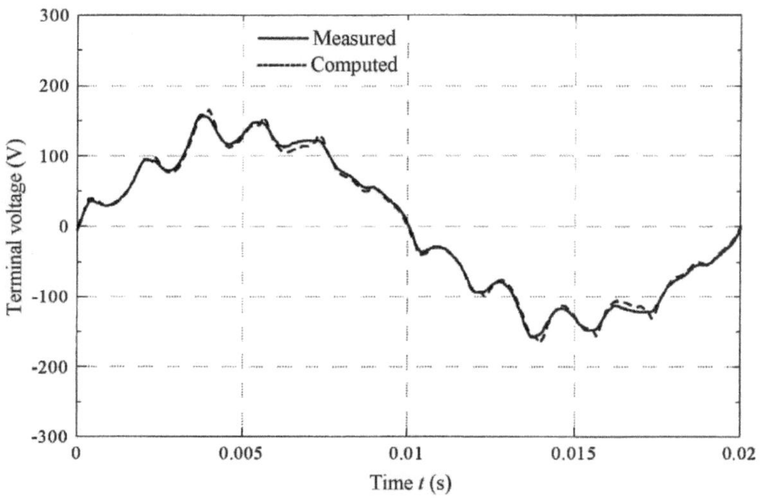


**Figure 18.** Steady-state terminal voltage versus time in IPM generator

Figures 20-22 show the steady-state load characteristics.

Figure 20 shows the measured and computed results of the terminal voltage versus the output. It can be seen that the good agreement between the measured and computed values exists except near maximum output.

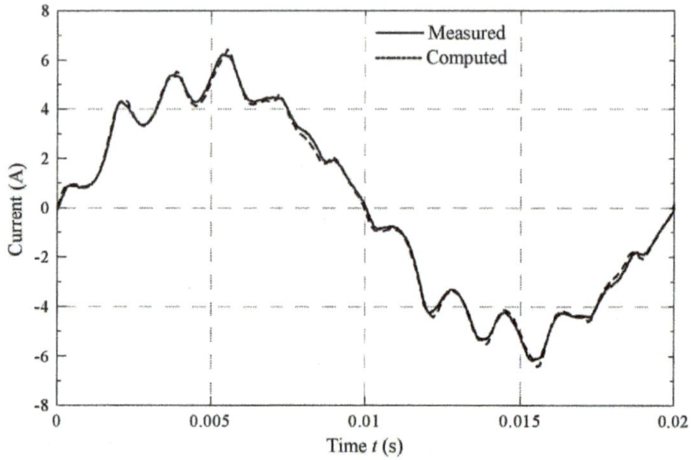


**Figure 19.** Steady-state line current versus time in IPM generator

Figure 21 shows the measured and computed results of the line current versus the output. It can be seen that the good agreement between the measured and computed values exists except near maximum output.

Figure 22 shows the measured results of the efficiency versus output. The efficiency was 85.8% at 600 W and 90% at 100W of light load. It is found that the efficiency is very high.

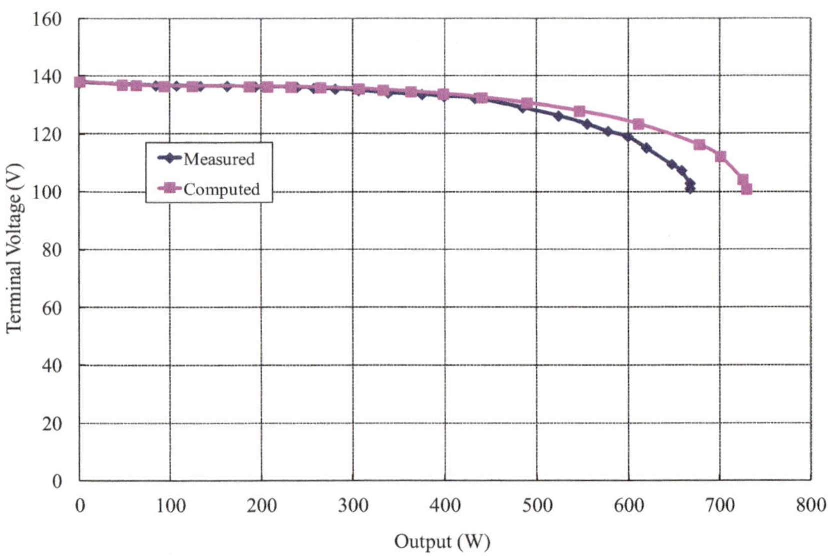


**Figure 20.** Measured and computed results of terminal voltage versus output in IPM generator

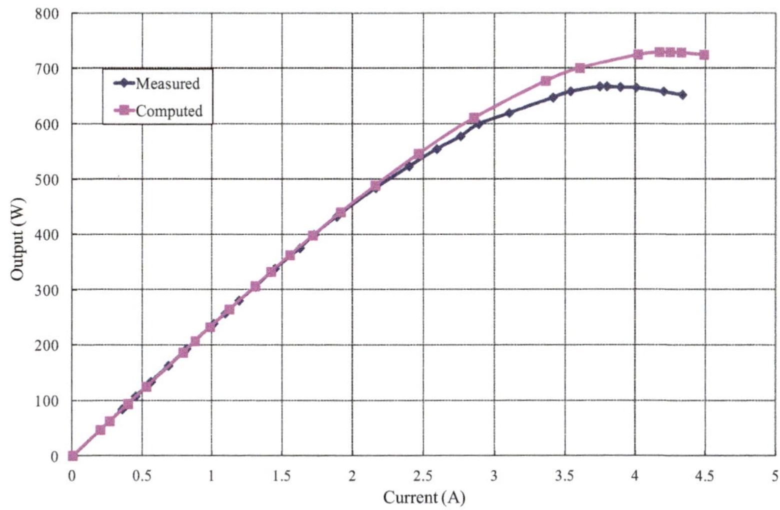


**Figure 21.** Measured and computed results of line current versus output in IPM generator

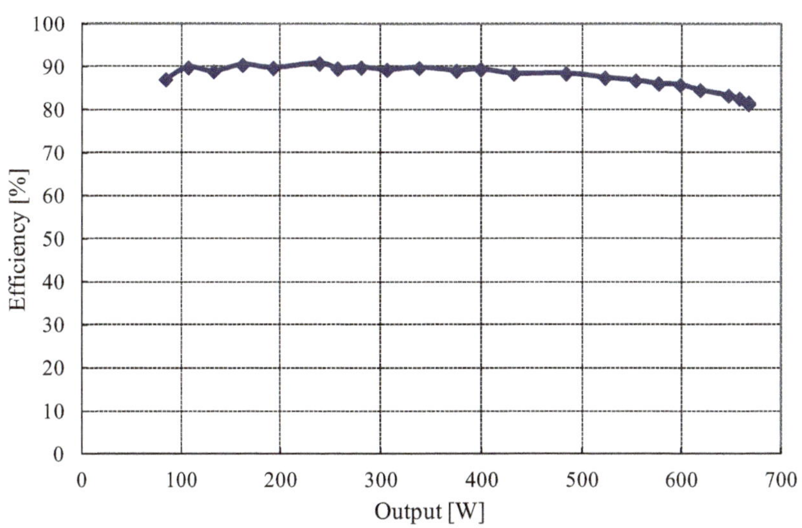


**Figure 22.** Measured results of efficiency versus output in IPM generator

## 5. Conclusion

A successful design of a high-efficiency small but novel IPM machine with cage bars was developed and tested. It is designed to operate with both high-efficiency line-start IPM motor and generator with damper bars. The IPM motor can start and synchronize fully with large load inertia. Beside, the effects of the damper bars on stability during load change and efficiency were investigated. Time-stepping finite element analysis has been used to successfully predict the steady-state and transient performances of the prototype IPM machines. It is clear that cage bars are used effectively to start up in a line-start IPM motor, and to operate stably in the IPM generator with damper bars. It has been found that the proposed design has yielded successful simulation and experimental results.

## Author details

Kazumi Kurihara
*Department of Electrical and Electronic Engineering, Ibaraki University, Hitachi, Japan*

## Acknowledgement

The author wish to thank Dr. Marushima of the Oriental Motor Company Ltd, for technical support and T. Kubota, T. Yasui, and T. Igari of the Ibaraki University for experimental support.

## 6. References

[1] Binns K. J, Jabbar M.A (1981) High-field self-start permanent magnet synchronous motor. Proc. IEE. B. 128:157-160.
[2] Rahman M. A, Osheiba A. M (1990) Performance of a large line-start permanent magnet synchronous motor. IEEE Trans. Energy Conversion 5: 211-217.
[3] Kurihara K, Rahman M. A (2004) High-efficiency line-start interior permanent-magnet synchronous motors. IEEE Trans. Ind. Applicat. 40: 789-796.
[4] Kurihara K (2010) Effects of Cage-Bars for Stability of Interior Permanent-Magnet Synchronous Generators. Int. Conf .Elect. Mach. and Systems.
[5] Kurihara K, Wakui G, Kubota T (1994) Steady-state performance analysis of permanent magnet synchronous motors including space harmonics. IEEE Trans. Magn. 30: 1306-1315.
[6] Kurihara K, Monzen T, Hori M (2006) Steady-state and transient performance analysis for line-start permanent-magnet synchronous motors with skewed slots. Int. Conf.Elect. Mach.
[7] Takeuchi J (1993) Design of Electrical Machines, Tokyo Ohmsha.
[8] Binns K.J, Riley C.P, Wong M (1985) The efficient evaluation of torque and field gradient in permanent-magnet machines with small air-gap. IEEE Trans. Magn. 21: 2435-2438.

Chapter 11

# High Sensitive Piezoresistive Cantilever MEMS Based Sensor by Introducing Stress Concentration Region (SCR)

Sh Mohd Firdaus, Husna Omar and Ishak Abd Azid

Additional information is available at the end of the chapter

## 1. Introduction

Micro-Electromechanical System (MEMS) device has become a hallmark technology for the 21$^{st}$ century. Its capability to sense, analyse, compute and control all within a single chip provide many new and powerful products. MEMS device is an emerging device in several areas of science and technology such as engineering structure, electronics and life sciences field such as chemistry, physic, biology and health sciences [Chollet and Liu (2007), Chivukula et al. (2006), Madou (1997)]. The two main key features for MEMS based device are mechanical structure that can be equated to motion and electrical signal. The addition of mechanical structure to an electronic chip gives a great enhancement to the functionality and performance. These devices have been dominantly used in the current market for computer storage system and automobiles [Madou (1997), Beeby et al. (2004), Hsu (2002)]. Smart vehicle are based on the extensive use of sensors and actuators. Various kind of sensors are used to detect the environment or road conditions and the actuators are used to execute any action are required to deals with conditions happen such as accelerometer for airbag system and Global positioning System (GPS) [Madou (1997), Hsu (2002)]. Most MEMS device are basically base on mechanical structure like cantilever beam, gears, pump and motor as shown in Fig. 1.

## 2. MEMS and finite element analysis

MEMS devices deal with nanofabrication process which related to microelectronics fabrication technology. This fabrication involves a series of high tech and high cost process such as ultraviolet lithography and doping. Due to expensive cost of fabrication, finite element analysis (FEA) has been used to characterize the MEMS structure behaviour during

DNA binding, through a water flow and vibration testing [Chollet and Liu (2007), Chivukula et al. (2006)]. FEA software helps MEMS designers to identify potential problem at early stage in design cycle before proceed on fabrication or production line, its help reducing working time to market. In design cycle, MEMS devices need to be check design intent, working operation, collision avoidance/detection and package stack-up. FEA capability scaling down from sub-micron to angstroms level features help designers come up with lower scale device design towards lead nano sensor/ actuator. Some MEMS base sensor devices is an assemblies of several parts and packaging, by using FEA, collision and contact surface can be determine [Hsu (2002), Liu (2006)].

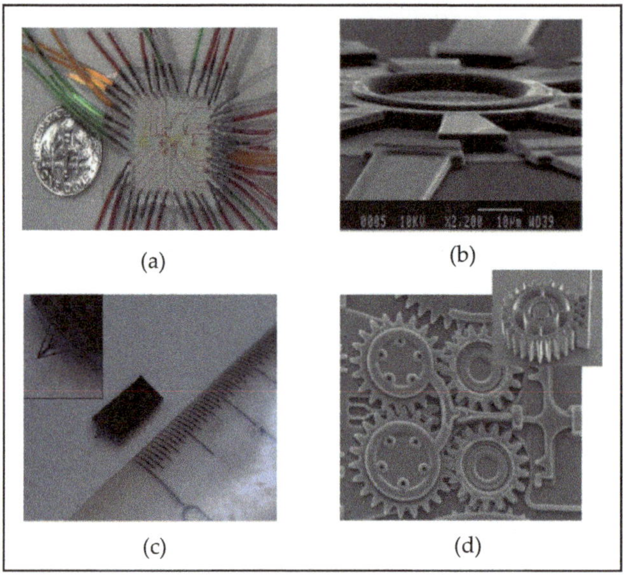


**Figure 1.** Example of MEMS devices; (a) micropump, (b) micromotor, (c) microcantilever, (d) microgears [Madou (1997), Hsu (2002), Arik et al. (1999)]

There are many FE software available in the market that has been used for analyse MEMS device like ANSYS, Solidworks, and Abaqus etc. Besides that there are also special dedicated MEMS FE software that integrates with MEMS device fabrication process such as CoventorWare, and IntelliCAD. In both software the modelling and fabrication file were combined and can transferred the fabrication machine [Madou (1997)]. The fabrication will be based on the attachment or design modelling file. This will not only help the MEMS designers to analyse and optimize the MEMS device design but also the manufacturability of the designed device. Flexibility in creating multiple design variations covering a wide range of needs such as die-mounted, package assemblies up to device efficiencies of configurations lead researchers to develop new device without any fabrication or prototype cost [Madou (1997), Hsu (2002), Liu (2006)].

## 3. Cantilever MEMS based sensor and finite element analysis

Brugger et al. (1999) and Thundat et al. (1995) have pointed out that cantilever based sensors are the simplest devices among MEMS devices that offer a very promising future for the development of novel physical, chemical and biological sensors. They have also been proven to be very versatile devices and have been used in several fields such as accelerometer, chemical sensors, etc [Vashist (2007)].

Basically MEMS cantilever sensor relies on the mechanical deformation of the structure, or in other words the deflection of membrane or beam structure. When the cantilever is loaded, its stressed elements deform. The MEMS cantilever will bend. As this deformation occur, the structure changes shape, and points on the structure displace. The concept is that deflection occurs when a disturbance or loading is applied to the cantilever is free end or along the MEMS cantilever surface. Normally the disturbance or loading is a force or mass that is attached to the MEMS cantilever in which it will make the MEMS cantilever bending Fig. 2 illustrates MEMS cantilever deflection working principal [Madou (1997), Hsu (2002), Lee et al. (2007)].

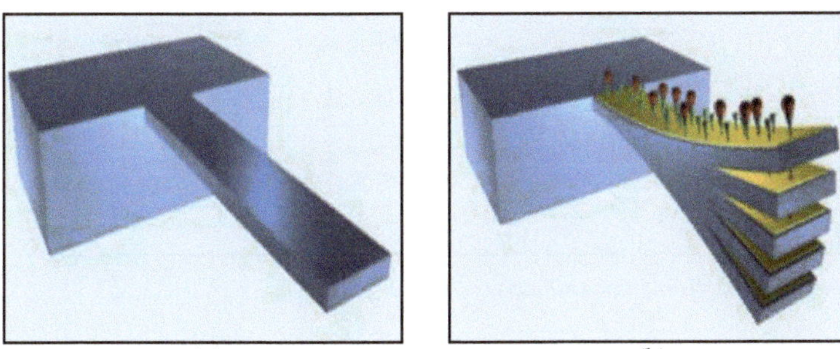

(a) (b)

**Figure 2.** MEMS Cantilever Sensor; (a) cantilever without binding mass, (b) cantilever deflects due to binding mass [Guillermo (2006)]

As the MEMS cantilever deflects, the resulted deformation is termed bending. External applied loads which cause bending will result in reactions at the free end, consisting of displacement or deflection, $\delta_{max}$ as shown in Fig.3. Maximum deflection during force applied for a beam that has constant cross section can be calculated using equation (1) [Cheng (1998), Benham et al. (1996)]. Fig.3 shows the schematic of cantilever deflection where it has one fixed end and one free end with force/mass applied.

$$\delta_{\max} = {}^{Fl^3}/_{3EI} \tag{1}$$

where $\delta_{max}$ is the maximum deflection, $F$ is force applied, $l$ is the cantilever length, $E$ is the Young's Modulus for the cantilever material which in this research is silicon and $I$ is the moment inertia for the cantilever.

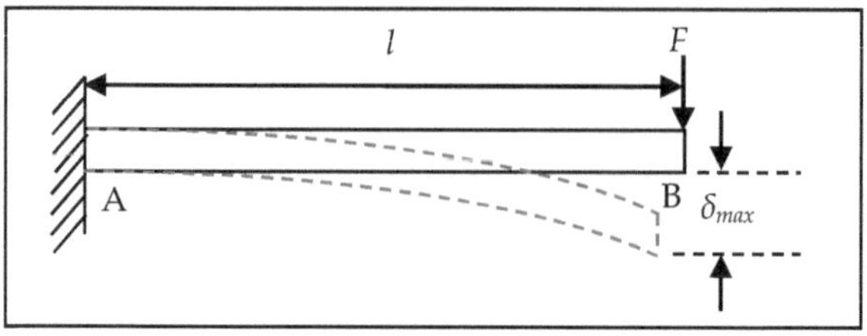


**Figure 3.** Schematic of MEMS cantilever deflection.

In the meanwhile, the cantilever will also sense stress that occurred during deflection. There are two type of stress occurred: tensile and compressive stress where tensile occurs at the top of cantilever and compression acts at the bottom of cantilever as illustrated in Fig.4. Since the piezoresistors are located at the top surface, research will be focuses at top surface of the cantilever.

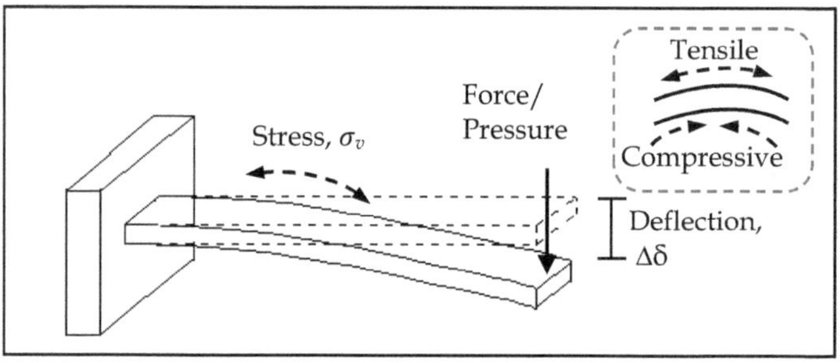


**Figure 4.** Stress occurred during force applied

Maximum stress can be calculated using equation (2) for a constant cross section beam.

$$\sigma_{max} = {}^{6Fl}/_{bh^2} \tag{2}$$

where *M, moment* = *F, force* x *l, cantilever length,* $\sigma_{max}$ is the maximum stress, c is the height from the center axis to the top surface of the cantilever and *I*, moment of inertia.

## 3.1. Piezoresistive effect in silicon and MEMS cantilever relationship

Piezoresistive effect describes the changing electrical resistance of a material due to applied mechanical stress. The effect causes a change in resistance value. This effect has been used for semiconductor based sensor such as germanium, silicon and polycrystalline silicon. Silicon offers remarkable piezoresistive effect and it has controllability for electronic circuits [Madou (1997), Streetman and Banerjee (2006)]. Semiconductor silicon is the most common material in the MEMS field. Naturally, the electrical and mechanical properties of silicon are of great interest which differs from conductor (e.g. metals) and insulator (e.g. rubbers). It has a conductivity which lies between a perfect insulator and a perfect conductor. Liu (2006) states that the resistivity of semiconductor changes as a function of deformed mechanism. Therefore, silicon is a true piezoresistor. Liu (2006) also mentioned that piezoresistive effect refers to piezoresistor or resistor which changes during applied force or mass. The change in piezoresistance is linearly related to the applied stress and strain according to Bhatti et al. (2007) and Liu (2006). These related expressions are shown in equation (3) and (4) below [Chu et al. (2006)]:

$$\frac{\Delta R}{R} = \pi\sigma_l + \pi\sigma_t = \pi(\sigma_l - \sigma_t) \tag{3}$$

$$\frac{\Delta R}{R} = G.\frac{\Delta l}{l} \tag{4}$$

where $\Delta R/_R$ is resistance change, $\sigma_l$ and $\sigma_t$ are the longitudinal and transverse stress components, $\pi$ is the piezoresistive coefficient, $G$ is gauge factor of piezoresistor (G=121, (Eklund and Shkel, 2007) , $\Delta l/_l$ is strain component. From equation (3) above, it shows that resistance change increases by maximizing the differential stress $(\sigma_l - \sigma_t)$.

Resistance change, $\Delta R/_R$ is often read using the Wheatstone bridge circuit configuration [Liu (2006)]. Wheatstone bridges consist of four resistors connected in a loop as shown in Fig. 5. An input voltage, $V_{in}$ is applied across two junctions that are separated by two resistors. Voltage drop across the other two junctions forms the output [Hsu (2002), Boleystad (2003), Cook (1996)]. By locating the piezoresistive on the surface of a cantilever beam structure, a piezoresistive response can be correlated to the stress occurred as the MEMS cantilever deflect. Stress that occurs will be converted into voltage output, $V_{out}$.

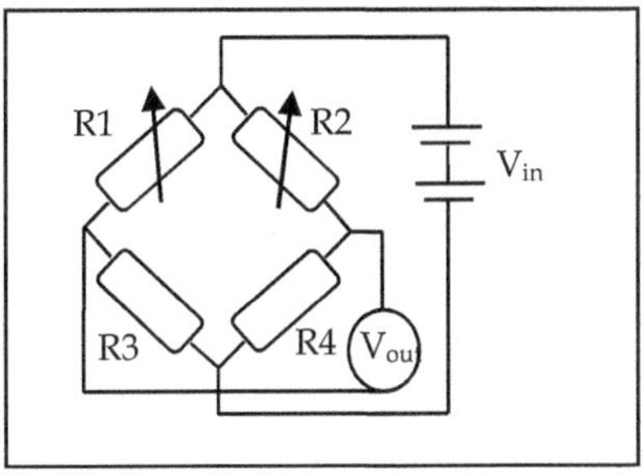


**Figure 5.** Wheatstone bridge circuit configuration; circuit consists of four piezoresistors in a loop [Cook (1996); Boylestad (2003); Chu et al.(2006)]

### 3.2. Piezoresistive MEMS cantilever design

In order to suit intended applications of MEMS cantilever, there are many available designs for MEMS cantilever. These designs vary in terms of the shape and parameter of the MEMS cantilever such as length, width, and thickness. In some published literatures, different designs at certain section of the MEMS cantilever are created where the shape is different from common MEMS cantilever design. Fig.6 shows the most common designs of piezoresistive MEMS cantilever available from literature studies such as rectangular shape, paddle pad and v-shape.

(a) (b) (c)

**Figure 6.** Type of shape for piezoresistive MEMS cantilever; (a) rectangle shape (b) paddle shape , (c) v-shape [Loui et al. (2008), Su et al. (1996), Saya et al. (2005)]

Additional designs or sections are proposed by some researchers to their fabricated device for the device protection according to the application purposes. Gel and Shimoyama (2004) have fabricated a protection head for their device to avoid the cantilever from easily being broken during handling as shown in Fig. 7(a). artificial hair cell (Fig. 7(b)) design are used for flow sensor as proposed by Fan et al. (2002).

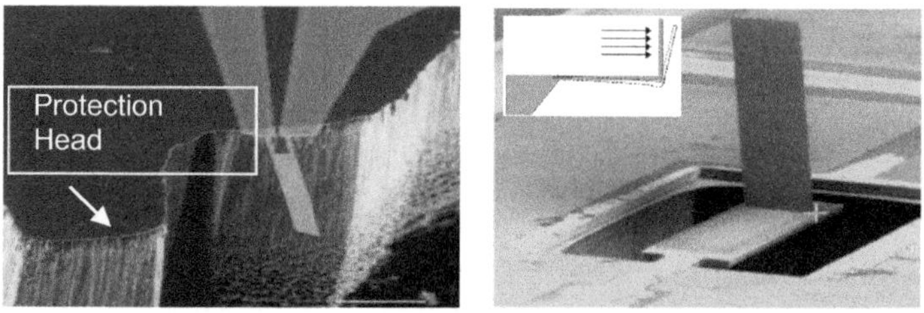


(a) (b)

**Figure 7.** Additional design for MEMS cantilever; (a) protection head , (b) artificial hair cell [ Gel and Shimoyama (2004), Fan et al. (2002)]

Table 1 summarizes MEMS cantilever designs shape, additional design, type of detection and also its applications. From the Table 1, it shows that a rectangular MEMS cantilever is a widely used for biosensor applications.

In this research, paddle pad type MEMS cantilever is chosen. The pad area is used as an area where force or mass can be applied or binding of biological mass. For rectangular type, the area for force or mass applied is smaller and it is difficult for the force to be applied. For safe handling during fabrication of the piezoresistive MEMS cantilever, the proposed design will also include the protection head.

| References | Design/Shape | Additional design | Type of detection | Applications |
|---|---|---|---|---|
| Gel & Shimoyama (2004) | Rectangular | Protection head | Piezoresistive | Force sensing |
| Loui et al. (2008) | Square & trapezoidal | - | Piezoresistive | Chemical sensor |
| Park et al. (2007) | Paddle type | - | Piezoresistive | Acceleration sensor |
| Peiner et al. (2008) | Rectangular | Tip | Piezoresistive | Force sensor |
| Sone et al. (2004) | v-type | Triangle shape | Piezoresistive | Biosensor |
| Yoo et al. (2007) | Rectangular type | - | Piezoresistive and optical | Biosensor |

**Table 1.** Summary of MEMS cantilever designs from literatures

### 3.3. Increasing the sensitivity of piezoresistive MEMS cantilever

There are several typical approaches to increase the sensitivity of piezoresistive MEMS cantilever as proposed in the published literature. The purpose of increasing the sensitivity for any MEM based device is to enhance the device capabilities to measure or detect small changes especially for biological mass detection which is to overcome low resolution of the read out system for piezoresistive detection method (Rosmazuin et al. (2008)).Table 2 summarizes the typical available approaches in order to increase sensitivity of piezoresistive MEMS cantilever. It looks like decreasing or making small dimension is the most popular approach in order to increase the sensitivity of piezoresistive MEMS cantilever. However, this approach requires high precision lithography and the equipment is very expensive, for example Micro/Nano Lithography machine. The same argument applied if the change to low Young's modulus material approach is taken. This approach needs deposition machine like LPCVD (low pressure chemical vapor deposition) or PECVD (plasma enhanced chemical vapor deposition) which is not available in many research labs. Another approach is by introducing stress concentration region (SCR).

| References | Approach |
|---|---|
| Chivukula et al. (2006); Li et al. (2007); Jiang et al. (2008); Brugger et al. (1999); Pramanik et al. (2006) | Decrease geometry & use low doping level |
| Calleja et al. (2005) | Material changes (use low Young's modulus) |
| Yu et al. (2007); Bhatti et al. (2007); He and Li (2008) | Introduce stress concentration region (SCR) |

**Table 2.** Summary of approaches taken to increase the sensitivity of MEMS cantilever

### *3.3.1. Stress concentration region (SCR)*

The main concept for this approach is to increase stress that occurred in the cantilever. SCR is an approach where defects or holes are made in order to increase stress. To produce SCR, no extra high tech equipment is needed because it just involves etching and mask design. So, this approach appears to be the most suitable for enhancing the sensitivity of piezoresistive MEMS cantilever since the piezoresistive material has good sensitivity to stress and no additional complicated equipment or process are required.

Yu et al. (2007) introduced holes to the beam in their finite element analysis to study the effect of surface stress on the sensitivity of MEMS cantilever. The result shows that by introducing holes, the sensitivity of the piezoresistive MEMS cantilever can be increased. Fig.8 shows their result using ANSYS® where the maximum stress occurred near to the fixed end and at the last two SCR holes.

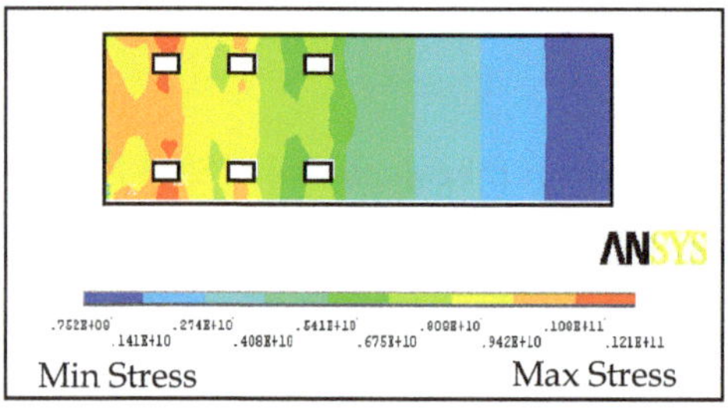


**Figure 8.** Surface stress effects along longitudinal distance cantilever with holes [Yu et al.(2007)]

Joshi et al. (2007) studied four types of SCR holes designs as shown in Fig.9 using Coventoreware2003. Long slit and staggered hole produced highest stress compared to other designs. It shows that more sharp corners can increase the stress occurred. This also agreed by He and Li (2006) which studied the surface stress effect on various types of SCR holes that are formed on the silicon cantilever using ANSYS®. Seven type of SCR holes shape have been analyzed such as rectangular, square, hexagonal, octagonal, circular and elliptical.

Table 3 summarizes the analysis result of surface stress or average stress difference for different types of SCR holes. The result shows that as the number of sides for SCR holes increases, the surface stress increases. The octagonal type of SCR holes gives the highest stress as it has the highest number of sides that creates surface stress.

He and Li (2006) also investigated the effect of adding more octagonal holes to the cantilever. Table 4 shows the surface stress occurred at SCR holes when different numbers of SCR holes of the same size are added along the length of the cantilever with the same spacing between the SCR holes. It shows that adding more SCR holes to the cantilever does not help to enhance the surface stress. Hence one octagonal SCR hole is enough to maximize the surface stress.

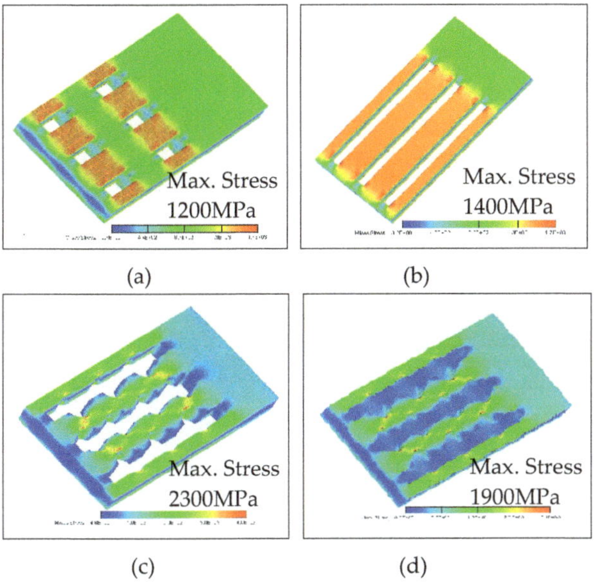


**Figure 9.** Coventorware 2003 analysis, (a) six rectangular hole type, (b) long rectangular slit, (c) long slit and staggered holes, (d) Partially-etched SCR [Joshi et al. (2007)]

| Shape of SCR holes | Maximum Stress (MPa) |
|---|---|
| Cantilever without any hole | 439 |
| Rectangular | 589 |
| Square | 563 |
| Hexagonal | 591 |
| Octagonal | 690 |
| Circular | 621 |
| Elliptical | 590 |

**Table 3.** Maximum stress for different shape of SCR holes [He and Li (2006)]

| No. of Holes | Maximum Stress (MPa) |
|---|---|
| 0 | 439.00 |
| 1 | 689.76 |
| 2 | 686.48 |
| 3 | 686.86 |
| 4 | 686.91 |

**Table 4.** Maximum stress when adding number of octagonal SCR holes [He and Li (2006)]

Bhatti et al. (2007) also simulated piezoresistive MEMS cantilever with paddle pad with rectangular SCR holes by adding the number of SCR holes. Table 5 shows the summary of the result on the effect of adding the number of rectangular SCR holes to the piezoresistive MEMS cantilever with paddle pad as shown in Fig.10. From the table, it shows that surface stress increases when one rectangular SCR holes is introduced. When adding more SCR holes to the cantilever, the surface stress does not change much. Compared with He and Li (2006), Bhatti et al. (2007) have the same surface stress behaviour when adding more SCR holes as shown in Fig.10. Both studies agreed that adding more SCR holes does not affect the surface stress; this happen because when a cantilever deflects, the bending moment is maximum at the fixed end. Hence the stress only shows significant increment at the first hole which is near to the fixed end because the sensitivity of piezoresistive MEMS cantilever cannot be further increased. In order to increase the sensitivity, one SCR hole is sufficient.

| No. of Holes | Maximum Stress (MPa) |
|---|---|
| 0 | 72.47 |
| 1 | 159.66 |
| 2 | 154.93 |
| 3 | 154.19 |
| 4 | 153.89 |

**Table 5.** Maximum stress with increasing the no. of SCR holes [Bhatti et al. (2007)]

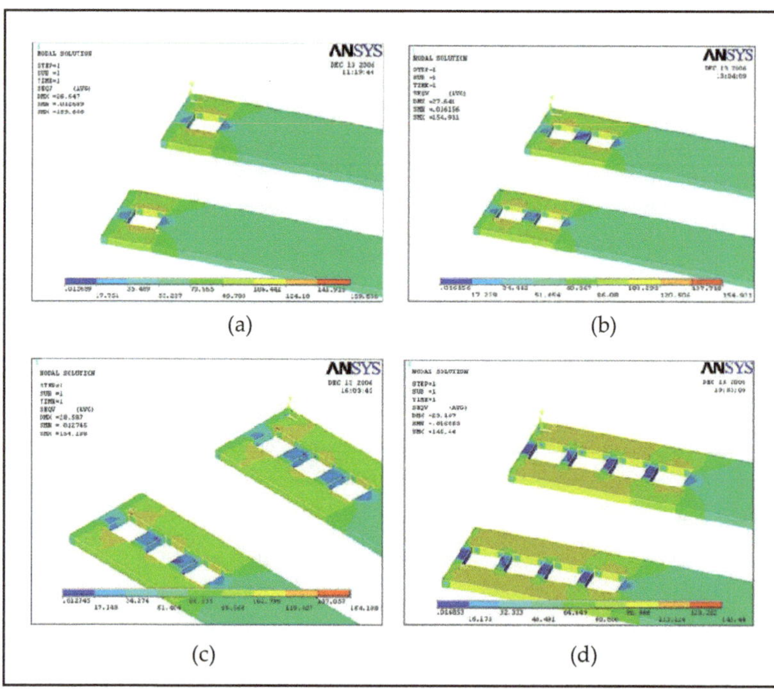


**Figure 10.** Cantilever stress distribution when adding SCR holes, (a) 1 hole, (b) 2 holes, (c) 3 holes, (d) 4 holes [Bhatti et al. (2007)]

In this research, all rectangular, hexagon, octagonal and decagonal types of SCR holes proposed by He and Li (2006) are selected for fabrication in order to increase the sensitivity and also to select which design is suitable with FEA and fabrication process. However the main problem of past literatures that used SCR method was most of the designs only did FEA simulation only and the study did not compared with the fabricated designs.

### 3.4. Location of piezoresistors to form Wheatstone bridge circuit

Conventionally, the piezoresistors are placed on the MEMS cantilever as close as possible to its clamped edge or fixed end. Fuller (2007) mentioned that the location of piezoresistor is best suited wherever the maximum stress occurs. Thus, the maximum stress that piezoresistors will sense is the maximum stress on the MEMS cantilever structure. Wheatstone bridge circuit configuration, there are two type of piezoresistors; the active type will be located at the high stress area whereas the passive type will be located at near zero stress area as illustrated in Fig.11 [Behren et al. (2003), Chu et al. (2007)]. The location of piezoresistor can be determined using FEA which will be discussed further in next section.

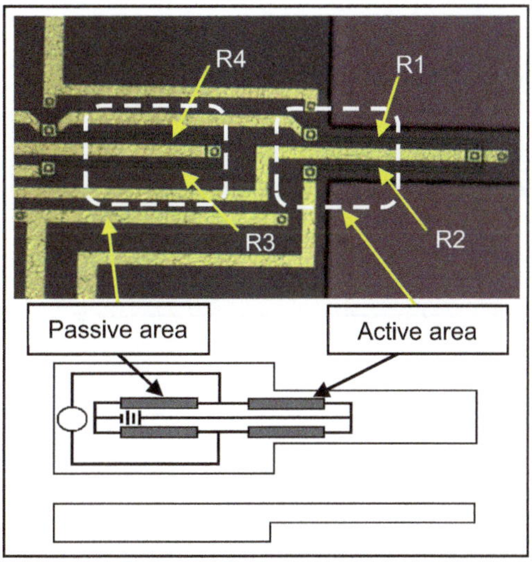


**Figure 11.** Passive and active area for piezoresistors location [Chu et al. (2007)]

### 3.5. Modelling and design of piezoresistive MEMS cantilever

Paddle type rectangular cantilever is chosen for this research since it has large area for mass binding or to apply external load onto its paddle pad. The selection of SCR designs base on past literature, then simulate to determine the stress characteristics. From the FEA results, the SCR designs are fabricated and not all designs suit with the fabrication process. The polygon SCR tend to be circular shape due to low SCR dimension and etching process interaction that over-etch the side SCR shape.

Two paddle type piezoresistive MEMS cantilevers were modeled using computer aided design (CAD) software Solidwork®. Those types are:

- Solid piezoresistive MEMS cantilever paddle pad model as shown in Fig.12a
- Piezoresistive MEMS cantilever paddle pad with stress concentration region (SCR) model as shown in Fig.12b

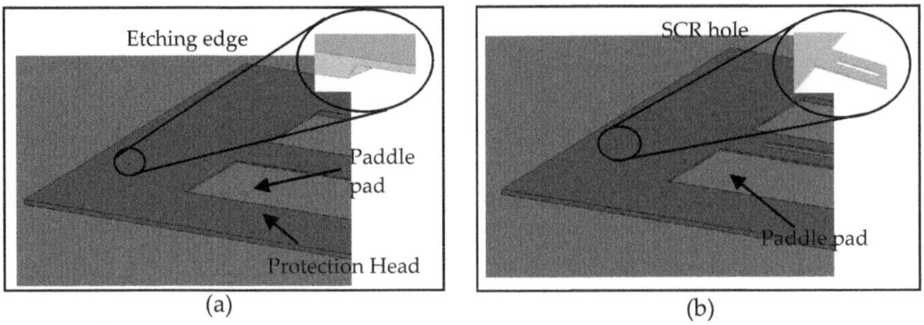


**Figure 12.** Piezoresistive MEMS cantilever model using Solidwork®;

Fig.13 shows the detailed drawing of piezoresistive MEMS cantilever model that is analyzed using ANSYS®. The dimensions were taken from the successful model of piezoresistive MEMS cantilever that was fabricated in the cleanroom for this research. The red line represents path line location that has been used for detail stress distribution and deflection profile plot.

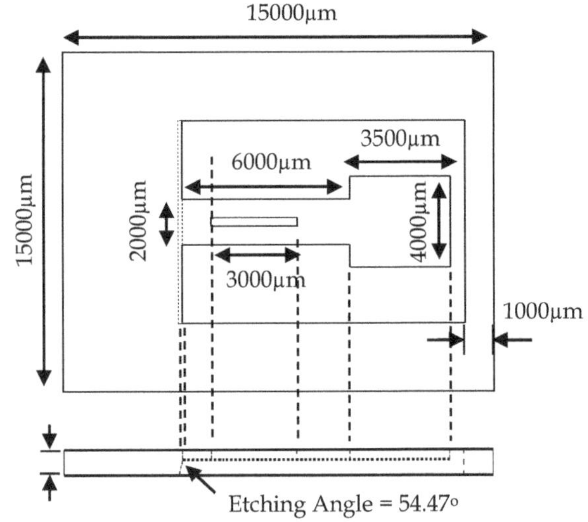


**Figure 13.** Detailed drawing for piezoresistive MEMS cantilever with paddle pad model

### 3.6. Finite element analysis (FEA) using ANSYS®

The piezoresistive MEMS cantilever model analysis is carried out by using ANSYS® version 9.0. The analysis is carried out to investigate and understand the stress and deflection of the piezoresistive MEMS cantilever when external pressure or load is applied. First, the model files were imported from Solidworks® into ANSYS® software so that there will be no error during analysis. The model of interest must be prepared in a manner where the solver will understand.

Then pre-processing is the second step and it is an important step when using ANSYS® prior to any solution execution. Some pre-processing procedures involved during analysis of piezoresistive MEMS cantilever models will be discussed here, including:

- **Element type**

  The proper selection of element is important to ensure desired analysis is carried out. The chosen element must be an elastic element with constant performance and suitable with the computer performance. Several types of element have been tested in order to suit the piezoresistive MEMS cantilever models, with the result verification and also along with computer performance so that the analysis would be finely carried out. Behrens et al. (2003) mentioned that tetrahedral element SOLID187 fits best to the shape of structure fabricated by anisotropic etching. After some verification with the available cantilever models in the literature [Bahtti et al. (2007), Yu et al. (2007)], the SOLID187 element was chosen as the element type.

- **Material properties**

  For this analysis, material properties used throughout both models are called linear properties. Linear properties are chosen because the analysis with these properties requires only a single iteration and not temperature dependent. The material is also defined as isotropic which means the same mechanical properties are applied in all directions. Silicon material will be used during fabrication of piezoresistive MEMS cantilever. Therefore, silicon properties are applied for ANSYS® models. Table 6 lists the material properties of silicon used for piezoresistive MEMS cantilever models.

| Properties | Value | Reference |
|---|---|---|
| Young's Modulus | 150GPa | Yu et al. (2007) |
| Poison's ratio | 0.22 | Yu et al. (2007) |
| Density | 2280 kg/m$^3$ | Yu et al. (2007) |

**Table 6.** Material properties used in ANSYS®

- **Meshing**

  The piezoresistive MEMS cantilever models are meshed by free meshing. Arik et al. (1999) have study the meshing effect for fine and coarse mesh structure. The results show that there is no major difference in both solutions. In this research, the analysis used coarse mesh in order to save time and to avoid crash during analysis. Table 7 lists the number of elements and node counted for both piezoresistive MEMS cantilever design. An element is a form of several nodes. For piezoresistive MEMS cantilever with SCR holes, number of nodes and elements are higher than solid cantilever because at the SCR holes the element are more refine. Fig.14 illustrates free meshed for both piezoresistive MEMS cantilever models.

| Design | No. of Nodes | No. of Elements |
|---|---|---|
| Solid Cantilever | 9365 | 4212 |
| Cantilever with SCR hole | 9813 | 4389 |

**Table 7.** Number of elements and nodes

- **Boundary conditions**

  Before solutions can be initiated, constraints or boundary conditions need to be imposed. Boundary conditions are a selected area or body that will be fixed with no displacement in any degree of freedom or any direction (DOF). When load is applied, the selected boundary condition area will remain constant which mean no deflection or movement occurred. In ANSYS®, boundary conditions or constraints are usually referred to as loads where the scope includes setting of boundary conditions (constraints, supports or boundary field specification) as well as other externally and internally applied loads. Most of these loads can be applied on the solid model (keypoints, lines, areas, and volume) or the finite element models (nodes and elements).

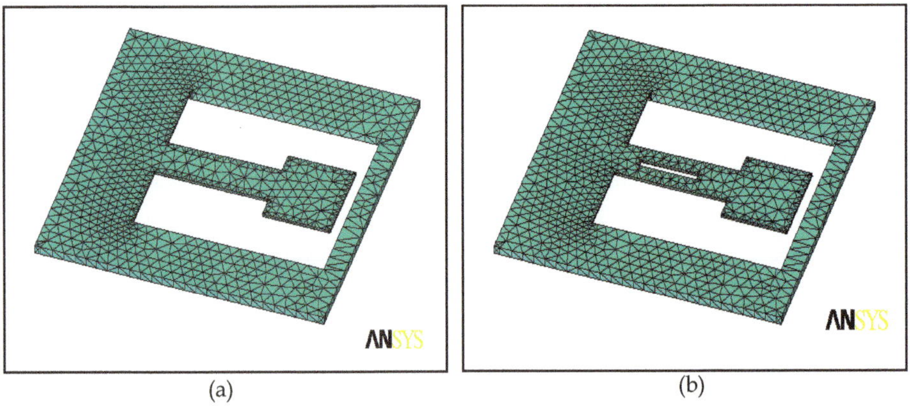


**Figure 14.** Element plot after meshing for piezoresistive MEMS cantilever: (a) solid cantilever, (b) cantilever with SCR hole

For this research, both models are constrained (zero DOF) in x, y, and z direction on the area as shown in Fig.15. Only the MEMS cantilever structure will reflect to the applied load.

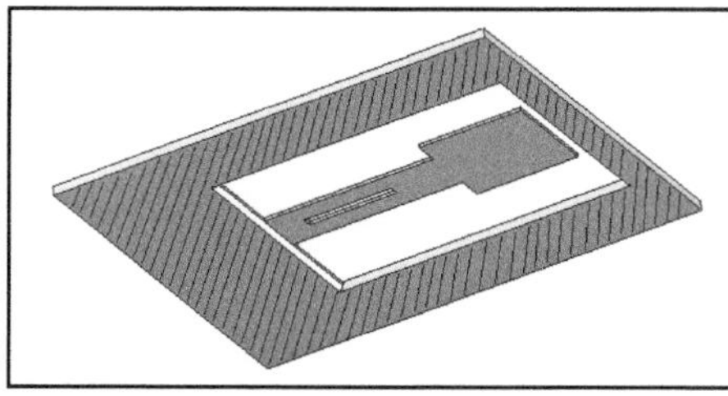

**Figure 15.** Selected area for boundary condition or constraint

- **Pressure applied and Contact area**

  In order to make the piezoresistive MEMS cantilever deflect, external force or mass should be applied at the free end area. From the literature, the external force or mass value depends on the limitation of the cantilever itself which means the smaller the cantilever geometry the lower is the force or mass it can detect or be applied. For this research, force or mass applied represents biological mass that is commonly applied for biosensor/ cantilever application [Yu et al. (2007), Vashist (2007)]. The mass value is converted to pressure so that it can suit with ANSYS®. The pressure will be applied on the paddle pad area only which is at the free end. Mass from 0 up to 5gram has been chosen for this analysis.

The conversion of pressure value is carried out using equation (5).

$$P = {}^{mg}/_{A} \tag{5}$$

where $P$ is pressure, mg is force and A is the applied area.

Table 8 lists the converted mass applied to pressure for ANSYS® analysis.

| Mass (g) | Area ($m^2$) $x10^{-5}$ | Pressure (Pa) |
|---|---|---|
| 0 | 1.5 | 0 |
| 0.2 | 1.5 | 130.67 |
| 0.4 | 1.5 | 261.33 |
| 0.6 | 1.5 | 392.67 |
| 0.8 | 1.5 | 523.33 |
| 1.0 | 1.5 | 654.00 |
| 1.5 | 1.5 | 981.33 |
| 2.0 | 1.5 | 1308.00 |
| 2.5 | 1.5 | 1635.33 |
| 3.0 | 1.5 | 1962.00 |
| 3.5 | 1.5 | 2289.33 |
| 4.0 | 1.5 | 2616.00 |
| 4.5 | 1.5 | 2943.00 |
| 5.0 | 1.5 | 3270.00 |

**Table 8.** Converted mass applied to pressure for ANSYS® analysis

The pressure is applied on the area at the cantilever free end. Fig.16 illustrates the area where the pressure is applied on the piezoresistive MEMS cantilever models for ANSYS® analysis.

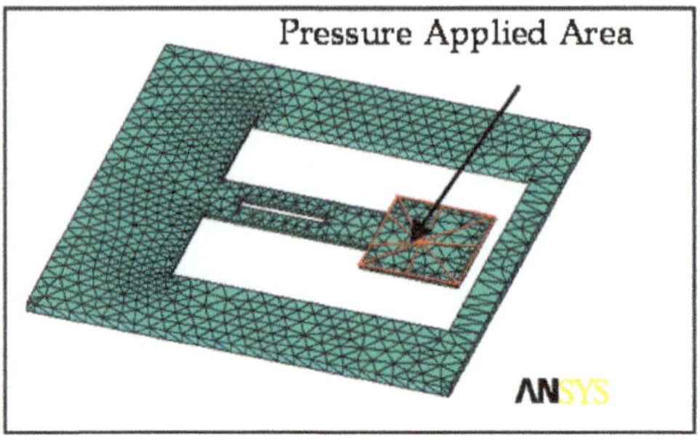


**Figure 16.** Pressure applied area for ANSYS® analysis

## 3.7. Wheatstone bridge circuit (piezoresistive circuit) analysis

The Wheatstone bridge has been used extensively in the literature to determine the output voltage for piezoresistive MEMS cantilever. Cook (1996) mentioned that Wheatstone bridge is commonly used for gathering and measuring the electrical signal generated from gauges. It consists of four resistors connected together and one of the resistors will be acting like the strain gauge. Fig.17 shows the Wheatstone bridge configuration in schematic diagram.

R1 is the active resistor and measuring gauge and the other three resistors are the passive resistors. Any variation in the current in the middle resistor will cause a change in output voltage from the circuit. In this research, software for circuit analysis named MultiSIM8® is used in order to study the circuit characteristic. Fig.17 shows the Wheatstone bridge circuit configuration using MultiSIM®. At ease, the circuit multimeter will show zero voltage as shown in Fig.17(a) and when the piezoresistive MEMS cantilever senses any stress at R1, the output voltage will change as shown in Fig.17(b). The resistance values are not fixed at any values. Hence any value can be taken as long as the measured output voltage is zero.

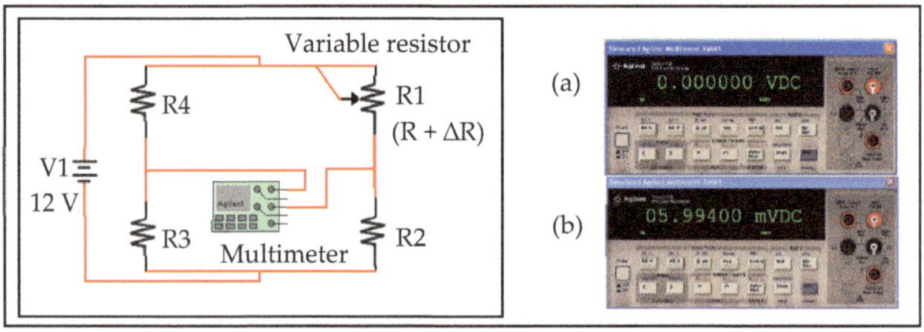


**Figure 17.** Wheatstone bridge configuration using MultiSIM®

### 3.8. Summary of fabrication and testing

In this research, two types of piezoresistive MEMS cantilever will be fabricated; solid design and 3mm SCR design piezoresistive MEMS cantilever. Fig.18 exemplifies the processes sequence schematically for piezoresistive MEMS cantilever fabrication.

Successful fabricated piezoresistive MEMS cantilever is shown in Fig.19. In this figure, two types of piezoresistive MEMS cantilever; without SCR and with rectangular SCR design had been fabricated which basically has 9500μm length, 2000μm width (4000μm X 3500μm for paddle pad) and 100μm thick. Other than these two cantilevers, protecting heads have also been successfully fabricated for handling safety during fabrication processes and testing.

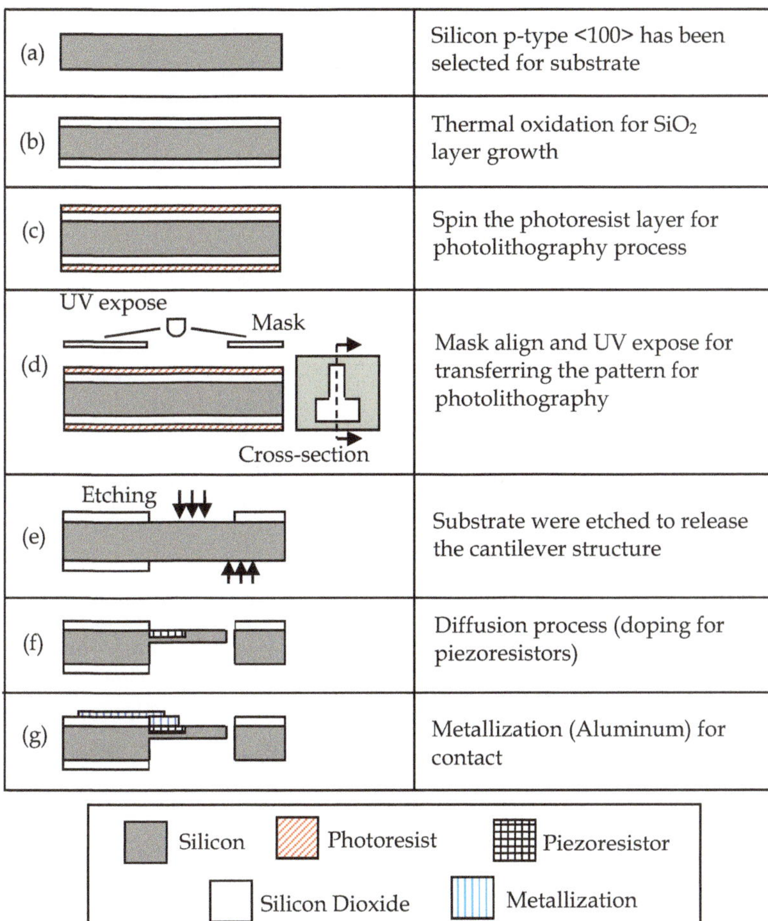


**Figure 18.** Fabrication of piezoresistive MEMS cantilever

Further Current-Voltage test (I-V test) was used to characterize the fabricated piezoresistors for piezoresistive MEMS cantilever as illustrated in Fig.20. Each piezoresistor was tested by applying voltage from -10volt to 10volt across piezoresistor and the resulting current was measured.

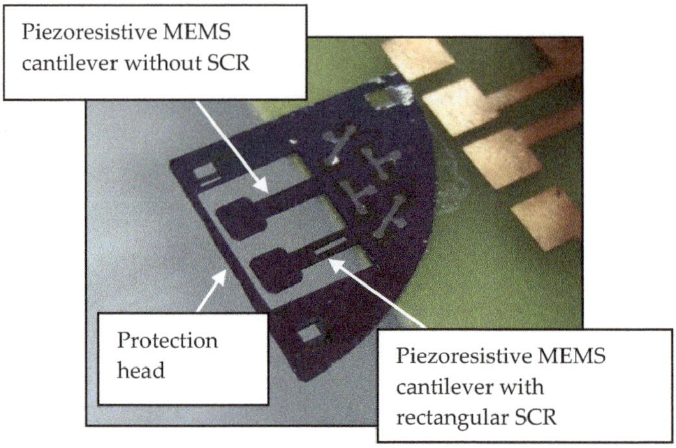


**Figure 19.** Successful fabricated solid and SCR piezoresistive MEMS cantilevers.

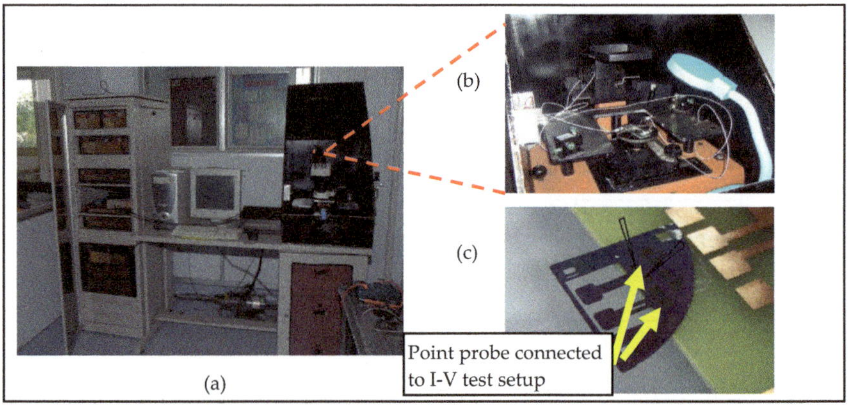


**Figure 20.** I-V test setup: (a) test setup system, (b) Point probe setup and magnifying glass, (c) Point probe connected to piezoresistor

## 4. Results and discussions

Results from the methodologies given in the earlier section are presented and discussed accordingly.

- Analysis on piezoresistive MEMS cantilever without SCR
- Selection of SCR designs for piezoresistive MEMS cantilever
- Analysis on cantilever without SCR and cantilever with selected SCR design
- Analysis on fraction change of piezoresistor using MultiSIM®
- Analysis of fabricated piezoresistive MEMS cantilever

## 4.1. Analysis on piezoresistive MEMS cantilever without SCR

As mentioned in the previous section, the mass applied from 0 to 5gram is converted to pressure during FE analysis throughout this research. From the results of the analysis, it can be deduced that the stress increases with the increase of the mass applied in a linear fashion of the maximum stress when mass is applied compared with calculation were not more 10% different as shown in Fig.21(a). Stress contour obtained when 1g mass is applied on the paddle pad area of the piezoresistive MEMS cantilever without SCR is shown in Fig.21(b). Maximum stress occurs at the cantilever fixed end. It shows that the cantilever has high bending moment at the fixed end.

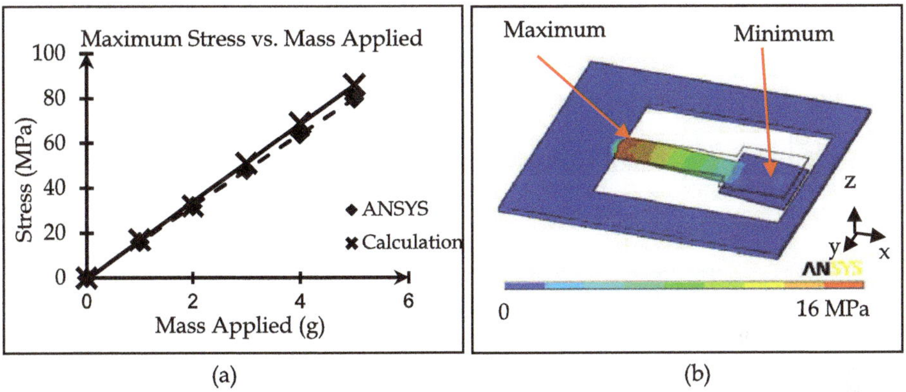


**Figure 21.** Maximum stress for piezoresistive MEMS cantilever without SCR when varying mass is applied; (a) max. stress plot, (b) stress contour when 1gram mass applied

Liu (2006) mentioned that the maximum stress associated with the individual cross sections changes linearly with respect to the distance to the free end. Stress occurs at the top and bottom surface then decrease when approaching to the middle of cantilever thickness. Fig.22 illustrates stress distribution along the cantilever length when force/mass is applied at the free end. Fig.22 also illustrates stress distribution along the cantilever thickness when force/mass is applied at the free end. This verifies that maximum stress is high at the fixed end and it is this reason piezoresistors are commonly fabricated on the surface of the cantilever and near to the fixed end.

Results from equation (1) and ANSYS® simulation were plotted in Fig.23. Both results show linear trend results where the increase of mass applied would increase the maximum deflection of solid piezoresistive MEMS cantilever.

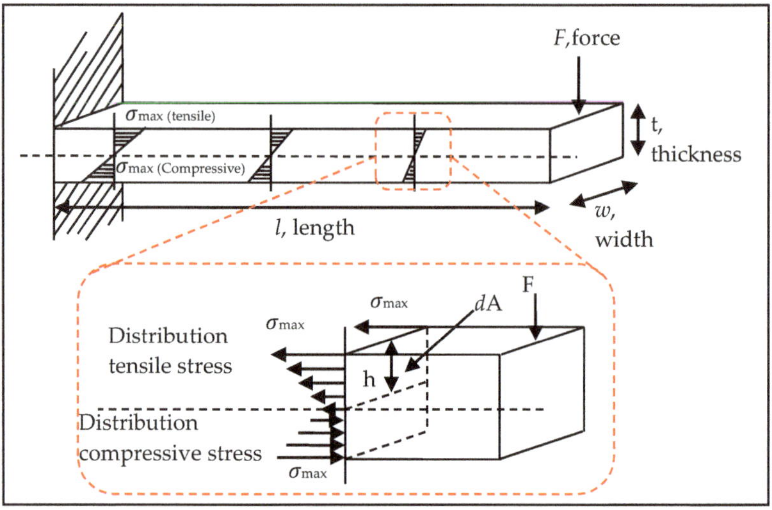


**Figure 22.** Stress distributions along the cantilever thickness when force or mass applied [Liu (2006)].

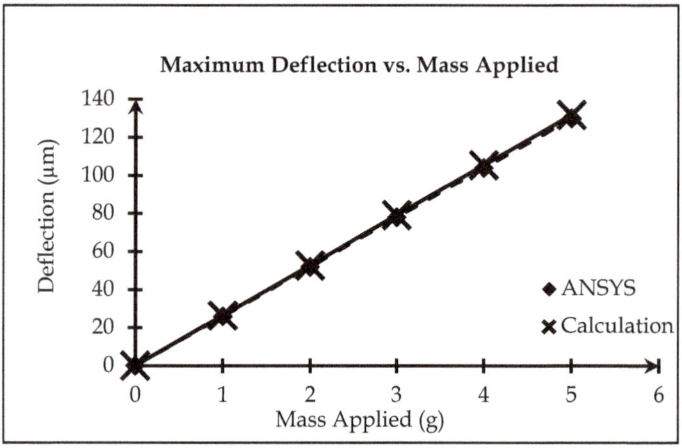


**Figure 23.** Maximum deflection plot with varying mass is applied.

### *4.1.1. Stress distribution and deflection profile along a path line for piezoresistive MEMS cantilever without SCR*

When plotting stress along the selected path, a detailed stress and deflection analysis can be studied. Fig.24 shows the stress distribution along the path when a 1g mass is applied at the free end of the piezoresistive MEMS cantilever without SCR. It also shows the stress occurs throughout the path line where the maximum stress of 16.2MPa occurs at the fixed end as shown in Fig.24(a). From the path plot the maximum stress occurs at the fixed end which is in good agreement with Liu (2006) in previous section.

Fig.24(b) illustrates deflection plot along the path line for 1g mass applied. Combining the path stress and deflection plot, the piezoresistive MEMS cantilever starts to deflect at the location of maximum stress. From the path plot, the maximum deflection of 70.2μm occurs at the free end of the cantilever as shown in Fig.24(b). The same plot pattern has been obtained by Behrens et al. (2003) for their piezoresistive MEMS cantilever model along the longitudinal distance when a 20μN load is applied at the free end. In this research, MEMS cantilever deflection was not the main consideration because the piezoresistive method is highly depend on stress occurred. Hence the research is focused more to stress characterization.

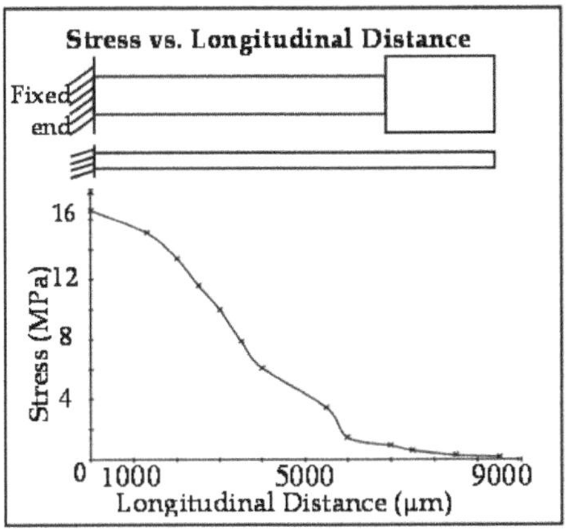


(a)

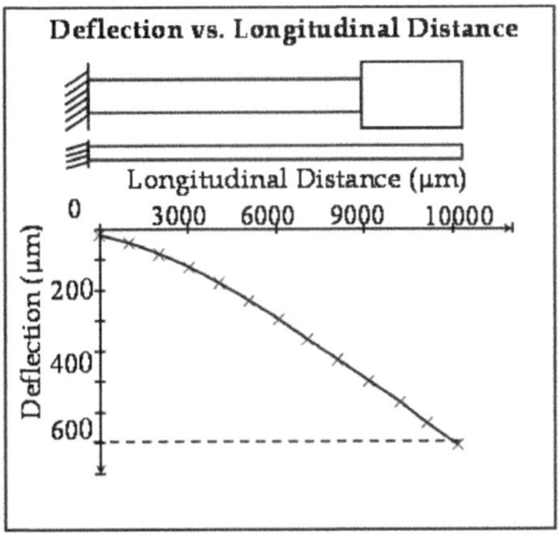


(b)

**Figure 24.** Stress through along the path line plot for piezoresistive MEMS cantilever without SCR when 1g mass applied.

## 4.2. Analysis on rectangular, hexagonal and octagonal SCR designs of piezoresistive MEMS cantilever

Three types of SCR designs have been choose from the past literature to study their stress characteristic when the mass is applied. Fig.25 shows the cantilever models with SCR designs dimensions.

Fig.26 shows the comparison between MEMS cantilever without SCR and MEMS cantilevers with SCR designs when varying mass is applied. From the plot, all SCR designs successfully increase the stress occurred at the cantilever. As the number of sides increase, the stress occurred also increases except rectangular SCR since its length of 1000μm will remain constant. Rectangular SCR designs have the highest stress plot but between these two designs, rectangular SCR design is selected due to its suitability with photolithography and etching process.

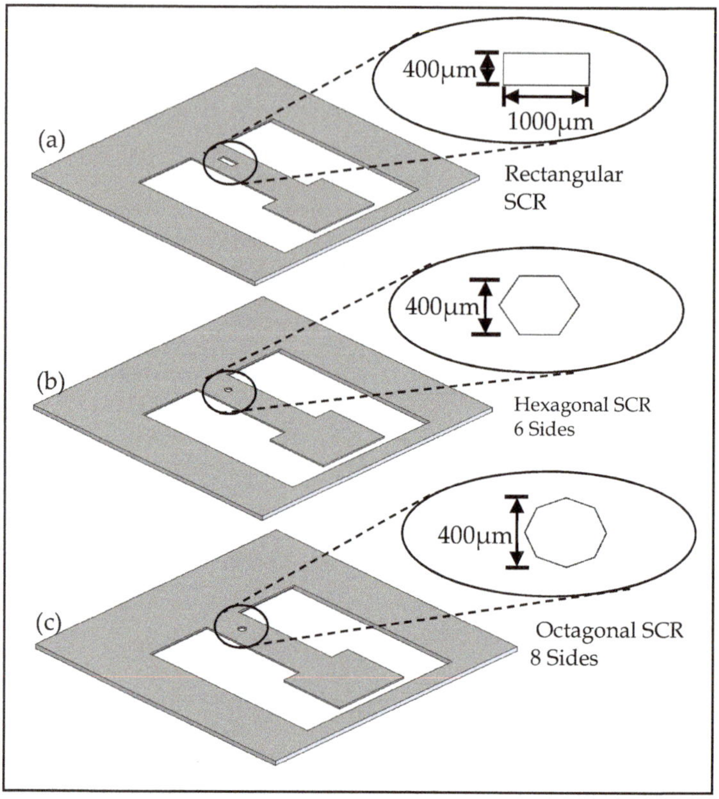


**Figure 25.** Piezoresistive MEMS cantilever with SCR designs: (a) rectangular, (b) hexagon, (c) octagonal

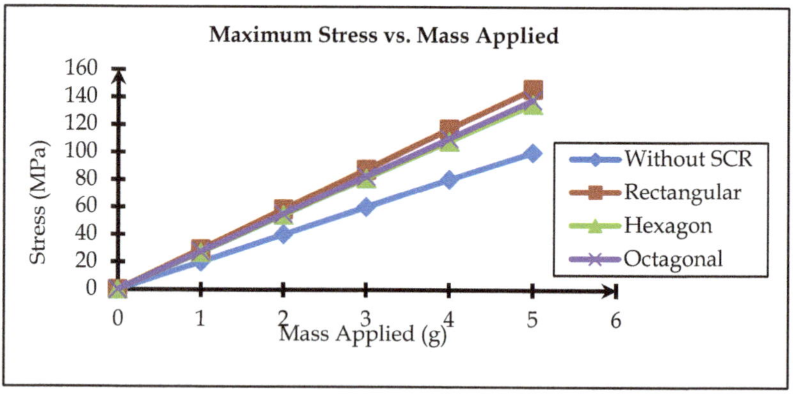


**Figure 26.** Maximum stress for piezoresistive MEMS cantilever without SCR and with various SCR designs for varying mass applied

## 4.3. Analysis on piezoresistive MEMS cantilever with various rectangular SCR dimensions

As mentioned earlier in the previous section, rectangular SCR design has been selected for detailed stress study where the length of rectangular SCR hole with constant width (400μm) is increased from 1000μm to 3000μm. All piezoresistive MEMS cantilever with rectangular SCR designs are compared with the MEMS cantilever without SCR in order to determine which rectangular SCR design will develop the highest stress. The selected design will be fabricated along with piezoresistive MEMS cantilever without SCR for characterization and functionality testing.

Fig.27 show the stress distribution plot along the path line is plotted for piezoresistive MEMS cantilever without SCR and all piezoresistive MEMS cantilever with rectangular designs when 1g mass applied. All piezoresistive MEMS cantilever with rectangular SCR has higher stress at the fixed end compared to piezoresistive MEMS cantilever without SCR. The stress occurred also increase along the length of rectangular SCR as the length increase. This shows that by increasing the length of rectangular SCR, the stress along the cantilever is also increased. It shows that piezoresistive MEMS cantilever with rectangular SCR A3 is the most suitable to fabricate since the increase in stress is more compared to rectangular SCR A1design and rectangular SCR A2 design.

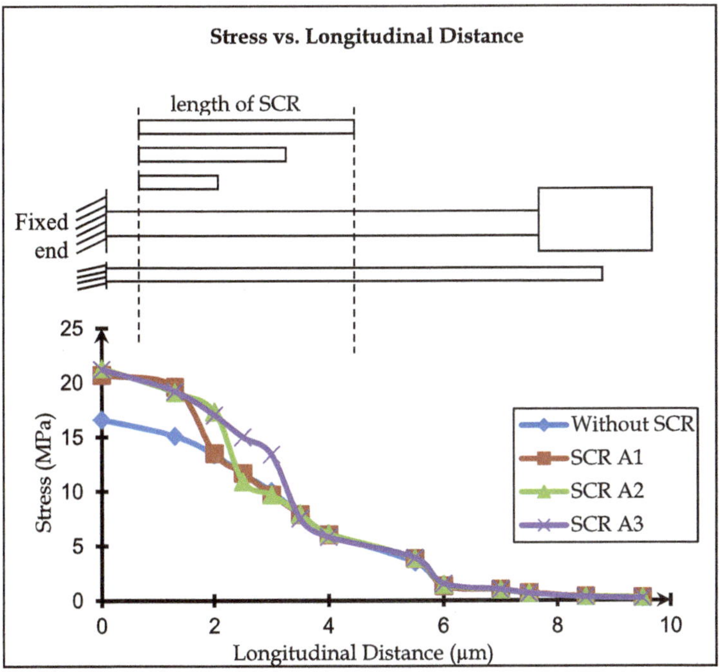


**Figure 27.** Stress along the path line plot for piezoresistive MEMS cantilever without SCR and piezoresistive MEMS cantilever with rectangular SCR designs when 1g mass applied.

Hence, the piezoresistive MEMS cantilever with rectangular SCR A3 design and piezoresistive MEMS cantilever without SCR will be used for Wheatstone bridge circuit analysis using MultiSIM®. Both piezoresistive MEMS cantilever designs will also be fabricated and tested

## 4.4 .Analysis on output voltage of piezoresistors using MultiSIMS®

Data from FEA were used for the piezoresistor circuit testing using MultiSIM® software. Fig.28 shows the graph of output voltage when mass is applied at the free end. The output voltage increases with the increase of mass applied. The piezoresistive MEMS cantilever without SCR seems to be less sensitive than the piezoresistive MEMS cantilever with rectangular SCR A3. This shows that by introducing rectangular SCR A3 to the piezoresistive MEMS cantilever can increase the sensitivity compared to the piezoresistive MEMS cantilever without SCR. Chu et al. (2007) also obtained a linear relationship for the output voltage versus applied force. Results shows that by introducing rectangular SCR A3 to the piezoresistive MEMS cantilever, the sensitivity is enhanced by almost about 2 times.

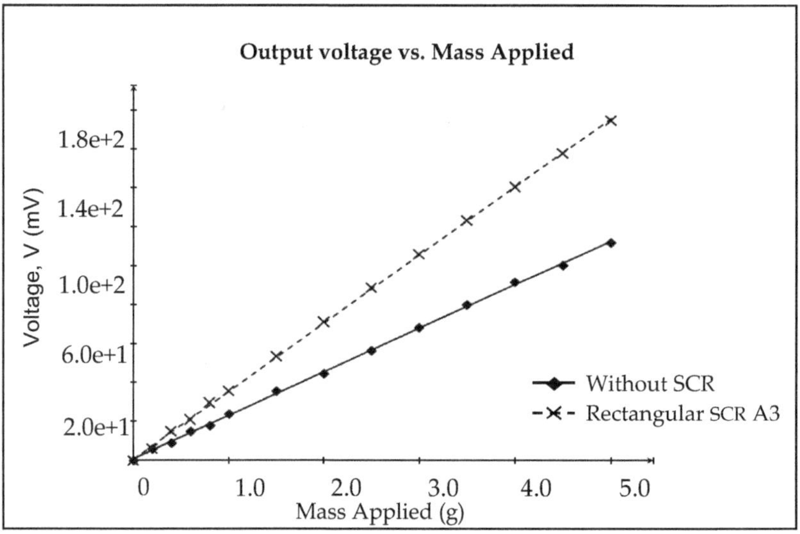


**Figure 28.** Output voltages when varying mass applied

## 4.5. Fabricated piezoresistive MEMS cantilever testing

Both fabricated piezoresistive MEMS cantilevers have been tested with 1, 2 and 3 gram of mass. Mass was applied at the paddle pad free end of the piezoresistive MEMS cantilever without SCR and with rectangular SCR A3 design. The output voltage change with varying mass applied was measured by Wheatstone bridge (WB) test setup as shown in the previous section. Fig.29 shows the output voltage from the Wheatstone bridge circuit when varying mass is applied for the fabricated piezoresistive MEMS cantilever without SCR and with rectangular SCR A3 design. Results show that both cantilever have polynomial relation, hence trend line is added for both curve fitting. From the $R^2$ values, both were near to 1 which mean the trend line almost balance through all plotted data. The output voltage not linear due to environment effect since the fabricated Wheatstone bridge circuit is sensitive to any changes in temperature. Chu et al. (2007) obtained linear relationship for output voltage versus applied force. From plotted results, the average sensitivity for piezoresistive MEMS cantilever with rectangular SCR A3 was $0.063mVg^{-1}$ and $0.032mVg^{-1}$ for the piezoresistive MEMS cantilever without SCR. When comparing both values, the piezoresistive MEMS cantilever with rectangular SCR A3 has successfully enhanced the sensitivity by 1.97 times from the piezoresistive MEMS cantilever without SCR.

Comparing the sensitivity value determined from MultiSIM® analysis, 2 times and fabricated device, 1.97 times, both results show that by introducing rectangular SCR A3 to the piezoresistive MEMS cantilever the sensitivity has successfully enhanced. The difference between both sensitivity is 1.5%. The different results obtained for both fabricated piezoresistive MEMS cantilever are due to uncontrollable fabrication processes such as etching rate, doping concentration, photolithography effect and the difference in the value of Young's Modulus etc. Besides that, environmental conditions like cleanroom temperature and vibration disturbance are also among enforcing effect because low dimension sensor has high sensitivity with the surrounding.

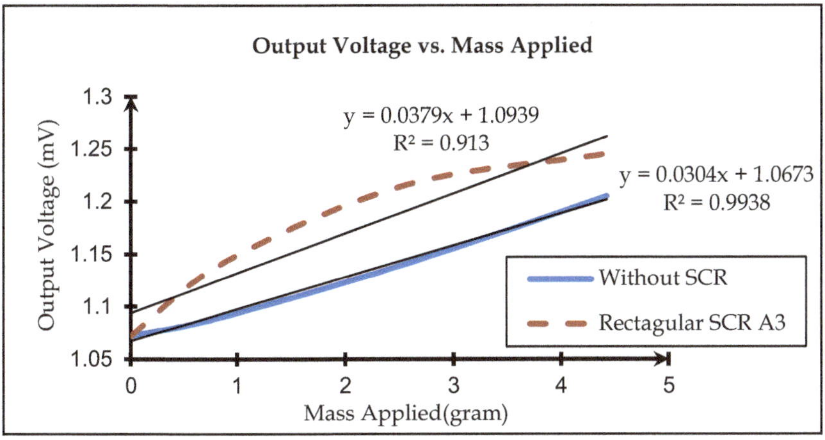


**Figure 29.** Output voltage when varying mass applied for piezoresistive MEMS cantilever without SCR and with rectangular SCR A3.

## 5. Conclusions

The piezoresistive MEMS cantilever has been design, analysed and fabricated in this research. FEA analysis results show that rectangular SCR design has the highest stress developed during force applied compared to other designs. The piezoresistive MEMS cantilever with rectangular SCR and without SCR were fabricated and characterised. Results show that the fabricated piezoresistive MEMS cantilever has successfully enhanced the sensitivity by 1.97 times compared to fabricated piezoresistive MEMS cantilever without SCR. The difference between FEA sensitivity analysis value and the fabricated device sensitivity value is 1.5%.

## Author details

Sh Mohd Firdaus*
*Faculty of Mechanical Engineering, University Technology MARA (UiTM), Penang, Malaysia*

Husna Omar
*Faculty of Applied Science, University Technology MARA (UiTM), Penang, Malaysia*

Ishak Abd Azid
*Faculty of Mechanical Engineering, University Science of Malaysia (USM), Penang, Malaysia*

## Acknowledgement

The authors wish to thank University Techology MARA, University Science of Malaysia and the Government of Malaysia for the offer of the e-science grant (Ref:6013205), fundamental research grant scheme (600-RMI/FRGS 5/3 (39/2012)) and Excellent Fund UiTM (600-RMI/ST/DANA 5/3/Dst (113/2011) that have enabled this research to be carried out.

## 6. References

Arik M, Zurn S.M, Yigit K.S and Bar-Cohen A, (1999). Design, Fabrication and Experimental-Numerical Study of PZT Sensors, University of Minnesota, MN 55455.

Arscott S., Legrand B., Buchaillot L. and Ashcroft A.E. (2007). A Silicon Beam-Based Microcantilever Nanoelectrosprayer. Sensors and Actuators B Vol.125, pp. 72–78.

Beeby S., Ensell G., Kraft M. and White N. (2004). MEMS Mechanical Sensors. Artech House,Canton Street, Norwood.

Behrens I., Doering L. and Peiner E. (2003). Piezoresistive Cantilever as Portable Micro Force Calibration Standard. Journal of Micromechanical Microengineering, Vol. 13, pp. 171-177.

Benham, P.P. Drawford, R.J. and Armstrong, C.G. (1996). *Mechanics of Engineering Materials.* Prentice Hall Ltd., UK

Bhatti M. A., Lee C. X., Lee Y. Z. and Ahmed N. A. (2007). Design and Finite Element Analysis of Piezoresistive Cantilever with Stress Concentration Holes. 2nd IEEE Conference on Industrial Electronics and Applications

Boylestad, R. L. (2003). Introductory Circuit Analysis. Pearson Pr entice Hall, Upper Saddle River, New Jersey.

Brugger J., Despont M., Rossel C., Rothuizen H., Vettiger P. and Willemin M. (1999). Microfabricated Ultrasensitive Piezoresistive Cantilevers for Torque Magnetometry. Sensors and Actuators A Vol.73, pp. 235-242.

* Corresponding Author

Calleja M., Nordstrom M., Alvarez M., Tamayo J., Lechuga L.M. and Boisen A. (2005). Highly Sensitive Polymer-based Cantilever-Sensors for DNA Detection, Journal of Ultramicroscopy, Vol. 105, pp. 215-222.

ChatrathiI K., Packirisamy M., Stiharu I., Nerguizian V. (2006). Effect of Curvature on Dynamic Behavior of Cantilever. MEMS IEEE ISIE 2006, Quebec, Canada.

Chollet Frank and Liu Haobing. (2007). A short introduction to MEMS. Micromachines Centre, School of MAE, Nanyang Technological University, Singapore.

Chivukula V., Wang M, Ji Hai-Feng, Abdul Khaliq, Ji Fang, Kody Varahramyan, (2006). Simulation of $SiO_2$ based Piezoresistive Microcantilevers. Sensors and Actuators A, Vol. 125, pp. 526-533.

Chu Duc T., Creemer J.F. and Sarro Pasqualina M., (2007), "Piezoresistive Cantilever Beam for Force Sensing in Two Dimensions", IEEE Sensors Journal, Vol.7, No.1.

Chu Duc T., Creemer J.F. and Sarro Pasqualina M. (2006). Lateral Nano-Newton Force-Sensing Piezoresistive Cantilever for Microparticle Handling. Journal of Micromechanical Microengineering, Vol. 16, pp. 102-106.

Cook Nigel P. (1996). Practical Electricity. Pearson Prentice Hall, Upper Saddle River, New Jersey.

Don Klaitabtim and Adisorn Tuantranon, (2005). Design Consideration and Finite Element Modeling of MEMS Cantilever for Nano-Biosensor Applications. Proceedings of 2005 5th IEEE Conference on Nanotechnology Nagoya, Japan.

Eklund E. J. and Shkel A. M., (2007). Single-mask fabrication of high-G piezoresistive accelerometers with extended temperature range. Journal of Micromechanics and Microengineering. Vol. 17 (4), pp. 730-736

Fan Z., Chen J., Zou J., Bullen D. 1, Liu C. and Delcomyn F. (2002). Design and Fabrication of Artificial Lateral Line Flow Sensors. Journal of Micromechanical Microengineering. Vol. 12, pp. 655–661.

Fuller L.F. Microelectronic Engineering Department of Rochester, Institute of Technology (2007).Available from World Wide Web: http: www.rit.edu/~iffee/[Accessed 8 January 2008].

Gel M., and Shimoyama I.(2004). Force Sensing Submicrometer Thick Cantilever with Ultra-Thin Piezoresistors by Rapid Thermal Diffusion. Journal of Micromechanical Microengineering, Vol.14, pp. 423-428.

Guillermo V. T. L. (2006). Development of Cantilever for Biomolecular Measurement. Ph.D. Dissertation, Universitat Autonoma de Barcelona, Barcelona, Spain.

Cheng F.H., (1998). Statics and Strength of Materials. Mc Graw-Hill International, Civil Engineering Series.

He J. H. and Li Y. F. (2006). High Sensitivity Piezoresistive Cantilever Sensor for Biomolecular Detection. Journal of Physic: Conference Series, Vol. 34, pp. 429-435.

Hsu, Tai-Ran. (2002). MEMS & Microsystems – Design and Manufacture. McGraw Hill International Series, Singapore Madou M. (1997). Fundamental of Microfabrication. CRC Press, Boca Raton.

Jiang Y.G., Ono T. and Eshashi M. (2008). Fabrication of Piezoresistive Nanocantilevers for Ultra-Sensitive Force Detection. Measurement Science and Technology, Vol.19, 84011

Joshi B.P., Joshi Aditee, and Gangal S.A. (2007), Performance Improvisation of Cantilever type Silicon Micro Acceleration Sensors using Stress Concentration Regions Technique, Defense Science Journal, Vol. 57, pp. 271-279.

Lee H. J., Young S. C., Lee Y.P., Jeong K.H. and Kim H.Y. (2007). Deflection of Microcantilever by Growing Vapor Bubble. Sensors and Actuators A, Vol. 136, pp. 717–722.
Li M, Tang H.X. and Roukes M.L.(2007). Ultra-Sensitive NEMS-based Cantilevers for Sensing, Scanned Probe and Very High-Frequency Applications. Nature Nanotechnology, Vol. 2, pp. 114-120.
Liu Chang. (2006). Foundation of MEMS. Illinois ECE series, Pearson Education, Upper Saddle River, New Jersey.
Loui A., Goericke F.T., Ratto T.V., Lee J., Hart B.R. and King W.P. (2008). The Effect of Piezoresistive Microcantilever Geometry on Cantilever Sensitivity during Surface Stress Chemical Sensing. Sensors and Actuators A, Vol. 147, pp. 516-521.
Park D. S., Yun D.J., Cho M. W. and Shin B.C. (2007). An Experimental Study on the Fabrication of Glass-based Acceleration Sensor Body Using Micro Powder Blasting Method. Sensors 2007, Vol. 7, pp. 697-707.
Peiner E., Doering L. and Balke M. (2008). Silicon Cantilever Sensor forMicro/Nanoscale Dimension and Force Metrology. Technical Paper of Microsystems Technology, Vol. 14, pp. 441-451
Pramanik C., Saha H. and Gangopadhyay U., (2006). Design Optimization of a High Performance Silicon MEMS Piezoresistive Pressure Sensor for Biomedical Applications, Journal of Micromech. Microeng. Vol. 16, pp. 2060-2066.
Rosmazuin A.R., Badariah B., and Burhanuddin Y.M., (2008), Design and Analysis of MEMS Piezoresistive $SiO_2$ Cantilever-based Sensor with Stress Concentration Region for Biosensing Applications, ICSE 2008 Proceeding, Johor Bahru, Malaysia.
Saya D., Belaubre P., Mathieu F., Lagrange D., Pourciel J.B., Bergaud C.(2005). Si-Piezoresistive Microcantilever for Highly Integrated Parallel Force Detection Applications. Sensors and Actuators A, Vol. 123-124, pp. 23-29.
Sone H. , Okano H. and Sumio H. (2004). Picogram Mass Sensor using Piezoresistive Cantilever for Biosensor. Japanese Journal of Applied Physics, Vol. 43(7b), pp. 4663–4666.
Streetman B. G. and Banerjee S. K. (2006). Solid State Electronic Devices 6thEdition. Pearson Prentice Hall, Upper Saddle River, New Jersey
Su Y., Evans A.G.R. and Brunnshweiler.(1996). Micromachine Silicon Cantilever Paddles with Piezoresistive Readout for Flow Sensing. Journal Micromechanical Microengineering, Vol.6, pp. 69-72.
Thundat T., Chen G.Y., Warmack R.J., Allison D.P. and Wachter E.A. (1995). Vapor Detection using Resonating Microcantilever. Anal. Chemical. Vol. 67, pp. 519-521.
Vashist K. S. (2007). A Review of Microcantilever for Sensing Applications. Journal of Nanotechnology Online, Vol. 3, June 2007.
Yoo K.A., Kim J.H., Nahm B.H., Kang C.J. and Kim Y.S. (2007) Fabrication and Characteristics of Microcantilever-based Biosensor for Detection of the Protein-Ligand Binding, Journal of Physics: Conference Series Vol. 61, pp. 1308–1311.
Yu X., Zhang H., Li X., Li T. and Zhang D. (2007). Design of High-Sensitivity Cantilever and Its Monolithic Integration with CMOS Circuits, IEEE Sensor Journal, Vol. 7, pp. 489-494.

Chapter 12

# Integrated FEA and Raytracing in Instrumentation for Astronomy: Smart Structures Evaluation and Structural Optimization

Marco Riva

Additional information is available at the end of the chapter

## 1. Introduction

The astronomical instrumentation needs high level of image quality and stability. The quality of images processed by an optical instrument can be referred to the size of the spot and/or the point spread function (p.s.f.), while the stability is related to the displacement of the spot centroid during the observations.

The opto-mechanical elements are designed and manufactured in order to have enough stiffness to minimize shape deformations and flexures due to thermo-gravitational loads. Old traditional design philosophy answered to the problem with high thicknesses and related high masses. Heavier glasses means heavier supports and high gravitational dependent misalignment. The technological research is nowadys devoted to the light weighing of opto-mechanical systems either keeping enough stiffness, or actively correcting optical surfaces and/or positions. Complementary to the technological research, the development of powerful numerical tools added to an huge enlargement of CPU computing capacity have been offered a significant improvement into the engineering design enhancing the complexity and efficiency of the design phase.

Optical lens design refers to the calculation of lens construction parameters (variables) that will meet a set of performance requirements and constraints, including cost and schedule limitations. Construction parameters include surface profile types (spherical, aspheric, holographic, diffractive, etc.), and the parameters for each surface type such as radius of curvature, distance to the next surface, glass type and optionally tilt and decenter. The optical design exploits numerical raytracing techniques to maximize the design efficiency. Ray tracing is a method for calculating the path of waves or particles through a system with regions of varying propagation velocity, absorption characteristics, and reflecting surfaces. Under these circumstances, wavefronts may bend, change direction, or reflect off surfaces, complicating analysis. Ray tracing solves the problem by repeatedly advancing idealized narrow beams called rays through the medium by discrete amounts. Simple problems can be analyzed by propagating a few rays using simple mathematics. More detailed analyses can be performed by using a computer to propagate many rays.

On the other hand, structural design is nowadays mainly based onto The finite element method (FEM). It is a numerical technique for finding approximate solutions of partial differential equations (PDE) as well as of integral equations. The solution approach is based either on eliminating the differential equation completely (steady state problems), or rendering the PDE into an approximating system of ordinary differential equations, which are then numerically integrated using standard techniques such as Euler's method, Runge-Kutta, etc.

The optimization procedure refers to choosing the best element from some set of available alternatives. In the simplest case, this means solving problems in which one seeks to minimize or maximize a real function by systematically choosing the values of real or integer variables from within an allowed set. This formulation, using a scalar, real-valued objective function, is probably the simplest example; the generalization of optimization theory and techniques to other formulations comprises a large area of applied mathematics. More generally, it means finding "best available" values of some objective function given a defined domain, including a variety of different types of objective functions and different types of domains.

In this chapter we present a possible simplified application of the optimization theory to opto-mechanical design that has been integrated into a multipurpose combined opto-mechanical numerical design process. In particular we will show a single variable optimization routine oriented to minimize mass while keeping the optical displacement below a certain value [12]. In addition considering the general purpose of this book will be briefly shown the modeling techniques used in some application to simulate the performances of functional materials like SHape Memory Alloys and Piezoelectrics. This techniques has been implemented in the optimization algorithm to maximize the efficiency of this devices in the actuation of active Mirrors based onto composite materials.

## 2. Framework

The integrated design procedure proposed exploits the huge power of numerical computation for the design of instrumentations for astronomy. The framework can be ideally seen as "server to client" communication, where a managing server code feeds input data to computing clients and extracts the desired results. The adopted software are:

- Matlab® is the server software: it adapts the inputs for the client codes and evaluates their outputs.
- MSC-Nastran® is the FEA client code: it receives from the server code proper models and computes mechanical results (thermo-gravitational displacements, eigen-frequencies, ...)
- ABAQUS® is a FEA client code used alternatively to MSC-Nastran® in case of user defined contitutive laws.
- Zemax® is the Raytracing client code: it receives the optical perturbations properly adapted and evaluates the optical performance of the system (image quality, image stability, ...).

The procedure obtains image quality and motion of an opto-mechanical system under thermal and/or gravitational loads. A simplified mesh (1d or 2d elements) of a possible opto-mechanical configuration is prepared starting from the optical design. Then Matlab® implement physical properties into the input file and feeds the MSC-Nastran® solver. The displacements are extrapolated and reorganized in order to be compliant with the Zemax® reference systems. Matlab® runs the Zemax® solver and extract the desired data (Spot radius, EE80, p.s.f., ...). In this way a first order estimation of the Instruments mechanical stability is obtained and can be easily implemented an optimization process based onto the smart modification of the physical and mechanical properties depending onto the Zemax® results.

Easy modifications can be introduced to improve the accuracy of the algorithm's results. A "qr" based Zernike fitting function has been implemented in order to allow the modeling of optical surface deformations introducing surface errors into the Zemax optical layout. This configuration can be used both for overall system analysis and for detailed object analysis like active optical element performances evaluation[11, 13].

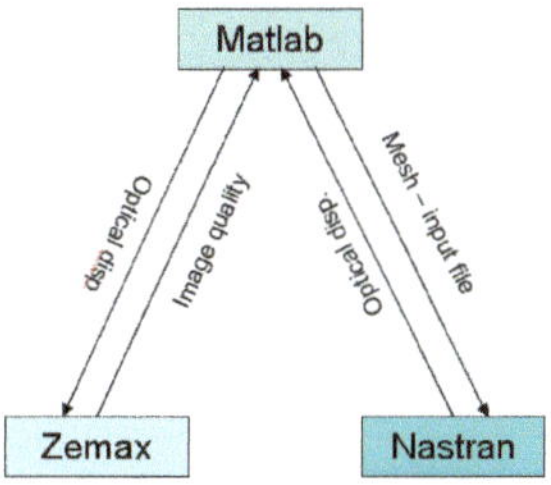


**Figure 1.** Numerical Framework

## 2.1. FEA

### *2.1.1. Mesh generator*

The opto-mechanical Finite Element modeling starts from the optical model. Masses, Center of Gravity and Optical Centers of each element are extracted from the optical file. A simplified Matlab® discretization function can be used when the instrument is very simple (i.e. two or three elements); in case of complex geometries the preliminary mesh can be prepared through dedicated software (Femap® , Hypermesh® , ...); this raw discretization usually models a bench or a box that profiles the optical systems through 2D elements. Due to the high interaction level required we decide to omit the automation of the whole bench meshing.

The opto-mechanical elements are then modeled through semi-rgid elements and concentrated masses. The weight of opto-mechanical subsystem is simplified considering double the optical element mass to include the contribution of mechanical mountings. The semi-rigid element (80%) is used instead of the rigid one in order to simulate the finite stiffness of the mountings; the master node is the optical vertex and the slave are the system's CoG and the connection points. Depending onto the geometry and the interface of the instruments, sometimes it is necessary to introduce some reinforcement beams or support trusses that can be easily modeled through 1d beam elements. In some cases, in particular with large optics or active optical systems, it is important to model the whole optical surface through 2D elements connected to the main bench via 1D beams.

Matlab® can read the model mesh and writes the input file for the Finite Element processor modifing both the geometry and the properties (thickness of 2D and section of 1D elements) if needed. This feature is the crucial point that helps the designer to optimize the performances of the system.

Different codes can be used for different application. The interaction with ABAQUS[1] is necessary for smart structures applications thanks to the higher performances of the code in presence of user defined constitutive laws. The routines interacts with NASTRAN[6] for opto-mechanical performances prediction and optimization.

*2.1.2. Fast extrapolation algorithm*

To evaluate the whole performances of an instrument is necessary to verify the performances within all the loading conditions. In particular it should be possible to find situations where the errors tends to auto-compensate or, on the other hand, are magnified by the interaction with other perturbations.

Here is presented a simple extraction algorithm that can be used to reduce the number of FEA analysis required. The basic idea is to perform less analyses as possible and extracts all the displacement fields in an analytical way. Doing this, one can also predict the displacements in every condition of the gravitational load with or without the thermal expansion.

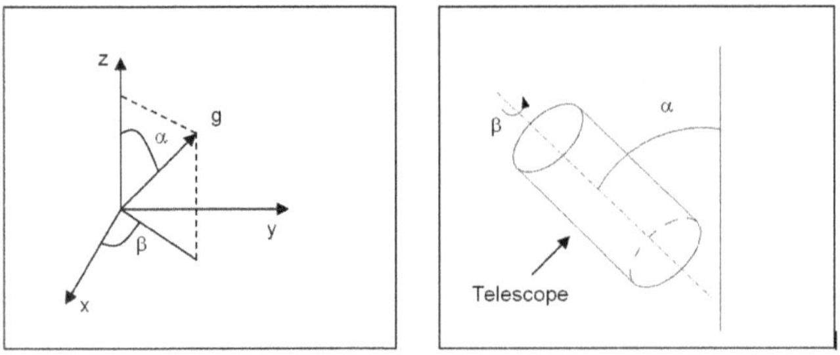


**Figure 2.** Reference systems

If the $g$ vector rotates in the whole 3D space the analytical definition of the displacements is more complex. First of all it is necessary to define how decompose the $g$vector in the three directions $x, y$ and $z$. We have decided to use that shown in Figure 2. In this way we can simulate the real rotations of the telescope, the $\alpha$ angle is the declination and $\beta$ angle represents the rotation of the telescope around the optical axis. Doing this, the components of the g-vector are:

$$\begin{cases} g_x = & g \cdot sin(\alpha) \cdot cos(\beta) \\ g_y = & g \cdot sin(\alpha) \cdot sin(\beta) \\ g_z = & g \cdot cos(\alpha) \end{cases} \tag{1}$$

The equation that ties force and displacements can be written as:

$$D = K^{-1} \cdot F \tag{2}$$

Where $D$ is the displacements vector, $K$ the stiffness matrix, and $F$ is the loads vector. If we combine Equations 1 ans 2, through some algebraical operations, we obtain the general form:

$$D = A_1 \cdot sin(\alpha) \cdot sin(\beta + \phi) + A_2 \cdot cos(\alpha) \tag{3}$$

Where $A_1$ is the vector of the maximum amplitudes of the displacements when the $g$ vector is in the $xy$ plane, $A_2$ is the vector of the maximum amplitudes of the displacements when the $g$ vector is orthogonal at the $xy$ plane, $\alpha$ and $\beta$ are the angles of equation 1 and $\phi$ is a phase vector of the sinusoids. In order to have the whole displacement we need three known points to determinate the three unknown variables: the amplitudes and the phase.

Equation 2 can be further detailed introducing the thermal loads simply by adding the $C \cdot T$ term, where C C is a vector constant which ties the temperature $T$ at the displacement $D$. If we consider $T_d$ as the the vector of the displacements due to the thermal loads, Equation 3 becomes:

$$D = A_1 \cdot sin(\alpha) \cdot sin(\beta + \phi) + A_2 \cdot cos(\alpha) + T_d \tag{4}$$

Now with only four known points (for four unknown variables: the two amplitudes, the phase and the thermal displacement) we can have the overall displacements and the possibility to decompose the thermal contribution.

In addition we should note that Equation 4 can be simplified if the known points are wisely taken. In fact if the sampling loading condition are:

1. $\alpha = 90$, $\beta = 0$, $\delta T = 0$: gravity vector along $X$ axis, no thermal load.
2. $\alpha = 90$, $\beta = 90$, $\delta T = 0$: gravity vector along $Y$ axis, no thermal load.
3. $\alpha = 0$, $\beta = 0$, $\delta T = 0$: gravity vector along $Z$ axis, no thermal load.
4. $g = 0$, $\delta T \neq 0$: only thermal load.

The algebraic simplification of Equations 1 and 2

$$D = X \cdot sin(\alpha) \cdot cos(\beta) + X \cdot sin(\alpha) \cdot sin(\beta) + Z \cdot cos(\alpha) + T_d \tag{5}$$

### 2.1.3. *Zernike fitting*

The surface Deformations are processed through a Zernike Fitting algorithm. In precision optical manufacturing, Zernike polynomials are used to characterize higher-order errors observed in interferometric analysis, in order to achieve desired system performance. They are commonly used in active and adaptive optics where they can be used to describe wavefront aberrations.

The most general way to express the Zernike polynomials is in the form:

$$R_n^m(\rho)e^{im\theta} = \begin{cases} R_n^m(\rho)cosm\theta \\ R_n^m(\rho)sinm\theta \end{cases} \tag{6}$$

Where the $n$ index defines the order of the radial power so an $n$ value of 5 would indicate all polynomials whose maximum radial power is $\rho^5$. Only certain values for $m$ are allowed once $n$ is chosen; $n + m$ must be even, and $0 \leq m \leq n$. The surface error is defined as[2]:

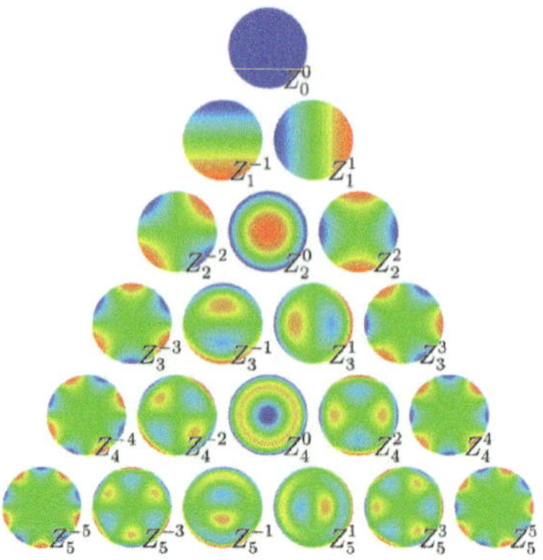


**Figure 3.** Zernike polynomial basis

$$E = \Sigma(\delta_i - D_i)^2 \tag{7}$$

where $i$ is the node number, $\delta_i$ is FE displacement of $i_{th}$ node, $D_i$ is the polynomial displacement of the $i_{th}$ node. The series of $D$ can be symbolically written as $D_i = \Sigma c_j f_{ij}$ where $c_j$ are the coefficients of the polynomial. The best fit coefficients can be found minimizing the error $E$ respect to $c_j$:

$$\frac{dE}{dc_j} = 2\Sigma(\delta_i - \Sigma c_j f_{ij}) f_{ij} = 0 \tag{8}$$

This system is then solved finding the best fitting coefficients through an orthogonal-triangular decomposition (Matlab® *qr* function). The first 28 coefficients of the Zernike Polynomial have been taken into account in this application.

## 2.2. Raytracing

### *2.2.1. Zemax® and Matlab® data exchange*

The raytrancing software Zemax® is used to evaluate the optical performances. Zemax® has a very powerful feature which allows another software to establish a communication link to extract lens data. The idea is based onto a program that use Zemax® like a remote application to trace rays through the lenses, and then extracts the data to be sent to other programs for further analysis or computation[17].

The communication between the application and Zemax® is accomplished using Dynamic Data Exchange (DDE). DDE is a protocol defined within the Windows operating system for sharing data between programs. Two programs can establish a DDE link, with one program acting as the "server" and the other the "client". The client generally requests specific data from the server, and the server sends the data back to the client.

Two main function must be used when exchanging data with Zemax that are the link opening and closing. To establish a DDE link with Zemax® , the client program must broadcast a message to all top level windows that includes a reference to the application name, and the topic name. The topic name indicates to Zemax® what data is being requested.

Zemax® supports a number of capabilities under DDE. Each function/item is given a name, that is passed to the Zemax® server using the proper request protocol. Zemax® responds to each item request with requested data. Most of them are passed from Zemax® back to the application (Matlab® ) in a string that must be properly managed.

### 2.2.2. *Zemax® computations*

In the case of thermal and gravitational performances evaluation, it is necessary an adaptation of the optical model. Coordinate Breaks (CB) are introduced before and after each Optical element paying attention to the transformation sequence. This is crucial to realize a feature that allow the introduction of local optical displacements without modifying the remaining optical layout. Matlab® extracts optical displacements from Nastran (global coordinate system), optical coordinate systems from Zemax® and applies the proper transformation matrices to obtain optical displacements in Zemax® local reference systems. In case of surface deformations, the surfaces of Zemax® file are modified from "Standard" to "Zernike Fringe Sag" in order to allow Matlab® to updates the extra data tables with the Zernike coefficients obtained from the fitting algorithm.

After the uploading of mechanical data into the optical file, Zemax® evaluate spot and p.s.f. information through its raytracing engine. If necessary Matlab® request a focusing optimization after having stored the focal plane distance to evaluate the relative focal variation. Finally Matlab® extract from Zemax Spot dimensions (Max and R.M.S. radius) and centroid displacement, eventually of multiple fields.

The results defines respectively the image quality and stability of the overall optical layout and, if necessary can be passed to an optimization algorithm that manage the mechanical properties of the system.

## 2.3. Optimization

Whereas optimization methods are nearly as old as calculus, numerical optimization reached prominence in the digital age. Its systematic application to structural design dates to its advocacy by Schmit in 1960. The success of structural optimization in the 1970s motivated the emergence of multidisciplinary design optimization (MDO) in the 1980s. Here will be presented a simplified single variable optimization approach that has been used as starting point[14].

### 2.3.1. *Problem statement*

The general optimization problem (for example minimization) wants to minimize the objective function:

$$F(X) \tag{9}$$

subjected to inequality constraints:

$$p_j(X) \leq 0 \;\; j = 1:m \tag{10}$$

and equality constraints:

$$h_k(X) \leq 0 \ \ k = 1 : l \tag{11}$$

and side constraints (if applicable):

$$X_i^l \leq X_i \leq X_i^u \ \ i = 1 : n \tag{12}$$

where vector $X$ is the $n$ dimension design variable vector.

In opto-mechanical variable design $X$ may consists into the instrument properties (shell thickness, beam diameter, ...) while $F(X)$ is the overall instrument Mass function dependent to design variables. The inequality constraint $p(X) \leq 0$ is the optical displacement that must be kept within certain specifications.

*2.3.2. Iterative procedure*

The optimization algorithm requires an initial set of variables $X_0$. In the opto-mechanical optimization we consider as staring point the preliminary mesh defined by the user based onto its own experience. The whole integrated analysis is carried out as shown in this chapter extracting from MSC-Nastran® the overall system mass and from Zemax® the first optical displacement set. A small perturbation is applied onto the design variable ($X_1 = X_0 + \delta X$) and the procedure evaluates a second set of Mass and optical displacements. After the initialization the gradient of Mass and Optical displacement functions are evaluate respect to the design variable and the automatic procedure is started.

The optimization routine is launched:

1. Evaluate Mass function gradient as $\nabla M(X_{i-1}) = (M(X_{i-1}) - M(X_{i-2}))/(X_{i-1} - X_{i-2})$;
2. Update the design variable: $X_i = X_{i-1} - g_M \nabla M(X_{i-1})$ where $g_M$ is a gain factor properly set in order to manage iteration number;
3. Evaluate Mass and optical displacement with new variables and update step counter;
4. Is optical displacement below specification? Yes: go to point 5; No: go to point 6
5. Is $M(X_i) - M(X_{i-1}) \leq Toll$ (where $Toll$ is a reference value) i.e. the minimization converging? Yes: optimization ended; No: go to point 1
6. Evaluate Displacement function gradient as $\nabla D(X_{i-1}) = (D(X_{i-1}) - D(X_{i-2}))/(X_{i-1} - X_{i-2})$;
7. Update the design variable: $X_i = X_{i-1} - g_D \nabla D(X_{i-1})$ where $g_D$ is a gain factor properly set in order to manage iteration and go to point 3

## 3. Modeling techniques

In the attitude of this book, following will be presented the modeling technique developed for the build up of the Finite Element Models of the Smart Structures. In particular the techniques adopted for the modeling of PiezoComposites and Shape Memory Alloy will be briefly introduced. It will not be described here the phsyical behaviour of the materials, detialed description of the full work and approach can be found in [11] and in [13].

### 3.1. Shape memory alloy Finite Element modeling: the Turner micro-mechanical approach

To perform Finite Element analysis regarding SMA is necessary to define the proper constitutive law that describe the material behavior. In this paper will be presented the FE implementation of three different constitutive law. A further detailed description of the laws and their application can be found in the references [7].

In physics, a constitutive equation is a relation between two physical quantities (often described by tensors) that is specific to a material or substance, and approximates the response of that material to external forces. It is combined with other equations governing physical laws to solve physical problems, like the flow of a fluid in a pipe, or the response of a crystal to an electric field.

During the last ten years researchers have been developed several constitutive laws that can be classified considering the different approaches used in their formulation such as: micro-mechanical or macro-mechanical and phenomenological or thermodynamic.

The micro-mechanical formulation essentially takes into account the properties of single crystals of the material averaging their behavior over a Representative Volume Element (EVE). Micro-mechanical models have been developed using a thermodynamic approach and evaluating the energy involved during a phase transformation. These models also use homogenization techniques to derive the overall behavior of the SMA.

The real benefit of this class of model lies in their ability to predict the real response of the material starting from the lattice parameters for crystalline and crystal and grain level data derived from martensitic transformation. Nevertheless these models are very complex and require a large number of computational operations.

The Turner model has been successfully implemented into the commercial code ABAQUS® [10]. This model defines temperature dependent Young modulus and Thermal expansion coefficient. Exploiting some peculiar properties of the software it is possible to define a look up table mapping the variation of material characteristics. As obtained from the calibration of the model (Figures 4 and 5), it has been used the definition of Young modulus and Thermal Expansion coefficient as a function of temperature.

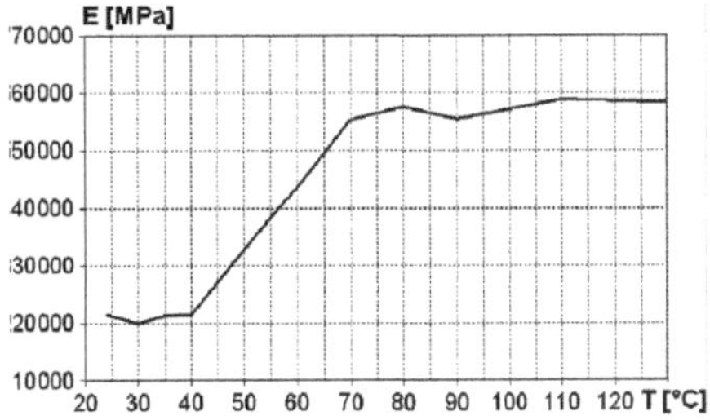


**Figure 4.** SMA $E(T)$

Different verification models were performed to select the type of the elements [10]. The comparison underlines how a shell model is precise enough with a high gain in terms of CPU time. Due to this consideration, the final comparison was conducted on a shell-based model

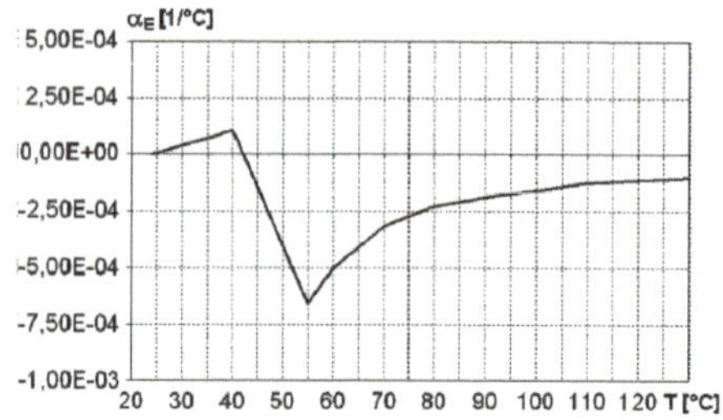


**Figure 5.** SMA $\alpha(T)$

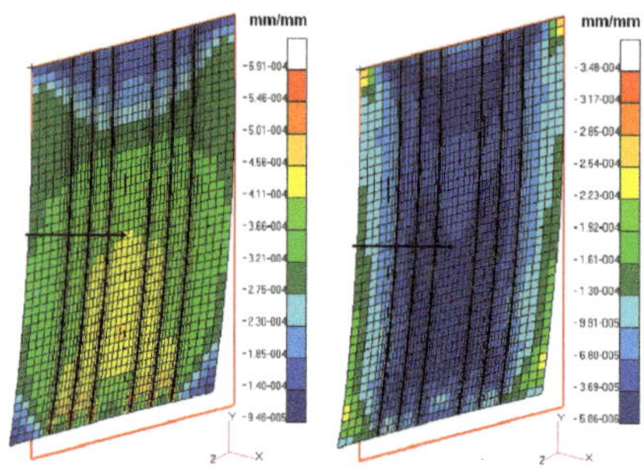


**Figure 6.** Turner Finite Element model

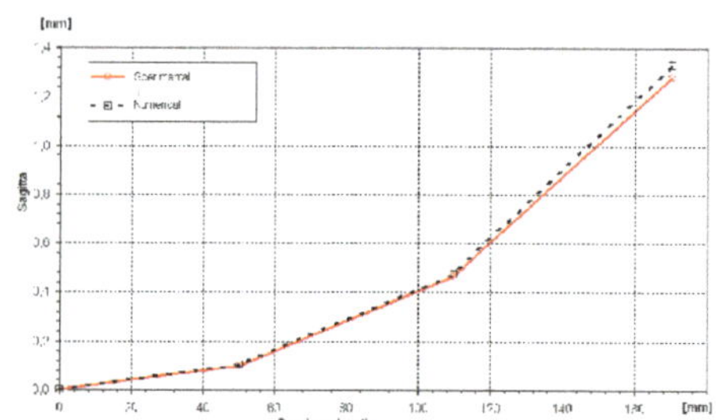


**Figure 7.** Displacement Turner correlation results

with composite laminate properties. The NiTiNOL actuators were modeled thickening the mesh and considering the material as a "ply" of the lamination sequence. The thermoelastic model receives as input the thermal load on the wires and derives the temperature behavior of all the nodes (Figure 6).

A carbon fiber-reinforced panel with six embedded NiTiNOL wires was modeled and analyzed using the ABAQUS commercial code. The actual panel was manufactured with 12 plies $[90/(0)_2/90/+45/-45]_s$ and its overall dimensions were: 30 x 170 x 1.2 mm. The actuators chosen were OWSME wires ($\phi = 0.38mm$) trained using standard heat treatments. They were embedded between the $2^{nd}$ and the and $3^{rd}$ plies with a 4% imposed strain. The manufacturing technolgoy is reported in [3, 10]. The panel was constrained on one side and activated via Joule effect. The numerical/experimental comparison shows that, after a short transition, the predicted displacement matches (max. error 20 $\mu m$) the experimental one (Figure 7).

## 3.2. Piezoelectric numerical modeling:"Structural-scale" modeling technique for MFC

Following will be presented the numerical technique adopted to model The Macro Fiber Composites (MFC) that have been developed by NASA Langley Research Center (LARC) in 2000 [16]. The components of the MFC are illustrated in Figure 9. The core is made by aligned piezoceramic fiber included in epoxy resin and joined between two groups of interdigited electrodes (IDE) supported by a Kapton film. The MFC have an overall thickness of $0.3mm$ and the dimensions of the region in which are placed PZT fibers called the active area, can vary from an area of $28 \times 14mm^2$ to $85 \times 28mm^2$. The spacing of IDE is $0.5mm$, while the fibers have a width of $350\mu m$ and a volume fraction above 85%.

**Figure 8.** MFC actuator

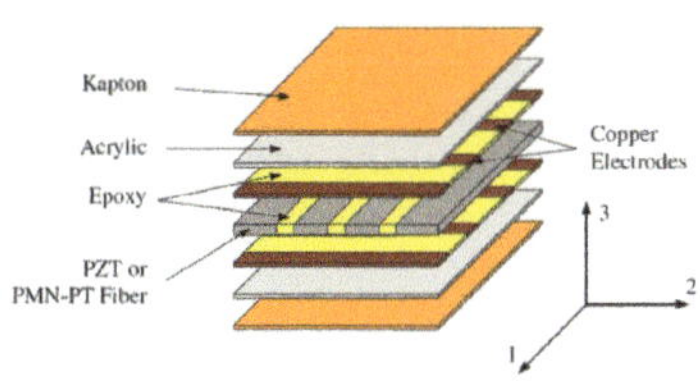


**Figure 9.** MFC layers

The set up of numerical procedure for the performance prediction as the selection of the best technology to couple the structure to the actuators are fundamental to design efficient smart structures. The numerical approach for the design of piezo based smart structures is dealt in this section . The design of smart structure is mainly oriented to the study of authority and then to the stress analysis. With the intent of reducing the design time, a technique able to predict the overall performances of the structure through "light-weight" numerical models was developed.

The overall smart structure can be conceptually divided into two sub-system: the structure itself consisting into the composite panel and the MFC actuators. The proposed technique is developed under the ABAQUS® commercial code [4] environment[1] and neglect the detailed analysis of stress state adopting essentially two type of elements: `S4R` four-node shell elements for the host material and `C3D20RE` twenty-node quadratic bricks (reduced

[1] And can be obviously ported to other Finite Element Codes

integration[2]) for the actuator. The two subsystems are modeled as separated mesh connected together through the `*TIE` algorithm which introduce a link between the nodes of the two different surfaces evaluating the distances between the two faces and obtaining an adhesion factor to be applied at each node (Figure 10).

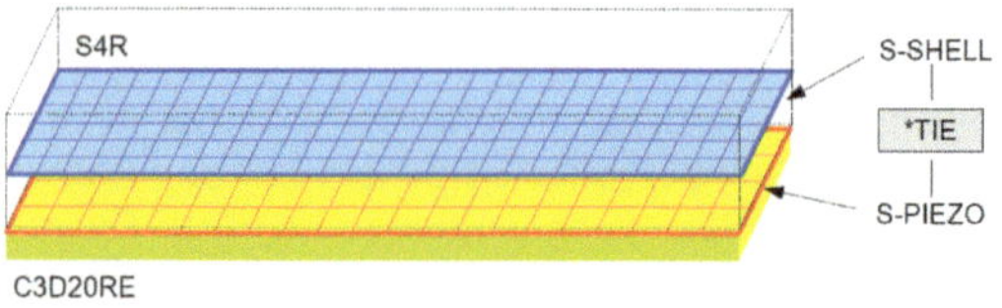


**Figure 10.** Sketch of the proposed structural scale technique.

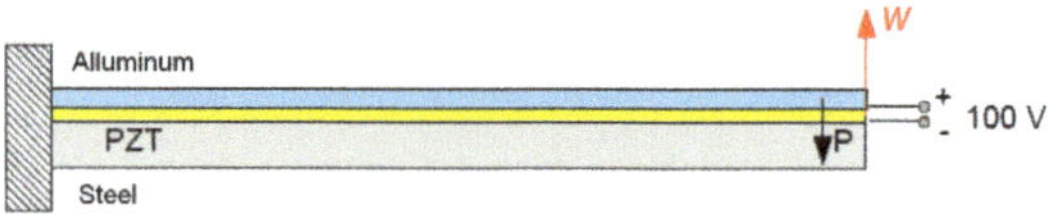


**Figure 11.** Layout for the structural scale validation.

A simple laminate (Figure 11) $200 \times 40mm$ made by three layers, with the piezo in the middle, has been considered to validate the proposed technique. A d.d.p. of $100V$ between the electrodes has been simulated and the transversal motion of the tip has been evaluated through the Classic Lamination Theory (CLT) [4]. Since the laminate is formed by three plates the `TIE` algorithm is used twice. Due to the fact that Piezo element are quadratic, while shell ones are linear different mesh densities must be adopted as shown in Figure 12

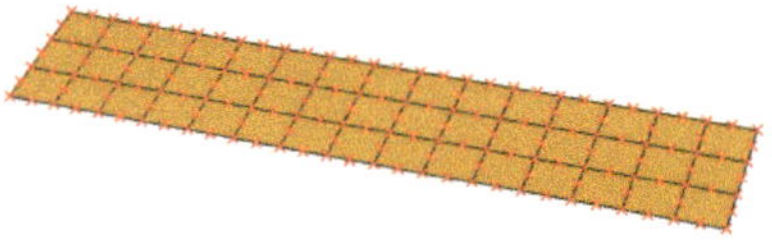

(a) Master

(b) Slave

**Figure 12.** Different mesh densities

It is important to pay attention to the constraint conditions, because those nodes are also influenced by the `TIE` algorithm which connect elements with six degrees of freedom (d.o.f.) per nodes to other with three d.o.f. per nodes. If compared with the CLT, the results coming from those analyses, strongly encourages the use of these techniques in fact the tip deflection error obtained is less then $1e - 3\%$.

The proposed technique can be also adopted for the modeling of smart structures with MFC actuators. For an overall performance evaluation is not necessary to model the whole

[2] the reduced integration is necessary to reduce the so called shear-locking numerical induced effect

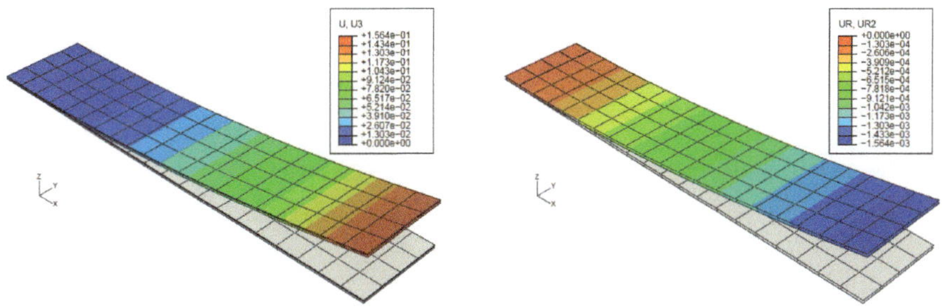


(a) Displacements (b) Rotations

**Figure 13.** Finite Element results of the test specimen

detailed MFC as a sequence of different layers and subcomponents. A previous comparison [4] between detailed models[3] confirm in fact, that the displacements resulting from the light-weighted technique are exceptionally close to the detailed ones.

The MFC have then been modeled through an homogeneous equivalent piezoelectric plates $MFC_{EQ}$ with the same effective coefficients derived from the data-sheets. The dimensions that have been taken into account are the ones of the active area, while the electrodes have been modeled as a voltage equivalent activation[4].

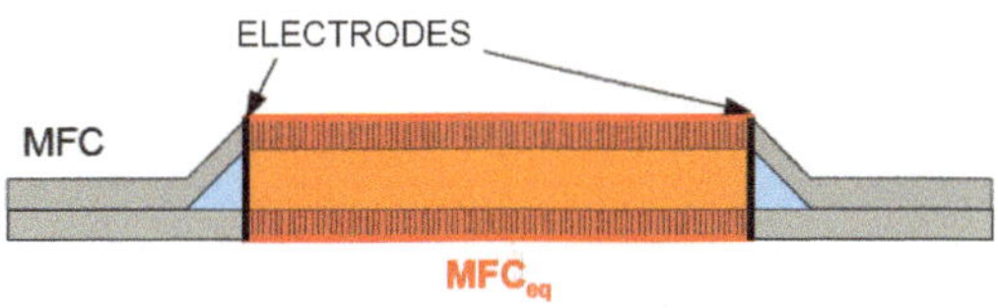


**Figure 14.** $MFC_{EQ}$ into its MFC counterpart

The $MFC_{EQ}$ has been then used for the modeling of simple smart panels with bonded or embedded actuators. The simply bonding of the piezo device onto a panel is simulated through the `TIE` algorithm and have been compared with the detailed one obtaining limited errors ($1e-2\%$).

## 4. Applications

### 4.1. Shape memory actuated deformable mirrors

This procedure has been used to evaluate optical performances deformable mirrors[7–9] based onto new technologies known as smart structures. Carbon fiber reinforced Mirrors actuated by Shape Memory Alloy wires have been modeled to verify their optical capabilities. Image

[3] The whole substructure was modeled in detail with $0.2mm$ mesh pitch

[4] the electric field between two digit times the number of digit gives the overall electric field and tension.

quality during refocusing oriented activation has been evaluated. Several runs of this code has been performed, varying the number/position of actuators and the structure lamination sequence in order to have the evaluation of the influence of each ingredient onto the efficiency of the overall system. In this particular application the routine interact with ABAQUS® that is a finite element solver more reliable with user defined Constitutive laws. The aim of the analysis was the evaluation of the authority of the actuators into the variation of the curvature radius of the shell keeping the image quality within acceptable limits.

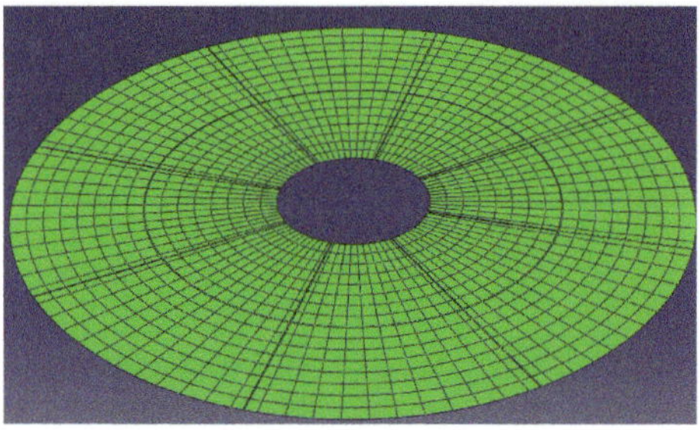

**Figure 15.** Undeformed SMA actuated Deformable mirror

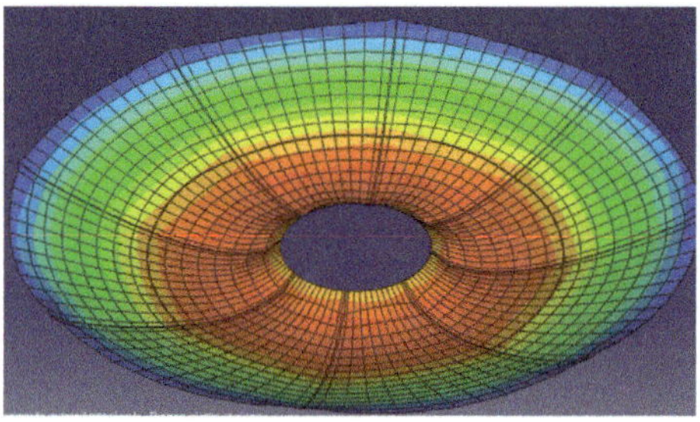

**Figure 16.** Deformed SMA actuated Deformable mirror

In Figure 17 is shown the overall activation sequence: the first part of the process is dominated by a contraction of the deformable mirror due to the mismatching of the CTE of NiTiNOL respect to CFRP; when the temperature raise up to the $A_S$ the phase transition starts and the actuators imposes the recovery strain deploying the deformable mirror. The image quality has been evaluated both in terms of focusing spot (Figures 19 and 21) and p.s.f. (Figures 18 and 20)

Several analyses have been performed to evaluare the focusing capabilities respect to the number of actuators and the stiffness of the substrate (i.e. number of composite plies). For the detail and the results of this analyses that are not purpose of this chapter we refer to [8]. As an example here we show how the actuators density modify the efficiency of the smart SMA actuated structure. Thus why thw he comparison between two different configurations that offer similar variations of focal positions is shown. In Figure 22 is displayed the overall focusing variation comparing a 48 plies substrate with 14 actuators (blue) and a 36 plies with 8 actuators (green). This behavior has been crosschecked with the RMS and max spot size variation (respectively Figures 23 and 24).

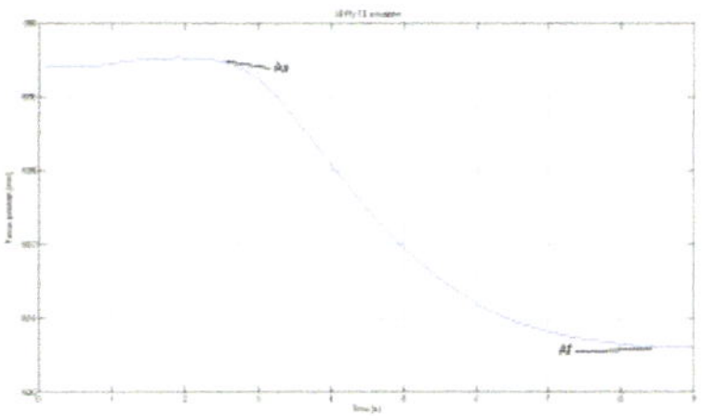

**Figure 17.** Example of variation of focus position during the activation of a SMA deformable mirror.

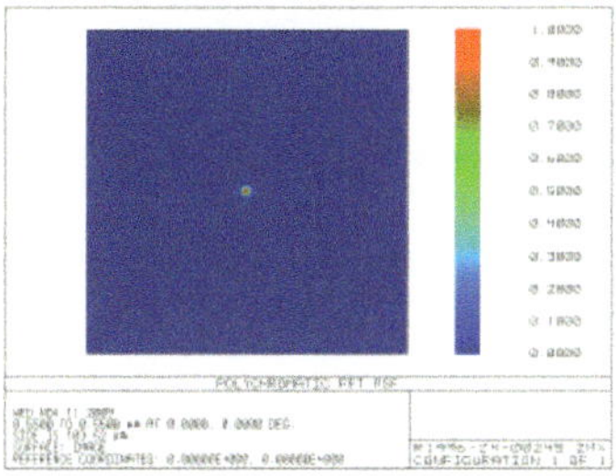

**Figure 18.** Starting p.s.f.

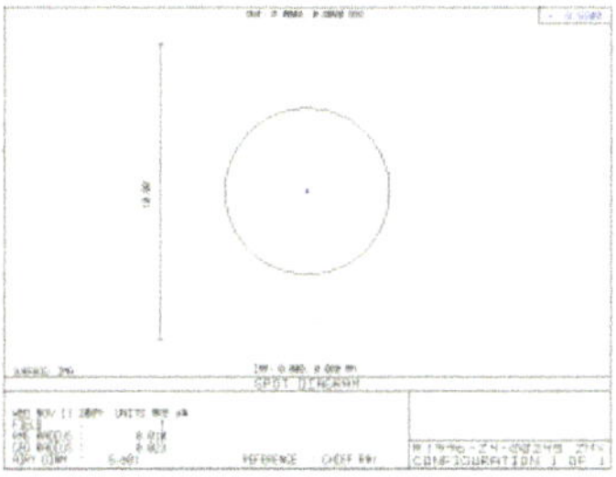

**Figure 19.** Starting focused spot

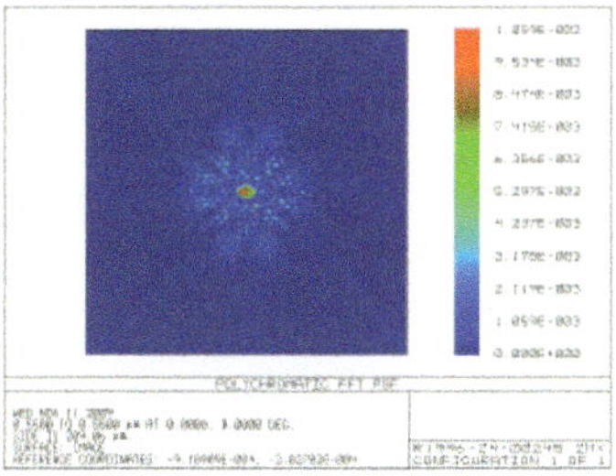

**Figure 20.** Example of Deformed p.s.f.

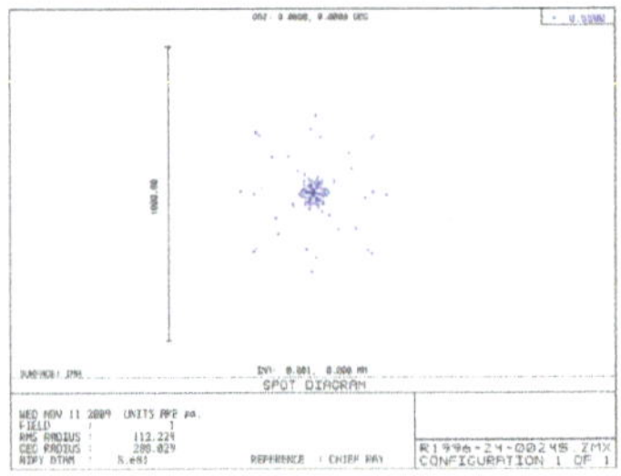

**Figure 21.** Example Deformed focused spot

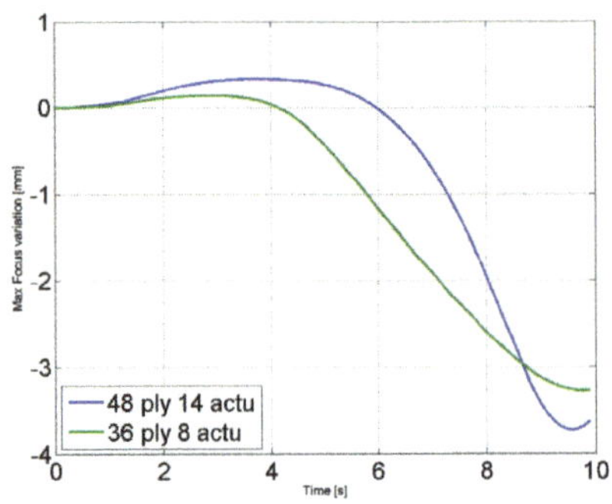


**Figure 22.** Focal position variation with similar focusing range

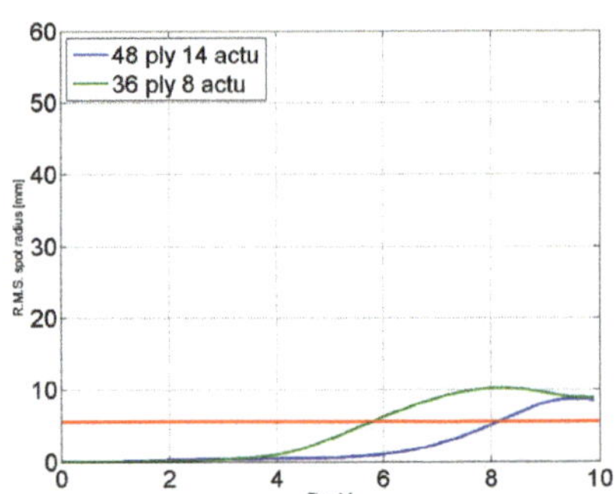


**Figure 23.** RMS spot radius with similar focusing range

## 4.2. MFC actuated deformable mirrors

The same approach (paragraph 4.1) has been used to evaluate Carbon fiber reinforced Mirrors actuated by Piezoelectric MFC actuators.

Analyses have been conducted based on the correlation data obtained by the validation of simplified modeling technique presented in Paragraph 3.2. In particular the surface quality

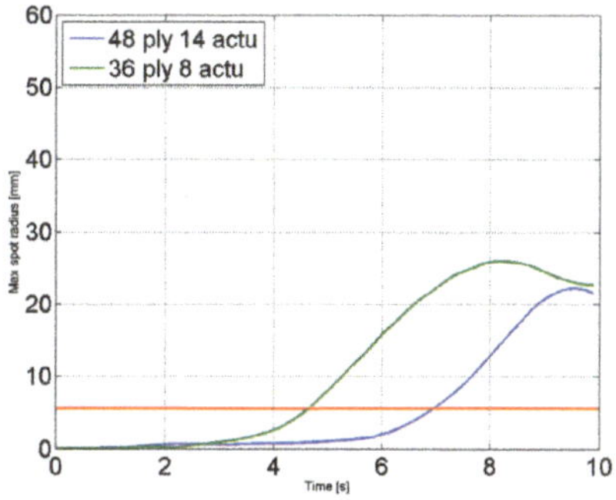


**Figure 24.** MAx spot radius with similar focusing range

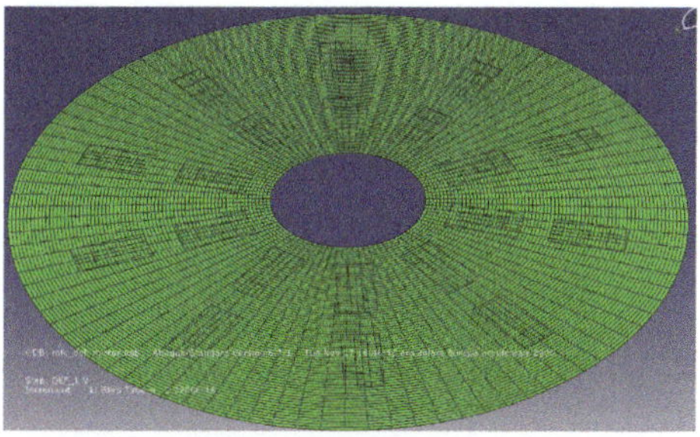

**Figure 25.** Undeformed Piezo actuated Deformable mirror

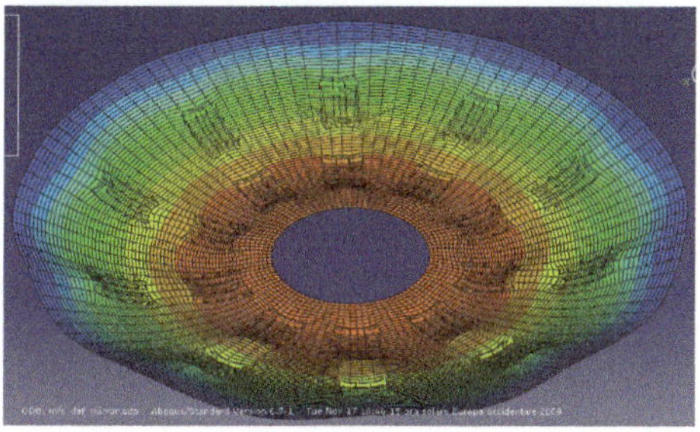

**Figure 26.** Deformed Piezo actuated Deformable mirror

and the curvature radius variation have been verified as functions of the number of actuators, the different configurations and the lamination sequence.

It has been considered the smallest type of actuator even if the analysis framework set up allow the implementation of several type and dimension of them, simply due to the fact that they were already available in the laboratory,

Considering this class of actuators the performances are not satisfactory when the MFC are placed onto a single concentric ring. In Figures 27, 28 and 29 the results obtained with a population of 15 actuators evaluated as a technological limit are plotted; the radial coordinates have been parametrically obtained searching for the best performances that have been obtained placing the actuators at 0.53% of the mirror radius. With this layout the spot size goes out of boundary limit with low radius variations (0.37$mm$) while the max spot radius is always unacceptable.

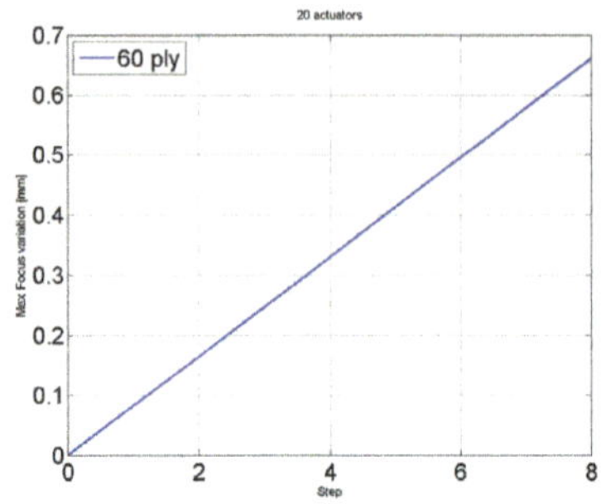


**Figure 27.** Focal position variation with 10 actuators along one ring (60 plies)

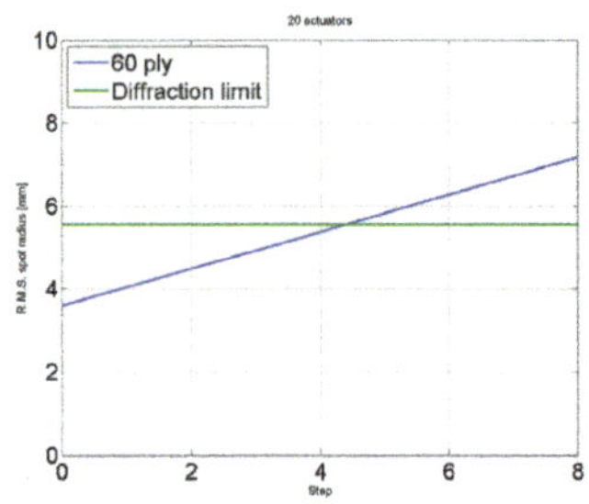


**Figure 28.** RMS spot radius with 10 actuators along one ring (60 plies)

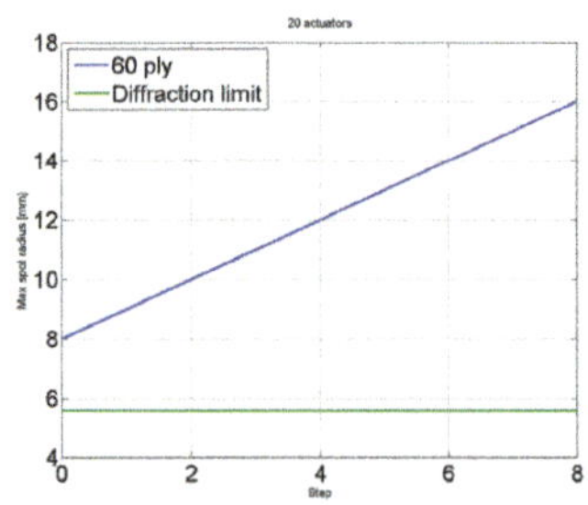


**Figure 29.** Max spot radius with 10 actuators along one ring (60 plies)

The performance of the piezo actuated smart deformable mirror are more interesting if we consider two ring of actuators, with the same angular coordinates (Figure 25). In Figure 26 is shown the contour of deformation. The results obtained comparing configurations with respectively 8 and 10 batteries of actuators (Figures 30, 31 and 32) shows the more interesting performances, both in terms of higher radius variation and in terms of better image quality processed, with the higher number of actuator that are then limited only by technological aspects, mainly related to the crowding of the surface.

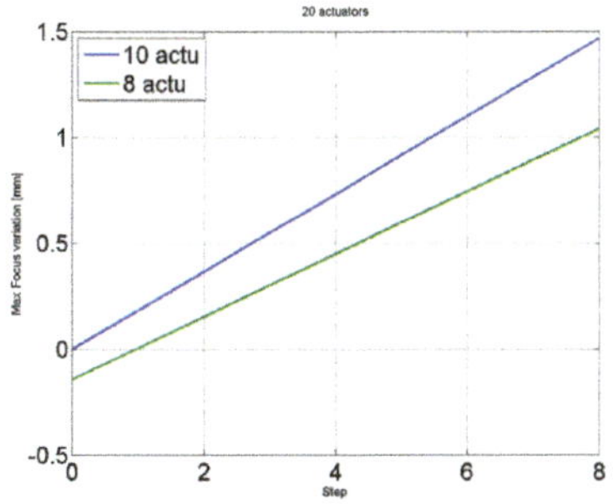


**Figure 30.** Focal position variation with 10 and 8 actuators per ring (two ring 60 plies)

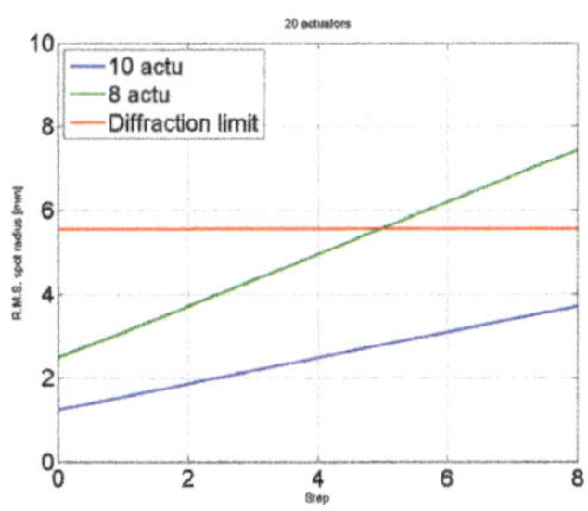


**Figure 31.** RMS spot radius with 10 and 8 actuators per ring (two ring 60 plies)

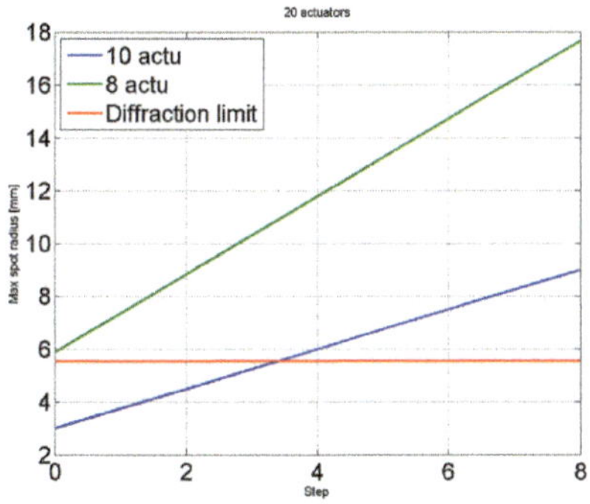


**Figure 32.** Max spot radius with 10 and 8 actuators per ring (two ring 60 plies

Also in this case have been performed analyses to evaluare the focusing capabilities respect to the number of actuators and the stiffness of the substrate. For the detail and the results of this analyses that are not purpose of this chapter we refer to [8].

### 4.3. Opto-mechanical design

Here we show a simple example of the optimization procedure. Within the framework of a feasibility study for a robotic 3m class telescope, equipped with VIS and NIR instruments, able to react to a satellite trigger in less than 50 sec (with a goal of 30 sec): CODEVISIR (Conceptual Design for a VISible and nIR telescope), we exploited the optimization procedure to define the thickness of the main bench of the instrumental suite[15].

The instrument includes seven camera able to cover the wavelength range from the Visible (VIS, $0.4 - 0.9\mu m$) to the Near Infrared (NIR, $1 - 2.5\mu m$) during the same exposure. This will be allowed by a multichannel imaging configuration that envisages a detector for each photometric band, delivered through a dichroic cascade along the optical path. Two ancillary instruments will complete the instrumentation suite, a visible spectrograph and a photo-polarimeter, fed by rotating the M3 mirror.

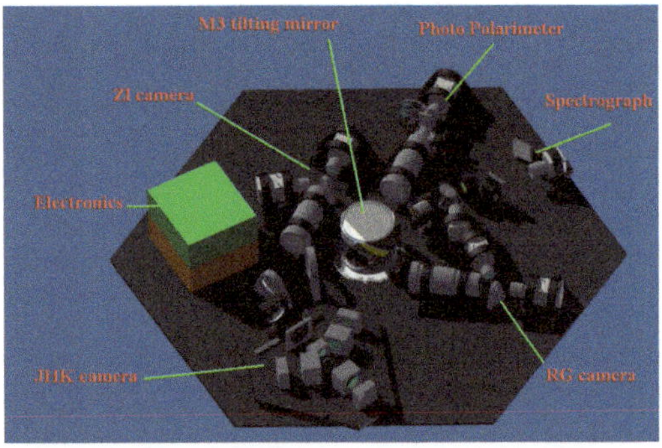


**Figure 33.** Codevisir Instrument suite

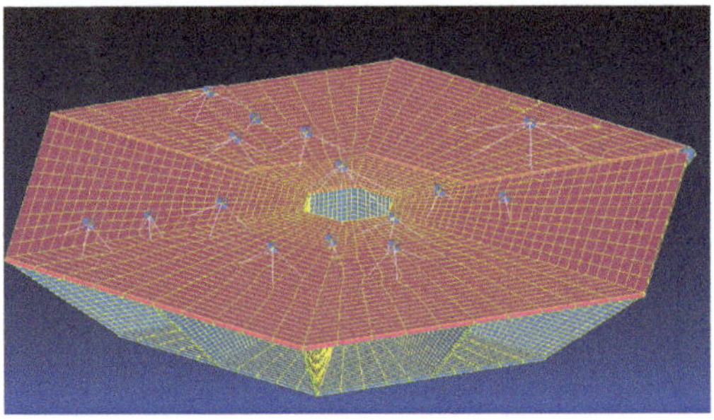

**Figure 34.** Finite Element simplified Model

The Optimization procedure has been oriented to weight minimizing of the overall system keeping the optical displacement[5] below proper requirements. The design variable considered is the thickness of the bench. In Figure 35 can be observed the image displacement behavior calculated as $D = \sqrt{\Delta X^2 + \Delta Y^2}$. The requirement was set at one fifth of a pixel (i.e. $3\mu m$) and the optimization variable has been modified starting from $35mm$. The optimization routine converges to a reasonable value (the thickness variation is less than a reasonable value

[5] of one of the seven arms that is also representative of the performances of all the others

($\Delta th = 0.016mm$)) within 10 steps. The final thickness is around $25mm$ and the optimized weight of the instrument is quite less than $1.2Ton$ and the final displacement is $2.88\mu m$. This procedure produce an overall weight saving of $270Kg$, starting from a very conservative value.

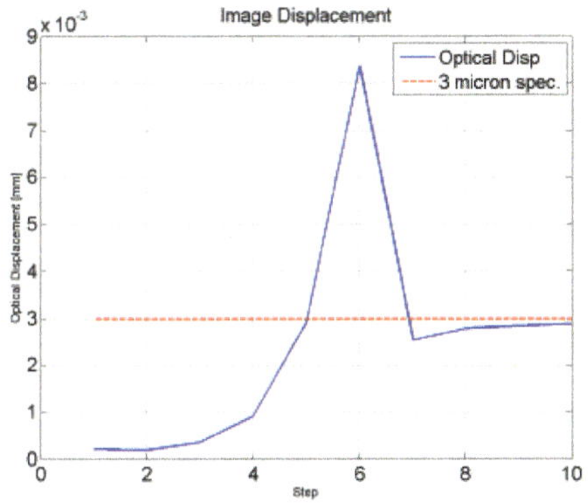


**Figure 35.** Image Displacement optimization

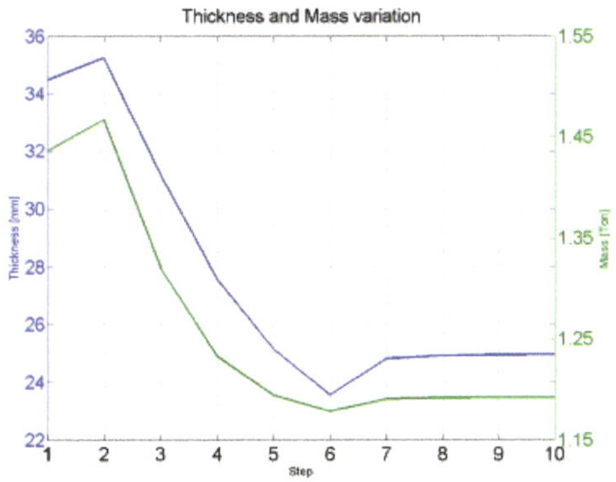


**Figure 36.** Thickness and Mass Variation

## 5. Conclusions

The aim of this chapter was to present the design procedure developed to the numerical capabilities of modern softwares and hardwares. The framework is very versatile and can be used simply for performance prediction of optical system or structres and/or for optimization strategies. The Matlab® code works as a "server" that interact with the "client" Finite Element (ABAQUS, Nastran, ...) solver managing the input file and extracting the results; the data are modified in order to be handled by the "client" raytracing software that evaluates the optical performances. The procedure is under development and integration in order to allow a multi-variable optimization. We are planning also to evaluates different optimization algorithms exploiting in particular the sensitivities analyses[5] capabilities of both Finite Element and raytracing softwares.

## Author details

Marco Riva

*I.N.A.F. - Osservatorio Astronomico di Brera, Italy*


## 6. References


[1] *ABAQUS keyword reference manual* [2007].

[2] Ahmad, A. [1999]. *Handbook of Optomechanical Engineering*, CRC Press.

[3] Bettini, P., Riva, M., Sala, G., DiLandro, L., Airoldi, A. & Cucco, J. [2009]. Carbon fiber reinforced smart laminates with embedded sma actuators-part i: Embedding techniques and interface analysis, *Journal of Materials Engineering and Performance* 18(5-6): 664–671.

[4] DiSanzo, D. [2009]. *Studio di pannelli attivati da mfc: modellazione numerica e prove sperimentali*, Master's thesis, Politecnico di Milano.

[5] Doyle, K., Genberg, V. & Michels, G. [2002]. *Integrated Optomechanical Analysis*, SPIE Press.

[6] *MSC.Nastran: User guide* [2004].

[7] Riva, M. [2010]. *Smart Structures in Instrumentation for Astronomy*, PhD thesis, Aerospace Engineering, Politecnico di Milano.

[8] Riva, M. [2011]. *Smart Structures in Instrumentation for Astronomy*, LAP LAMBERT Academic Publishing.

[9] Riva, M., Bettini, P., DiLandro, L. & Sala, G. [2009]. Shape memory composite deformable mirrors, Vol. 7288, SPIE, pp. 72881T–1/72881T–12.

[10] Riva, M., Bettini, P., DiLandro, L., Sala, G. & Airoldi, A. [2009]. Carbon fiber-reinforced smart laminates with embedded sma actuators-part ii: Numerical models and empirical correlations, *Journal of Materials Engineering and Performance* 18(5-6): 672–678.

[11] Riva, M., Bettini, P., DiLandro, L., Sala, G. & Zerbi, F. [2010]. Smart structures for deformable mirrors actuated by shape memory alloy, Vol. 7739, SPIE, pp. 77391M–1/77391M–20.

[12] Riva, M., DeCaprio, V., Spanó, P. & M.Tintori [2010]. Integrated fiĄnite element analysis and raytracing, oriented to structural optimization, for astronomical instrument design, Vol. 7738, SPIE, pp. 77380J–2/77380J–9.

[13] Riva, M., DiSanzo, D., Airoldi, A., Sala, G. & Zerbi, F. [2010]. Smart structures for deformable mirrors actuated by piezocomposites, Vol. 7739, SPIE, pp. 77391N–1/77391N–16.

[14] Vanderplaats, G. [2007]. *Multidiscipline design optimization*, VR&D Press.

[15] Vitali, F., Chincarini, G., M.Zannoni & all. [2010]. A path to the stars: the evolution of the species in the hunting to the grbs, Vol. 7730, SPIE, pp. 77330W–1/ 77330W–14.

[16] Wilkie, W., Bryant, G. & et all, J. H. [2000]. Low cost piezocomposite actuator for structural control applications, *7th Annual International Symposium on Smart Structures and Materials*, SPIE, Newport Beach.

[17] *ZEMAX: Optical Design Program User's Guide* [2005].

Section 3

# Recent Advances of Finite Element Analysis in "Civil & Structural Engineering"

# Chapter 13

# Finite Element Analysis for the Problem of Tunnel Excavation Successive Phases and Lining Mounting

Simona Roatesi

Additional information is available at the end of the chapter

## 1. Introduction

Numerical application to geomechanics is an area of research that has grown rapidly since its origins in the late 1960s. This growth led to new methodologies and analysis approaches that are nowadays commonly employed in geotechnical, mining or petroleum engineering practice.

The circular tunnel boundary problem is frequently encountered in rock and soil mechanics, geotechnics and generally in mining and petroleum engineering. Consequently, there is a great interest in solving boundary problems involving the excavation of an underground opening. Moreover, time dependent behavior of geomaterials benefits by a special attention in constitutive and numerical approach. Regarding the constitutive modeling and analytical approach [1] concerns both with vertical and horizontal underground openings excavated in an elasto-viscoplastic rock mass, governed by the constitutive laws which go by his name and for which analytical solution for the displacement and viscoplastic strain are derived; [2] provides analytical results for circular and non-circular openings with thermal effects as well; [3-5] present results in approaching the problem both analytically (convergence-confinement method) and numerically by FEM.

The viscoplastic approach in FEM has been considered by many authors, in terms either of numerical stability of schemes used in viscoplasticity, see [6-8], or practical applications using FEM solutions in different areas. For instance, tunneling using FEM for viscoplastic materials has been investigated widely, as well, e.g. [9] approaches the problem with outstanding results in computational geomechanics, with special reference to earthquake engineering, in numerical modeling of dynamic soil and pore fluid interaction and earthquake-induced liquefaction and multiphase pollutant transport in partially saturated porous media; Zienkiewicz provides outstanding numerically approaches in a wide range of problems, e.g., [10] is devoted to computational geomechanics; [11] focuses a great interest in the multi-disciplinarily aspect of the problem, taking into account also adjacent phenomenon occurring at the rock-support interaction; [12] deals with ductile damage and fracture FE modeling of viscoplastic voided materials for high strain rate problems,[13] provides a finite-strain viscoplastic law coupled with anisotropic damage both theoretical and numerical approach, [14] develops a FE procedure to model the tunnel installation and the liner to predict the likely extent of damage to surface structures caused by nearby shallow tunneling, [15] deals with FE modelling of excavation and advancement process of a shield tunnelling machine, [16] develops a FE micromechanical-based model for hydro-mechanical coupling for tunnelling application, [17] applies a model based on plastic damage evolution and permeability to excavation-disturbed zone simulation of the mudstone shield tunnel; [18] analyses tunnel depth effect on the stress and strain state around the tunnel; [19], [20] studies the face tunnel influence in the analysis of a circular tunnel with a time-dependent behaviour, etc.

This chapter deals with a FE code implementation of an elasto-viscoplastic constitutive law. The numerical calculations are performed with a finite element code called CESAR made in LCPC-Paris [21]. The viscoplastic module is coded and implemented in the FE code CESAR by the author. The validation of the numerical code is performed in different steps: with the boundary problem solution of the triaxial laboratory tests, with analytical solution of creep step, and of supported and unsupported underground openings in viscoplastic rock mass. This chapter is focused on further complex applications, such as the tunnel excavation successive phases and lining mounting which are approached using the viscoplastic constitutive module.

## 2. The elasto-viscoplastic constitutive equation

The constitutive law implemented in finite element code is proposed by Cristescu. The hypotheses for the constitutive equation formulation are (see [1]):

i. The material is considered homogeneous and isotropic. Thus, the constitutive functions depend only on the invariants of the stress and strain tensors. The stress tensor and the strain tensor are denoted $\boldsymbol{\sigma}$ and $\boldsymbol{\varepsilon}$, respectively and the stress tensor principal components are denoted as: $\sigma_1, \sigma_2, \sigma_3, \varepsilon_1, \varepsilon_2, \varepsilon_3$. Among the stress invariants, those with an important physical meaning are:

- the mean stress: $\sigma = \dfrac{(\sigma_1 + \sigma_2 + \sigma_3)}{3}$;

- the equivalent stress $\bar{\sigma}^2 = \sigma_1^2 + \sigma_2^2 + \sigma_3^2 - \sigma_1\sigma_2 - \sigma_2\sigma_3 - \sigma_3\sigma_1$ (or the octahedral shear stress $\tau = \sqrt{2}/3\,\bar{\sigma} = \sqrt{2}/\sqrt{3}\, II_{\sigma'}$ with $II_{\sigma'}$ being the second invariant of the stress deviator).

ii. The displacements and the rotations are assumed small, so that $\dot{\boldsymbol{\varepsilon}} = \dot{\boldsymbol{\varepsilon}}^E + \dot{\boldsymbol{\varepsilon}}^I$, where $\dot{\boldsymbol{\varepsilon}}^E$ and $\dot{\boldsymbol{\varepsilon}}^I$ being the elastic strain rate and the irreversible strain rate respectively.

iii. The component of the elastic strain rate satisfies the Hooke law

$$\dot{\boldsymbol{\varepsilon}}^E = \left(\frac{1}{3K} - \frac{1}{2G}\right)\dot{\sigma}\mathbf{I} + \frac{1}{2G}\dot{\boldsymbol{\sigma}}$$

with $K$, $G$ being the bulk and shear modulus, respectively and $\mathbf{I}$ being the identity tensor.

iv. The component of the irreversible strain rate satisfies:

$$\dot{\varepsilon}^I = k(\boldsymbol{\sigma}, d)\left\langle 1 - \frac{W^I(t)}{H(\bar{\sigma}, \sigma)}\right\rangle \frac{\partial F}{\partial \boldsymbol{\sigma}},$$

where: $k$ represents the viscosity coefficient, that can depend on the stress state and the strain state, and probably on a damage parameter $d$ describing the history of the micro cracking the rock was subjected to, and the bracket < > represents the positive part of respective function:

$$\langle A\rangle = \left(A + |A|\right)/2 = A^+$$

The irreversible stress work is used as a hardening parameter or internal state variable, split into volumetric and deviatoric parts and it is given by the quantity:

$$W^I(T) = \int_0^T \sigma(t)\cdot\dot{\varepsilon}^I(t)dt = W_v^I(T) + W_d^I(T). \tag{1}$$

We introduce the *damage parameter d*, see [1], defined by:

$$d(t) = W_v^I(t_{max}) - W_v^I(t) \tag{2}$$

describing the energy released by micro-cracking during the entire dilatancy period. In (2) $t_{max}$ represents the time for which $W_v^I$ is maximum. The failure threshold is considered to be the total energy released by micro cracking during the entire dilatancy process and it is characterized by the following parameter (constant):

$$d_f = W_v^I(t_{max}) - W_v^I(t_{failure}). \tag{3}$$

$H(\boldsymbol{\sigma})$ represents the loading function, generally a function of stress tensor $\boldsymbol{\sigma}$, with $H(\sigma, \bar{\sigma}) = W^I(t)$ the creep stabilization boundary equation, the function $H$ depending on the two stress invariants mentioned above.

$F(\boldsymbol{\sigma})$ represents a viscoplastic potential, that establishes the orientation of $\dot{\boldsymbol{\varepsilon}}^{\mathbf{I}}$ .

v. The initial yield stress of the material is zero or very close to zero.
vi. The applicable domain for the constitutive equation is considered for compressive stresses (positive) and bounded by the failure surface which may be incorporated in the constitutive equation.

We consider therefore, the following constitutive equation:

$$\dot{\boldsymbol{\varepsilon}} = \left(\frac{1}{3K} - \frac{1}{2G}\right)\boldsymbol{\sigma}\mathbf{I} + \frac{1}{2G}\boldsymbol{\sigma} + k(\boldsymbol{\sigma},d)\left\langle 1 - \frac{W^I(t)}{H(\bar{\sigma},\sigma)}\right\rangle \frac{\partial F}{\partial \boldsymbol{\sigma}} \tag{4}$$

For instance, for the model describing the *Borod coal* behavior (see [1]), the constitutive functions and material constants used are:

$$H(\sigma,\bar{\sigma}) := a_0 \frac{(\bar{\sigma}/\sigma_*)^2}{a_2\sigma/\sigma_* + a_1} + b_0\bar{\sigma}/\sigma_* + \begin{cases} c_0 \sin(\omega\sigma/\sigma_* + \varphi) + c_1 & \text{if } 0 \le \sigma \le \sigma_0 \\ c_0 + c_1 & \text{if } \sigma_0 \le \sigma \end{cases} \quad \text{(the yield surface)}$$

$$F(\sigma,\bar{\sigma}) := H(\sigma,\bar{\sigma}) \tag{5}$$

where $a_0 = 7.65 \times 10^{-4}$ MPa; $a_1 = 0.55$; $a_2 = 8.159 \times 10^{-3}$; $b_0 = 0.001$ MPa; $c_0 = 4.957 \times 10^{-4}$ MPa; $c_1 = 4.8955 \times 10^{-4}$ MPa; $\omega = 171.927^0$;$\sigma_0$=1,0996MPa; $k = 6 \times 10^{-6}$; $s^{-1}$ (the viscosity coefficient); $d_f$= $4.48 \times 10^3$ Jm$^{-3}$; $\sigma_* = 1$ MPa and the elastic constants are: $E = 798.385$ MPa, $\upsilon = 0{,}327$.

The relation (5) shows that the viscoplastic potential equals the loading function, fact that determines an *associated flow rule*.

For the model describing the *saturated sand* behaviour, the constitutive functions and material constants are:

$$H(\sigma,\bar{\sigma}) := a\frac{(\bar{\sigma})^7}{\left(\sigma - \frac{\bar{\sigma}^3}{3}\right)^5} + b\bar{\sigma} + c\sigma \quad \text{(the yield surface)}$$

$$F(\sigma,\bar{\sigma}) := -\frac{2h_1\sigma^{\frac{3}{2}}}{2f+\alpha} + 2fh_1\left(\frac{2}{3}\frac{\sigma^{\frac{3}{2}}}{2f+\alpha} + \frac{2\left(1+\frac{\alpha}{3}\right)^{\frac{3}{2}}}{(2f+\alpha)^3}\right) +$$

$$\left(h_1\bar{\sigma}\frac{\left[\left(1+\frac{\alpha}{3}\right)\bar{\sigma}\right]^{\frac{1}{2}}}{(2f+\alpha)^{\frac{3}{2}}} - 2fh_1\frac{\left[\left(1+\frac{\alpha}{3}\right)\bar{\sigma}\right]^{\frac{3}{2}}}{(2f+\alpha)^{\frac{5}{2}}}\right)\ln\frac{\left[(2f+\alpha)\sigma\right]^{\frac{1}{2}} + \left[\left(1+\frac{\alpha}{3}\right)\bar{\sigma}\right]^{\frac{1}{2}}}{\left[(2f+\alpha)\right]^{\frac{1}{2}} - \left[\left(1+\frac{\alpha}{3}\right)\bar{\sigma}\right]^{\frac{1}{2}}} +$$

$$g_0\bar{\sigma} + g_1\bar{\sigma}^3 \qquad \text{(the plastic potential)}$$

where $a$=$4.834\times10^{-7}$ (kPa)$^{-1}$, $b$=$1.33\times10^{-3}$, $c$=$1.058\times10^{-3}$, $h_1$=$2.34\times10$ (kPa)$^{-1/2}$, $g_0$=0.005, $g_1$=$0.62\times10^{-6}$(kPa)$^{-2}$, $f$=0.562, $\alpha$ =1.34, $k$=$3.6\times10^{-6}$ s$^{-1}$ (the viscosity coefficient) and the elastic constants are: $E$ = 205300 kPa, $\upsilon$ = 0.33.

For this model, as the viscoplastic potential differs to the loading function, determines a *non-associated flow rule*.

## 3. The numerical integration of the elasto-viscoplastic equation

In this paragraph the way to implement the elasto-viscoplastic law presented in previous paragraph in finite element code is described. The interface with the program is performed by the calling of the subroutine that carries out the numerical integration of the proposed constitutive law, at a Gauss integration point level. The implicit form for the constitutive law is used and two solving methods, namely: Euler semi implicit method ($\theta$ scheme) and Runge - Kutta method of fourth order.

The numerical integration of equation (4) is performed alternatively by two methods:

a. Semi implicit Euler method ($\theta$ scheme) in which the evaluation of the viscoplastic strain increment is done with the rule: $\Delta\boldsymbol{\varepsilon}_{vp} = (\dot{\boldsymbol{\varepsilon}}^1_{vp}(1-\theta) + \dot{\boldsymbol{\varepsilon}}^2_{vp}\theta)\Delta t, \theta \in [0,1]$ where $\dot{\boldsymbol{\varepsilon}}^1_{vp} = f(\boldsymbol{\sigma}, W)$, $\dot{\boldsymbol{\varepsilon}}^2_{vp} = f(\boldsymbol{\sigma} + \Delta\boldsymbol{\sigma}^1, W + \Delta W)$ and $\Delta\boldsymbol{\sigma}^1, \Delta W^1$ are the stress increment and associated irreversible work increment calculated with the standard scheme, corresponding to the viscoplastic strain increment $\Delta\boldsymbol{\varepsilon}^1_{vp}$; $\boldsymbol{\sigma}, W$ are the known values of stress and irreversible work at the beginning of the step and function $f(\boldsymbol{\sigma}, W) = k(\boldsymbol{\sigma}, d)\left\langle 1 - \frac{W^I(t)}{H(\sigma, \bar{\sigma})} \right\rangle \frac{\partial F}{\partial \boldsymbol{\sigma}}$.

b. Runge-Kutta method performs more evaluations of the function inside each time step *dt*, propagating thus on an interval a solution that is a combination of a few Euler type steps (each of it implies an evaluation of function *f*) and further using the obtained information to match an expansion in Taylor series of higher orders.

In particular, Runge-Kutta method of fourth order employed to integrate the proposed constitutive equation uses four evaluations of function *f* for the considered time step:

$$
\begin{aligned}
\dot{\boldsymbol{\varepsilon}}^1_{vp} &= f(\boldsymbol{\sigma}, W) \\
\dot{\boldsymbol{\varepsilon}}^2_{vp} &= f\left(\boldsymbol{\sigma} + \Delta\boldsymbol{\sigma}^1, W + \Delta W^1\right) \\
\dot{\boldsymbol{\varepsilon}}^3_{vp} &= f\left(\boldsymbol{\sigma} + \Delta\boldsymbol{\sigma}^2, W + \Delta W^2\right) \\
\dot{\boldsymbol{\varepsilon}}^4_{vp} &= f\left(\boldsymbol{\sigma} + \Delta\boldsymbol{\sigma}^3, W + \Delta W^3\right)
\end{aligned}
$$

with $\Delta\boldsymbol{\sigma}^1, \Delta\boldsymbol{\sigma}^2, \Delta\boldsymbol{\sigma}^3, \Delta W^1, \Delta W^2, \Delta W^3$ the intermediate evaluations of the stress increment, and the irreversible work increment respectively.

Runge-Kutta method of fourth order is used for the numerically integration of equation (4) to evaluate the viscoplastic strain increment and it is also used to evaluate the irreversible work (1) from the evolution differential equation of the hardening parameter $\Delta W^i$.

## 4. The numerical solution of nonlinear problems

The classical solution algorithms used in finite element method are displacement type, incremental and iterative ones that often present some convergence difficulties related to the applied loading and the searching of the limit loading.

Generally, the solution algorithms of nonlinear problems will depend on the specific problem under consideration. At the same time, the algorithms are made up of two levels: one global level, of the general scheme of solving, that enables the calculation of the displacement field at the nodal points of the discretized structure; and a local level, of the integration scheme of a nonlinear constitutive equation, which enables at a point, as starting with the stress and strain history at that point, a new state of stress to be calculated.

The two schemes have a great reciprocal influence upon the solving process. Concerning the integration of the constitutive law, it may be done through explicit or implicit schemes.

The treating of a nonlinear problem with finite element method leads to the solving of a system of equations that may be put in the form:

$$\psi(\mathbf{u},\lambda) = \mathbf{R}(\mathbf{u}) - \lambda(t)\mathbf{P} \tag{6}$$

where: **u** represents the displacement vector in the nodal points of the structure under consideration, **R(u)** is the nodal forces vector corresponding to the stresses at moment $t$, **P** designates the total loading applied to the structure and $\lambda(t)$ represents the loading factor applied at moment t.

The solution of the system of equations (6) is in fact the pair $(\mathbf{u}, \lambda(t))$ associated to the displacement response of the structure at the loading which it is subjected to. Generally it is impossible to obtain a solution with a direct solution technique, so an iterative process has to be adopted. Often, the used iterative process is based by the linearization of the nonlinear equations to be solved.

The iterative process assumes that the constitutive equation can enable the calculation of the exact value of the inequilibrium (residual) with respect to the only unknown entity, which is the present displacement $u_1$. For this reason it is necessary to perform an incremental loading of the structure.

The algorithm for solving a static nonlinear problem with the initial stress method used in the framework of the finite element program is:

$$\begin{cases} u_1 = 0 \\ \psi_1 = \mathbf{P} \end{cases} \quad \text{initialization}$$

$j = 1$ loop on the increments

The incrementation of the loading **P**

$i = 1$ loop on the iterations (We assumed $i$ state known and $i$+1 state has to be calculated).

Setting up the term $\Delta F_i = P_i^j - \int \mathbf{B}^t \sigma_i dv$ that has to be calculated as accurate as possible.

The calculation of the global stiffness matrix $\mathbf{K} = \int \mathbf{B}^t \mathbf{D} \mathbf{B} dv$ , where the matrix **D** needs not be quite accurate and can be:

- elastic matrix (the initial stress method)
- the constitutive matrix $\mathbf{D}^{evp}$ (the tangent stiffness method)

The solving of the algebraic system $\mathbf{K}\Delta u_i = \Delta \mathrm{F}_i$ (nonlinear if **K=K(u)**)

$\Delta u_i = \mathbf{K}^{-1} \Delta \mathrm{F}_i$ solving the system of equations
$u_{i+1} = u_i + \Delta u_i$ updating displacement
$\Delta \varepsilon_i = \mathbf{B} \Delta u_1$ displacement-strain relations
$\sigma_{i+1} = \sigma_i + \Delta \sigma_i$ calculation of the incremental stresses with the constitutive law; $\Delta\sigma$ can be calculated as:

- elasticity: $\Delta\sigma_i = \mathbf{D}^e \Delta\varepsilon_i$ .
- viscoplasticity: * $\Delta\sigma_i = \mathbf{D}^e (\Delta\varepsilon_i - \Delta\varepsilon_i^{vp})$, with $\Delta\varepsilon_i^{vp} = \Lambda \partial f / \partial \sigma$ ,proportional with the derivative of the viscoplastic potential.

* $\Delta\sigma_i = \mathbf{D}^{evp} \Delta\varepsilon_i$ (explicit form).

From the law $\Delta\varepsilon_i = f(\Delta\sigma_i)$ (implicit form), the numerical integration may be done with $\theta$ scheme, with Runge-Kutta method, the consistent matrix, etc.

$\mathbf{R}(u_{i+1}) = \int \mathbf{B}^t \sigma_{i+1} dv$ calculation of the residue

$\varphi_{i+1} = \mathrm{P}^j - \mathbf{R}(u_{i+1})$ equilibrium check
Convergence test:
no $i = i+1$ (next iteration)
yes $j = j+1$ (next step)

where **u** represents the vector of nodal displacements, **v** is the vector of out of balance nodal forces, **B** designates the matrix of the shape functions for strain, $\mathbf{P}^j$ represents the vector of the known nodal applied forces, $\mathbf{R}(u_i)$ is the vector of the equivalent nodal forces, due to the stresses $\sigma_i$. These nodal forces are consistent with the current value of unknown **u**.

Returning again on the fact that it can be also deduced from the algorithm above, namely that two calculation levels can be distinguished:

- local level: the application of the constitutive law for a material point (in fact, Gauss integration point in finite element analysis) for the calculation of the constitutive matrix **D**, $(\Delta\sigma = \mathbf{D}\Delta\varepsilon)$ and the stresses $\sigma_{i+1}$ .
- global level: the application of the iterative process upon the vector of nodal displacements.

In order to ensure the convergence of the iterative process, the non-equilibrium of the structure (the residual), the displacement variation and the work variation done during the iteration have to be tested at every iteration. So, the testing is done upon the following quantities:

- testing the residue $\frac{|\psi(u_i)|}{|\psi(u_0)|} \leq \varepsilon_R$;
- testing the displacement $\frac{|\Delta u_i|}{|u_i|} \leq \varepsilon_D$;
- testing the work $\frac{|\psi(u_i)^T.\Delta u_i|}{|\psi(u_0)^T.\Delta u_0|} \leq \varepsilon_W$.

It has to be mentioned the fact that an iteration has no physical meaning: in fact, at the beginning of the iteration, the equilibrium is satisfied, but, at the end of one iteration, the constitutive law is satisfied in preference to the equilibrium. Therefore, only at the end of an incremental stage, the solution can converge to a physical sense for the studied structure.

## 5. Comparison of the numerical and analytical solution for the creep step

In order to test the subroutine which performs the numerical integration of the elasto-viscoplastic constitutive equation, a comparison of the numerical solution with the analytical formula for the creep step is performed. For this purpose, an analytical formula is used for the calculation of the strain rate in the case of the application of a number of successive steps at constant stress, namely: it is assumed that at moment $t_0$ the stress state is increased suddenly to the value of $\boldsymbol{\sigma}(t_0)$ and it is kept constant.

### 5.1. Determination of the analytical solution for creep step

To establish the formulae that describe the creep deformation, we will write firstly the formula which supplies the variation of $W^I(t)$ when all stress components are constant; it is easily obtained by integration of the constitutive law (4) (see [1]):

$$1-\frac{W^I(t)}{H(\sigma,\bar{\sigma})}=\left(1-\frac{W^I(t_0)}{H(\sigma,\bar{\sigma})}\right)\exp\left[\frac{k}{H}\frac{\partial F}{\partial \boldsymbol{\sigma}}\cdot\boldsymbol{\sigma}(t_0-t)\right] \tag{7}$$

where $\boldsymbol{\sigma}(t)=\boldsymbol{\sigma}(t_0)=\text{constant}$ and $W^I(t_0)$ is the initial value of $W^I$ for $t=t_0$, denoted further $W^{IP}$.

This is the relation that describes the variation in time of $W^I(t)$ under a constant stress. The variation in time of $W^I(t)$ is longer or shorter depending especially on the value of the viscosity parameter $k$.

Introducing the left hand side of the relation (7) into the constitutive equation (4), we get the total strain under constant stress:

$$\boldsymbol{\varepsilon}=\boldsymbol{\varepsilon}^0+\frac{\left(1-\dfrac{W^I(t_0)}{H(\sigma,\bar{\sigma})}\right)\dfrac{\partial F}{\partial \boldsymbol{\sigma}}}{\dfrac{1}{H}\dfrac{\partial F}{\partial \boldsymbol{\sigma}}\cdot\boldsymbol{\sigma}}\left\{1-\exp\left[\frac{k}{H}\frac{\partial F}{\partial \boldsymbol{\sigma}}\cdot\boldsymbol{\sigma}(t_0-t)\right]\right\} \tag{8}$$

using the initial conditions:

$$t=t_0;\ \boldsymbol{\varepsilon}^I=0;\ \boldsymbol{\varepsilon}=\boldsymbol{\varepsilon}^0=\left(\frac{1}{3K}-\frac{1}{2G}\right)\sigma_0\mathbf{I}+\frac{1}{2G}\boldsymbol{\sigma}_0 \tag{9}$$

where $\boldsymbol{\sigma}(t)=\boldsymbol{\sigma}(t_0)$ is the initial stress state, reached instantaneously and taken with respect to the state at $t_0$ when the loading is applied (it represents the elastic response). It is observed that the value of irreversible strain rate during the creep is governed by the expression $1-\dfrac{W^I(t)}{H(\sigma,\bar{\sigma})}$.

If we wish to have a stress path made up of small stress increments $\Delta\boldsymbol{\sigma}$, followed by time intervals of constant stress, we get for the strain increments some formulae similar with (8), namely:

$$\begin{aligned}\varepsilon_1(t)&=\frac{\sigma_1(t)}{2G}+\frac{\left(1-\dfrac{W^I(t_0)}{H(\sigma,\bar{\sigma})}\right)\dfrac{\partial F}{\partial \sigma_1}}{\dfrac{1}{H}\dfrac{\partial F}{\partial \boldsymbol{\sigma}}\cdot\boldsymbol{\sigma}}\left\{1-\exp\left[\frac{k}{H}\frac{\partial F}{\partial \boldsymbol{\sigma}}\cdot\boldsymbol{\sigma}(t_0-t)\right]\right\},\\ \varepsilon_2(t)&=\frac{\sigma_2(t)}{2G}+\frac{\left(1-\dfrac{W^I(t_0)}{H(\sigma,\bar{\sigma})}\right)\dfrac{\partial F}{\partial \sigma_2}}{\dfrac{1}{H}\dfrac{\partial F}{\partial \boldsymbol{\sigma}}\cdot\boldsymbol{\sigma}}\left\{1-\exp\left[\frac{k}{H}\frac{\partial F}{\partial \boldsymbol{\sigma}}\cdot\boldsymbol{\sigma}(t_0-t)\right]\right\},\end{aligned} \tag{10}$$

where $\boldsymbol{\sigma}_1$ and $\boldsymbol{\sigma}_2$ being the principal stresses and where during the time interval $t\in(t_0,t]$ the state of stress is constant and $W^I(t_0)$ is calculated at moment $t_0$, just before of a new stress increment $\Delta\boldsymbol{\sigma}$ that occurs at $t_0$.

## 5.2. Comparison of the numerical results with the analytical solution

The numerical solution is obtained for the same loading path as in the calculation of the analytical solution. Since the calculation of the analytical solution uses the hypothesis of an instantaneous loading, for the numerical solution the loading is applied in a very small time interval $\Delta t$, namely $10^{-10}$ s. Similar to the calculation of the analytical solution in the hypothesis of a constant stress, for the numerical solution, at every time step, null loading increment is applied.

To perform a comparison between the analytical solution and the numerical results obtained using the Euler method ($\theta$ scheme) and Runge-Kutta method of fourth order, we present as example, one step loading path, with a null initial state of strain and an initial stress state of $\sigma_1^0 = 0.7\text{kPa}, \sigma_2^0 = 0.1\text{kPa}, \sigma_3^0 = 0.1\text{kPa}$. The numerical results obtained using Runge-Kutta method of fourth order shows the superiority of this method.

For instance, in the case of $\Delta t$ = 0.01 s, the analytical solution for the strain components gives:

$$\varepsilon_1^{analytic} = 3.08839 \times 10^{-6}, \varepsilon_2^{analytic} = -7.99371 \times 10^{-7}, \varepsilon_3^{analytic} = -7.99371 \times 10^{-7};$$

while the same components using $\theta$ scheme are:

$$\varepsilon_1^{Euler} = 3.08841 \times 10^{-6}, \varepsilon_2^{Euler} = -7.99383 \times 10^{-7}, \varepsilon_3^{Euler} = -7.99383 \times 10^{-7};$$

and using Runge-Kutta of fourth order methods, are:

$$\varepsilon_1^{R-K} = 3.08840 \times 10^{-6}, \varepsilon_2^{R-K} = -7.99378 \times 10^{-7}, \varepsilon_3^{R-K} = -7.99378 \times 10^{-7}.$$

## 6. Comparison of the numerical solution of the triaxial test boundary problem and the experimental data

The numerical solution is performed using the elasto-viscoplastic constitutive law presented above and represents the triaxial test for a cylindrical sample of *saturated sand* $(\sigma_1 \text{ increasing}, \sigma_2 = \sigma_3)$ under axi-symmetric hypothesis, see [20].

A quarter of a sample was considered because of the symmetries, requiring only one quadrilateral finite element with 8 nodal points. So, the mesh is a very simple one, like in Figure 1, the width being denoted with *a* and the length *l*. In this case, we consider *a* = *l* = 1.

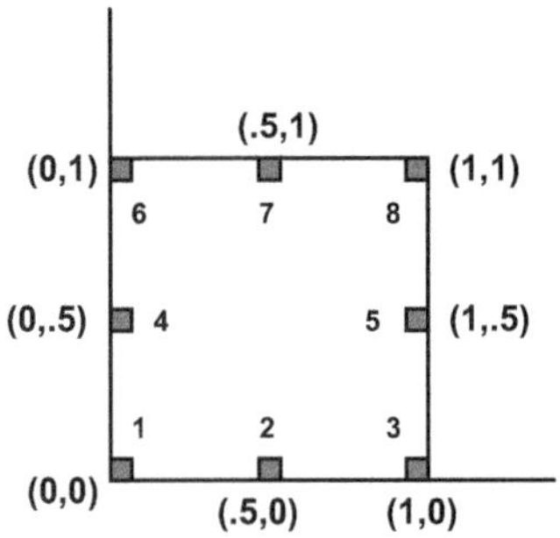


**Figure 1.** The mesh for the simulation of the triaxial test for a cylindrical sample

The stress state in element being uniform, the 8 nodal points of the element present the same state of stress and strain. For validation we deal only with 8th nodal point element, which has the advantage that since its height being equal with unity, its vertical displacement is equal with the longitudinal strain, and its horizontal displacement is equal with the radial strain.

The initial boundary problem is then:

$$\begin{aligned}
&(\text{inner face}) \quad z=0 \text{ and } r\in[0,a] \quad u_r(t)=0,\ \sigma_{rz}(t)=0,\ t\geq 0.\\
&(\text{left lateral surface}) \quad z\in[0,l] \text{ and } r=0 \quad u_r(t)=0,\ \sigma_{rz}(t)=0,\ t\geq 0.\\
&(\text{right lateral surface}) \quad z\in[0,l] \text{ and } r=a \quad \sigma_{rr}(t)=f(t),\ \sigma_{rz}(t)=0,\ t\geq 0.\\
&(\text{outer face}) \quad z=l \text{ and } r\in[0,a] \quad \sigma_{zz}(t)=g(t),\ \sigma_{rz}(t)=0,\ t\geq 0.
\end{aligned}$$

We consider for the functions *f* and g some particular forms that can simulate the triaxial test in two stages, namely:

- the hydrostatic stage when both lateral pressure and the vertical one are increased (in steps for this test) until a certain value is reached.
- the deviatoric stage, when the lateral pressure remains constant at while the vertical one is increased (in steps as well) until failure.

The following functions are used for the two stages:

$$f(t)=\begin{cases}\sigma_h \dfrac{t}{T} \text{ for } 0\leq t\leq T & (\text{the hydrostatic stage})\\ \sigma_h \text{ for } T\leq t & (\text{the deviatoric stage})\end{cases}$$

$$g(t)=\begin{cases}\sigma_h \dfrac{t}{T} \text{ for } 0\leq t\leq T & (\text{the hydrostatic stage})\\ \sigma_h+\Sigma_h\left(\dfrac{t-T}{\tau}\right) \text{ for } T\leq t & (\text{the deviatoric stage})\end{cases}$$

with a linear variation during the hydrostatic stage. $\Sigma_h, \tau, T$ are constants that characterize the loading.

We are going to consider as well a quadratic variation with respect to the time of the loading through the functions:

$$f'(t)=\begin{cases}\sigma_h \dfrac{t^2}{T'^2} \text{ for } 0\leq t\leq T' & (\text{the hydrostatic stage})\\ \sigma_h \text{ for } T'\leq t & (\text{the deviatoric stage})\end{cases}$$

$$g'(t)=\begin{cases}\sigma_h \dfrac{t^2}{T^2} \text{ for } 0\leq t\leq T' & (\text{the hydrostatic stage})\\ \sigma_h+\Sigma_h\left(\dfrac{t-T'}{\tau}\right)^2 \text{ for } T'\leq t & (\text{the deviatoric stage})\end{cases}$$

Let us consider now a hydrostatic stage of a classical triaxial test (we consider for the functions *f*, *g*, *f'*, *g'* the upper branch of the function form defined above). We will try to find again some of the important constitutive features of the material with the numerical solution.

The comparison of the results for a linear incremental in time loading (considering the functions *f*, respective *g*) and a quadratic incremental in time loading (considering the functions *f'*, respective *g'*) are made for two different duration of creep time step, 10s and 30s respectively. In both cases, the volumetric strain corresponding to the quadratic loading is bigger than in the case of the linear one in time, the difference being more marked at smaller loading rates.

A comparison with respect to the loading rates for 7 loading steps, and 14 loading steps hydrostatic test respectively is made too. In both cases one can observe grater strains in the case of smaller loading rates, the strains being as great as the test performed with more loading steps.

For the deviatoric stage of the classical triaxial test (the lower branches of the functions *f*, *g*, *f'*, *g'*) in which the stress state that was reached in the hydrostatic stage is maintained constant and only the vertical component of the stress is increased, a comparison between the numerical results and the experimental data is achieved for saturated sand (see [22]).

There are two sets of tests with the confining pressure of 14.7 kPa and 29.4 kPa respectively. Figures 2a,b represents the experimental data, while Figures 3 a, b presents the numerical results for the two tests mentioned above, respectively.

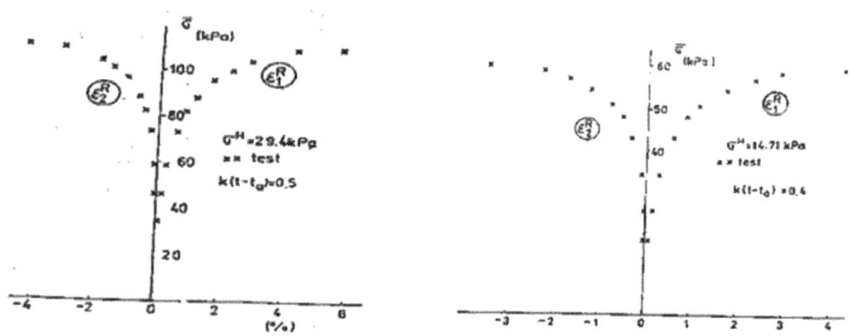


(a) (b)

**Figure 2.** a), b). Experimental data for 2 tests with confining pressure of 14.7 kPa , 29.4 kPa respectively

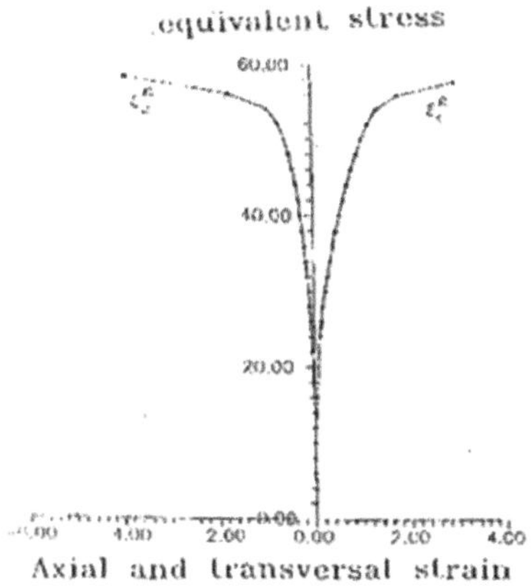


(a)

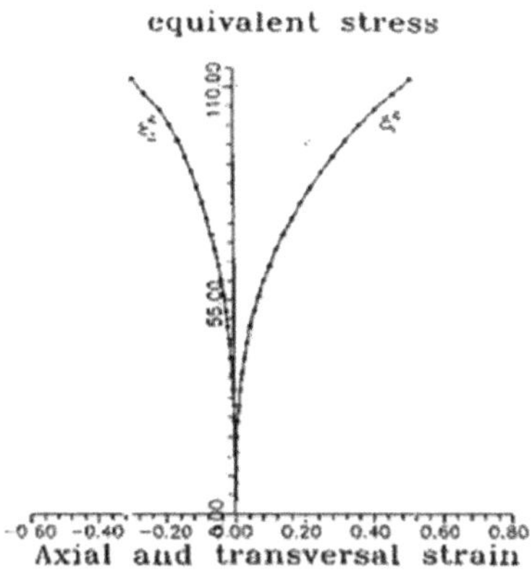


(b)

**Figure 3.** a), b). Numerical solutions for 2 tests with confining pressure of 14.7 kPa,29.4 kPa, respectively

## 7. Comparison of numerical and analytical solution for underground openings

In this paragraph, a comparison between the numerical results and an available analytical solution for the problem of underground openings in viscoplastic rock mass is performed and it represents the next step of validation of the numerical code. The approach presented here concerns the applications of supported and unsupported underground openings in viscoplastic rock mass with the assumption of constant primary stress in the whole domain.

### 7.1. Problem formulation

The proposed boundary problem is as follows: it is assumed that the rock mass is an infinite body in which circular opening is made. Therefore only plane strain condition is considered.

Assuming further that the underground opening is at a certain depth where the horizontal and vertical components of the primary (initial) stress $\sigma_h$ and $\sigma_v$ are known and, generally, distinct. It is also assumed that in the neighbourhood of the opening these components are constant and equal with their corresponding value for the opening axis depth. The influence of the ground surface is neglected, so that the opening is imagined as a cylindrical cavity of infinite length, excavated in an infinite space.

Cylindrical coordinate system is chosen for convenience with axis Oz being the symmetry axis of the opening (Figure 4).

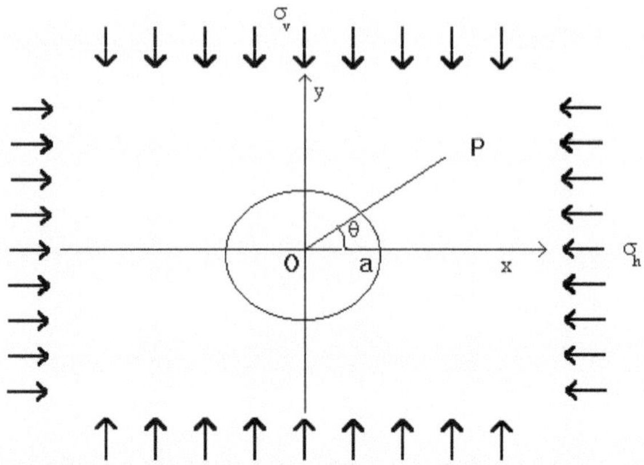


**Figure 4.** The domain of the boundary value problem

A pressure loading *p(t)* is considered on the surface $r = a$ of the opening, that could be due to different causes. Therefore, the boundary conditions are:

$$\left.\begin{array}{l} r=a \\ \text{any } z \end{array}\right\}: \ \sigma_{rr}^{S}=p, \sigma_{r\theta}^{S}=0, \quad \left.\begin{array}{l} r\to\infty \\ \text{any } z \end{array}\right\}: \ \sigma_{xx/\infty}^{S}=\sigma_{h}, \sigma_{yy/\infty}^{S}=\sigma_{v}, \tag{11}$$

with $Ox$ and $Oy$ the horizontal and the vertical axis respectively.

The conditions (9) can be written in the cylindrical coordinates as follows (see [1]):

$$\begin{aligned} \sigma_{rr/\infty}^{S} &= \sigma_{rr}^{P} = \tfrac{1}{2}(\sigma_h+\sigma_v)+\tfrac{1}{2}(\sigma_h-\sigma_v)\cos 2\theta, \\ \sigma_{\theta\theta/\infty}^{S} &= \sigma_{\theta\theta}^{P} = \tfrac{1}{2}(\sigma_h+\sigma_v)-\tfrac{1}{2}(\sigma_h-\sigma_v)\cos 2\theta, \\ \sigma_{r\theta/\infty}^{S} &= \sigma_{r\theta}^{P} = -\tfrac{1}{2}(\sigma_h-\sigma_v)\sin 2\theta. \end{aligned}$$

The superscripts *S, P, R* mean the secondary, primary and relative components of the stress, displacement and strain respectively.

Further, the fundamental equations of the problem are presented.

The equilibrium equations written in the relative components are:

$$\begin{aligned} &\frac{\partial\sigma_{rr}}{\partial r}+\frac{1}{r}\frac{\partial\sigma_{r\theta}}{\partial\theta}+\frac{\partial\sigma_{rz}}{\partial_z}+\frac{\sigma_{rr}-\sigma_{\theta\theta}}{r}=0; \\ &\frac{\partial\sigma_{r\theta}}{\partial r}+\frac{1}{r}\frac{\partial\sigma_{\theta\theta}}{\partial\theta}+\frac{\partial\sigma_{\theta z}}{\partial_z}+2\frac{\sigma_{r\theta}}{r}=0; \\ &\frac{\partial\sigma_{rz}}{\partial r}+\frac{1}{r}\frac{\partial\sigma_{z\theta}}{\partial\theta}+\frac{\partial\theta_{zz}}{\partial_z}+\frac{\sigma_{rz}}{r}=0. \end{aligned}$$

The components of the small strains:

$$\varepsilon_{rr} = \frac{\partial u_r}{\partial r}, \varepsilon_{r\theta} = \frac{1}{2}\left[\frac{1}{r}\frac{\partial u_r}{\partial \theta} + \frac{\partial u_\theta}{\partial r} - \frac{u_\theta}{r}\right];$$
$$\varepsilon_{rz} = \frac{1}{2}\left[\frac{\partial u_r}{\partial z} + \frac{\partial u_z}{\partial r}\right];$$
$$\varepsilon_{\theta\theta} = \left[\frac{1}{r}\frac{\partial u_\theta}{\partial \theta} + \frac{u_r}{r}\right], \varepsilon_{z\theta} = \frac{1}{2}\left[\frac{1}{r}\frac{\partial u_z}{\partial \theta} + \frac{\partial u_\theta}{\partial z}\right], \varepsilon_{zz} = \frac{\partial u_z}{\partial z}.$$

and the compatibility equations in cylindrical coordinates has to be added, as well:

$$2\frac{\partial^2 (r\varepsilon_{r_\theta})}{\partial r\ \partial \theta} - \frac{\partial^2 \varepsilon_{rr}}{\partial \theta^2} - r\frac{\partial^2 (r\varepsilon_{\theta\theta})}{\partial r^2} + r\frac{\partial\ \varepsilon_{rr}}{\partial r} = 0,$$
$$2r\frac{\partial^2 \varepsilon_{\theta z}}{\partial z\ \partial \theta} - r^2\frac{\partial^2 \varepsilon_{\theta\theta}}{\partial z^2} - \frac{\partial^2 \varepsilon_{zz}}{\partial \theta^2} + r\left(2\frac{\partial \varepsilon_{rz}}{\partial z} - \frac{\partial \varepsilon_{zz}}{\partial r}\right) = 0,$$
$$2\frac{\partial^2 \varepsilon_{rz}}{\partial z\ \partial r} - \frac{\partial^2 \varepsilon_{zz}}{\partial r^2} - \frac{\partial^2 \varepsilon_{rr}}{\partial z^2} = 0,$$
$$\frac{\partial^2 \varepsilon_{rz}}{\partial r\ \partial \theta} + 2\frac{\partial \varepsilon_{r\theta}}{\partial z} - r\frac{\partial^2 \varepsilon_{\theta z}}{\partial r^2} + \frac{1}{r}\left(\frac{\partial \varepsilon_{rz}}{\partial \theta} - \varepsilon_{\theta z}\right) + r\frac{\partial^2 \varepsilon_{\theta r}}{\partial r \partial z} - \frac{\partial^2 \varepsilon_{rr}}{\partial z \partial \theta} - \frac{\partial \varepsilon_{\theta z}}{\partial r} = 0,$$
$$r\frac{\partial^2 (\varepsilon_{r\theta})}{\partial z \partial \theta} - \frac{\partial^2 \varepsilon_{rz}}{\partial \theta^2} - r\frac{\partial^2 (r\varepsilon_{\theta\theta})}{\partial r \partial z} + r\frac{\partial \varepsilon_{rr}}{\partial z} + \frac{\partial \varepsilon_{\theta z}}{\partial \theta} + r\frac{\partial \varepsilon^2{}_{\theta z}}{\partial r \partial \theta} = 0,$$
$$r\frac{\partial^2 \varepsilon_{\theta z}}{\partial r \partial z} + \frac{\partial^2 \varepsilon_{rz}}{\partial z \partial \theta} - r\frac{\partial^2{}_{\theta r}}{\partial z^2} - \frac{1}{r}\left(r\frac{\partial \varepsilon_{\theta z}}{\partial z} - \frac{\partial \varepsilon_{zz}}{\partial \theta}\right) - \frac{\partial \varepsilon^2{}_{zz}}{\partial \theta \partial r} = 0.$$

Due to the plane strain hypothesis, we have $\frac{\partial}{\partial z} = 0$, $u_z^R = 0$, and then, according to the components of the small strains, zero values are founded for the relative strains $\varepsilon_{zz}^R = \varepsilon_{rz}^R = \varepsilon_{\theta z}^R = 0$.

A viscoplastic constitutive equation as presented in the previous paragraph is considered.

## 7.2. Analytical solution

It is assumed that the opening (or only a part of it) is excavated in a very short time interval $t \in (0, t_0)$ and then it is exploited in a much longer time interval $t \in (t_0, t_1)$. It is also assumed that during the first time interval the response of the rock is "instantaneous" and during the second one different time effects are possible, such as: creep and a slow variation in time of stress.

Consequently, if the tunnel is excavated at $t_0$, then immediately after excavation, the stress, the strain and the displacement of the rock are given by the elastic solution. Concerning the second interval $(t_0, t_1)$ a possible solution will be present, obtained under a number of assumptions that would simplify a lot the analytical solution thus making it amenable to analysis.

The main hypothesis is that the same distribution of stresses for the elasto-viscoplastic model is assumed as in the elastic model: by considering that in the interval $(t_0, t_1)$ the stress components $\sigma_{rr}$, $\sigma_{\theta\theta}$, $\sigma_{r\theta}$ remain constant and equal with those given by the elastic solution (see [1]).

Due to the fact of the plane strain hypothesis $\varepsilon_{zz}^{R} = 0$ it has to be accepted that $\sigma_{zz}$ varies during this interval and therefore it satisfies the differential equation:

$$\frac{1}{3}\left(\frac{1}{3K}+\frac{1}{G}\right)\dot{\sigma}_{zz} = -k\left(1-\frac{W^I}{H}\right)\frac{\partial F}{\partial \sigma_{zz}}.$$

This differential equation may be integrated numerically using the initial conditions $\sigma_{zz} = \sigma_h$. It was found that a very small variation in time (negligible) of $\sigma_{zz}$ was exhibited and a very fast stabilization of this variation. Therefore, it can be assumed that all stress components could be constant during the rock creep around the opening.

In order to arrive at the expression that describes the creep strain, the equation (7) that supplies the variation of $W^I(t)$ when all stress components are constant is used.

The time taken for $W^I(t)$ to reach the asymptotic value depends mainly on the value of the constitutive parameter *k*, the viscosity. A reasonable correct value for *k* can be obtained only observing the convergence of a tunnel wall in time.

Then one can obtain, similarly as in the previous paragraph, the formulae (10) for the strain variation by integrating the constitutive equation (4) taking into account (7) and using the initial conditions (9).

In the case of $\sigma_h = \sigma_v$ one can deduce a formula for the wall opening convergence as:

$$u = u^{R0} + \frac{\left(1-\dfrac{W^{IP}}{H(\sigma,\bar{\sigma})}\right)\dfrac{\partial F}{\partial \sigma_\theta}}{\dfrac{1}{H}\dfrac{\partial F}{\partial \boldsymbol{\sigma}}\cdot\boldsymbol{\sigma}}\left\{1-\exp\left[\frac{k}{H}\frac{\partial F}{\partial \boldsymbol{\sigma}}\cdot\boldsymbol{\sigma}(t_0-t)\right]\right\}r \quad \text{with } \mathbf{u}^{R} \xrightarrow{\infty} 0.$$

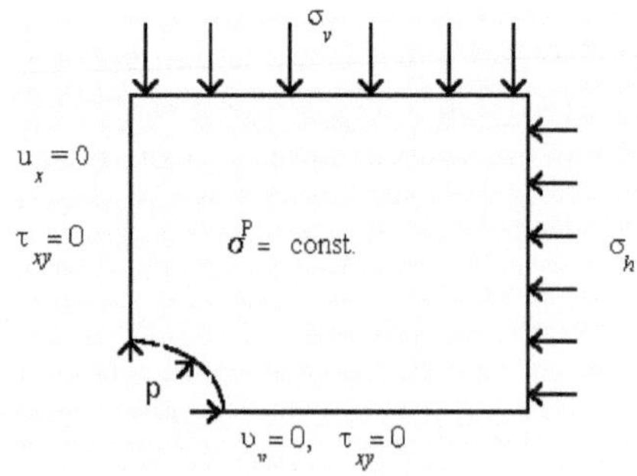


**Figure 5.** Domain used in numerical formulation

### 7.3. The numerical solution

In the case of numerical approach, the domain to be discretized and the boundary conditions, see Figure 5 and the FEM mesh, see Figure 6, are:

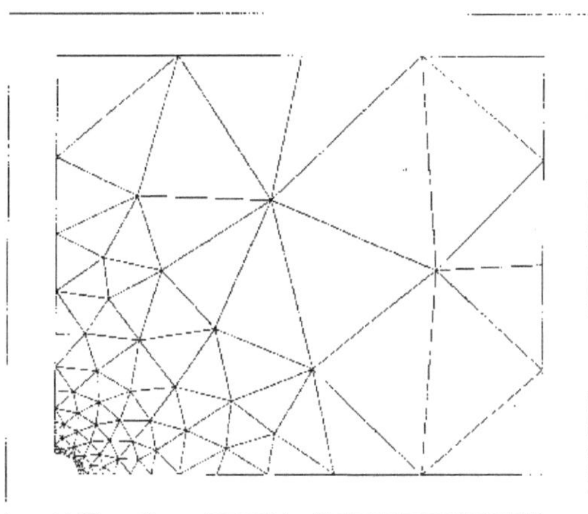

**Figure 6.** FEM mesh

In this case a quarter of the domain needs to be considered as the principal components of the primary stress $\sigma_h$ and $\sigma_v$ are assumed constant over the entire domain.

For this problem a mesh of 6 nodded triangles was used. The presence of the lining is simulated either by an internal pressure acting on the tunnel walls or by introducing some elements (8 nodded quadrilaterals) in contact with the tunnel walls, with the mechanical characteristics of the lining material (Figure 6).

In the following examples the lining was considered made of concrete with Young modulus E = 200000 kPa and Poisson coefficient $\upsilon = 0.3$.

There were considered the cases when $\sigma_h = \sigma_v = 2000$ kPa and the case $\sigma_h \neq \sigma_v$, namely $\sigma_h =$ 1500 kPa and $\sigma_v = 2000$ kPa. The tunnel radius is 1 m.

It is well known that the choice of the time step is quite important in viscoplasticity problems. This choice must be correlated with the viscosity parameter k as well, so that the value of $k\,dt$ to ensure a good convergence of the numerical scheme.

In the model under consideration of saturated sand it was observed that values for $k\,\Delta t$ which exceeds the magnitude order of $10^{-3}$ produce a divergence of the numeric calculation with the present method (the initial stress method and the initial stress method combined with different methods of acceleration). So, some stronger methods have to be implemented, in order to offer a faster convergence of the numerical calculus, e.g. the method of consistent tangential operator, backward Euler method, etc.).

The next Figures present other aspects of the numerical solution both in the case of an internal pressure acting on the tunnel wall, and for the case of concrete lining, that cannot be represented by the analytical solution.

### *7.3.1. Numerical solution for the case of internal pressure acting on the tunnel wall*

The following Figures represents the numerical solution for the case of a tunnel subjected to a hydrostatic primary stress of $\sigma_h = \sigma_v = 2000$ kPa and an internal pressure on the tunnel walls $p = 1000$ kPa which simulates the lining.

Figures 7a and 7b represent the evolution of the Euclidian norm of the viscoplastic strain $\left|\varepsilon_{vp}^{ij}\right|$ for the first time step and after 130 time steps with one time step 10000 s, respectively.

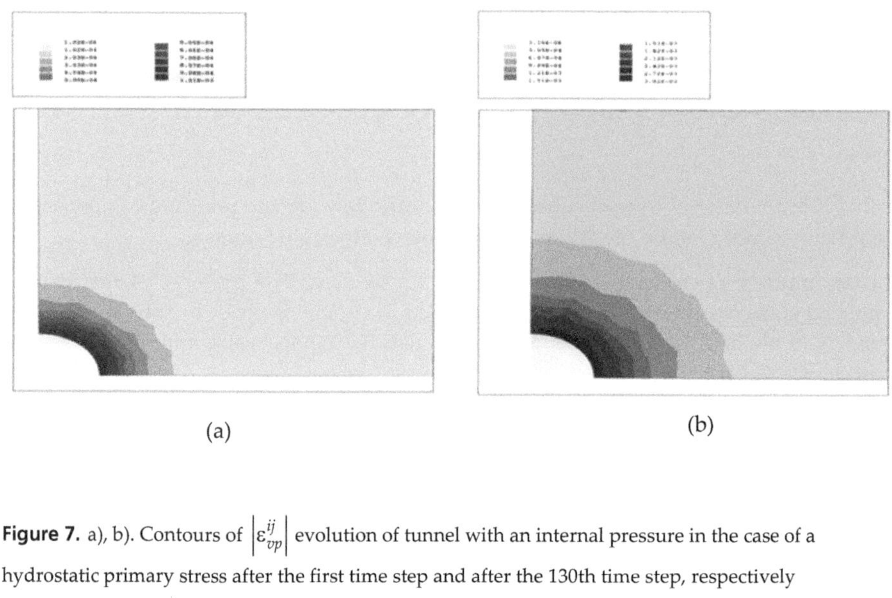

(a) (b)

**Figure 7.** a), b). Contours of $\left|\varepsilon_{vp}^{ij}\right|$ evolution of tunnel with an internal pressure in the case of a hydrostatic primary stress after the first time step and after the 130th time step, respectively

A tunnel driven in a rock mass with a non-hydrostatic primary stress of $\sigma_h = 1500$ kPa and $\sigma_v = 2000$ kPa, exhibits for instance, a development of a viscoplastic zone in the tunnel wall that increases from 1.13E-4 for $\left|\varepsilon_{vp}^{ij}\right|$ after the first time step Figure 8 to 5.13E-3 for $\left|\varepsilon_{vp}^{ij}\right|$ after 130 time steps.

The radial stress component $\sigma_{rr}$ presents a very slow variation in time and it is presented in Figure 9.

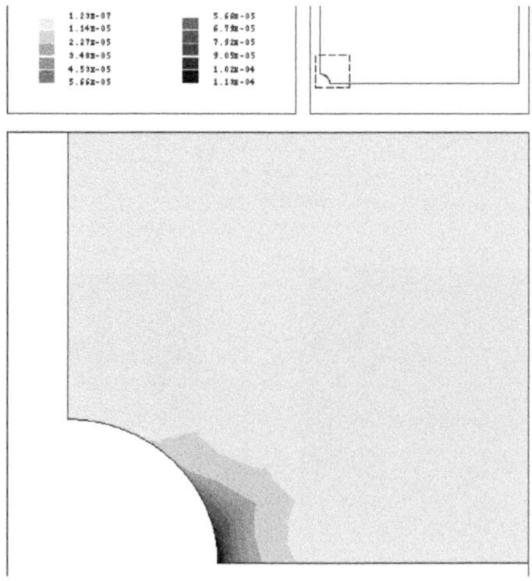


**Figure 8.** Contours of $\left|\varepsilon_{vp}^{ij}\right|$ after the first time step

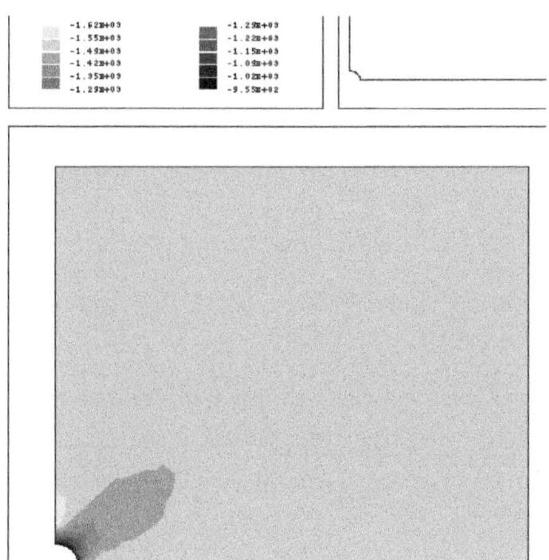


**Figure 9.** Contours of $\sigma_{rr}$ after the first time step

For the same boundary problem the evolution of the hoop stress $\sigma_{\theta\theta}$ is designed in Figures 10a for the first time step and 10b for the 130th time step.

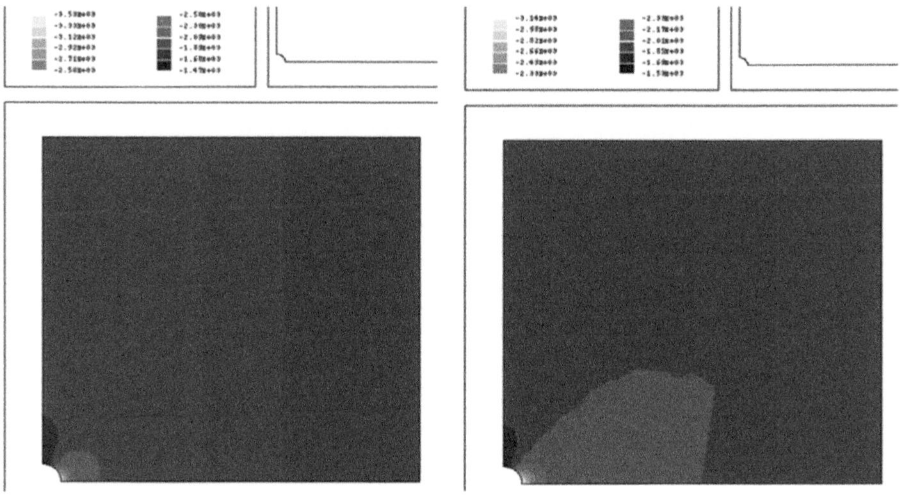

(a) (b)

**Figure 10.** a) Contours $\sigma_{\theta\theta}$ after the first time step, b) Contours $\sigma_{\theta\theta}$ after the 130th time step

### *7.3.2. Numerical solution for the case concrete lining*

In this subparagraph, the numerical solution for a concrete lining as a distinct material and mesh elements group is presented.

For a case of a lined tunnel by a concrete lining with Young modulus E = 200000 kPa and Poisson coefficient $\upsilon$ = 0.3 for the same type of rock mass subjected to a primary stress of $\sigma_h$ = 1500 kPa and $\sigma_v$ = 2000 kPa the diagrams of variation are presented as follows: the total displacement $u_{\text{total}} = \sqrt{u_x^2 + u_y^2}$ in Figures 11a and 11b, and the norm of the viscoplastic strain $\left|\varepsilon_{vp}^{ij}\right|$ in Figures 12a and 12b respectively.

The evolution of the equivalent stress $\overline{\sigma}$ and hoop stress $\sigma_{\theta\theta}$ is also studied. Both in the case of the equivalent stress and of the hoop stress $\sigma_{\theta\theta}$, an increasing of the smaller value zone from the roof, and a decreasing of the big value zone from the tunnel wall are observed.

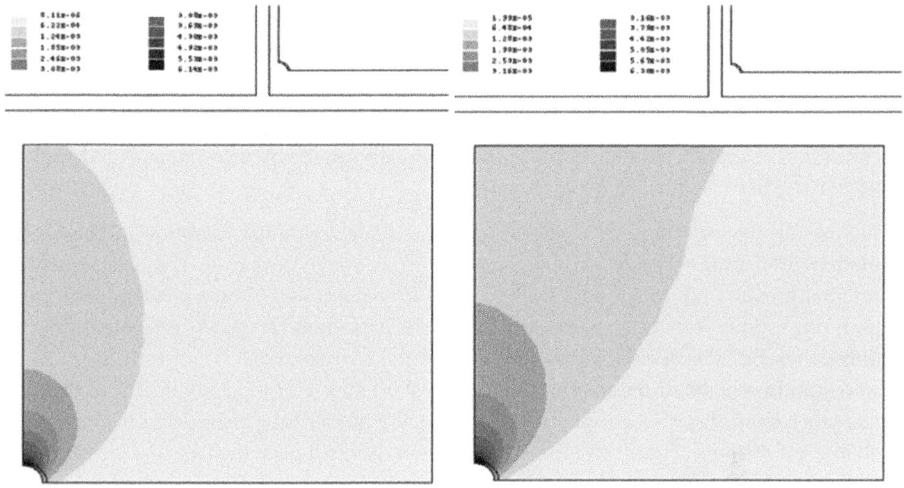

(a) (b)

**Figure 11.** a), b) Contours of $u_{total}$ evolution in a supported tunnel excavated in a rock mass with non-hydrostatic primary stress, after the first time step and after 130th time step, respectively

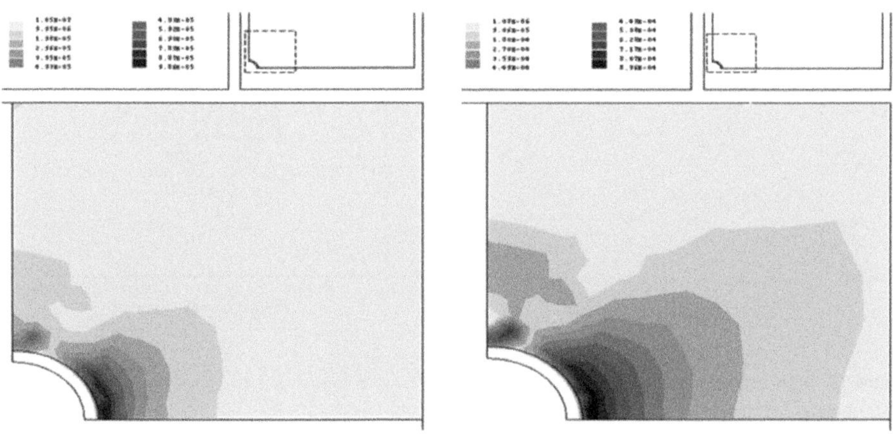

(a) (b)

**Figure 12.** a), b) Contours of $\left|\varepsilon_{vp}^{ij}\right|$ evolution in a supported tunnel excavated in a rock mass with non-hydrostatic primary stress, after the first time step and after the130th time step, respectively

### 7.4. Comparisons between the numerical and analytical solutions

Although the analytical solution was obtained under the assumptions of constant stress during the creep, it represents a good benchmark for the numerical calculation, supplying important information both quantitatively and qualitatively.

On the other hand, the numerical solution introduces certain truncation due to its discretisation errors. Nevertheless, the comparison between the two solutions could lead to some constructive conclusions for both solutions.

In Figure 13a the variation with respect to the distance in radial direction of the radial displacement in a tunnel excavated in condition of primary stress $\sigma_h = \sigma_v = 2000$ kPa, with an internal pressure on the tunnel wall $p = 1000$ kPa, at a chosen time, namely after 5 time steps ( one time step is considered 10000 ) is presented. A good agreement in the neighbourhood of the opening is observed, then the two solutions begin to differ, taking into account on one hand the assumptions of constant stress for the analytical solution and on the other hand the errors accumulated during the numerical process, due to the course mash at great distance. Figure 4b represents a comparison between the two displacements at two locations (the polar radius $r = a$ and the polar angle $\theta = 0$) at tunnel boundary, under the same condition as above, but with respect to time. One can observe that for small numbers of time steps, the numerical solution is superior, but for larger number of time steps (i.e. more than 10 time steps), the analytical solution becomes larger.

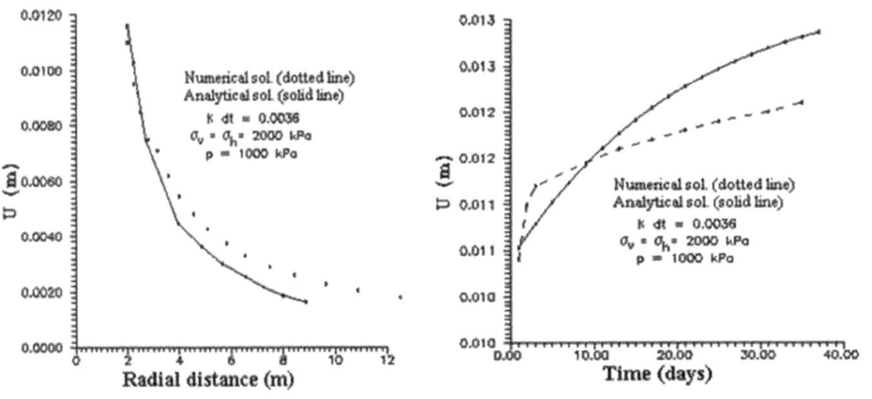


**Figure 13.** a), b). Tunnel outline total displacement with respect to distance and time, respectively, for the numerical and analytical solutions

In the case of the viscoplastic strain, the analytical solution predicts greater values than the numerical one, even from the beginning of time analysis. That can be observed in Figure 14, for the case of a tunnel excavated in a rock mass with a non-hydrostatic primary stress of $\sigma_h$ = 1500 kPa, $\sigma_v$ = 2000 kPa, and the internal pressure $p = 1000$ kPa.

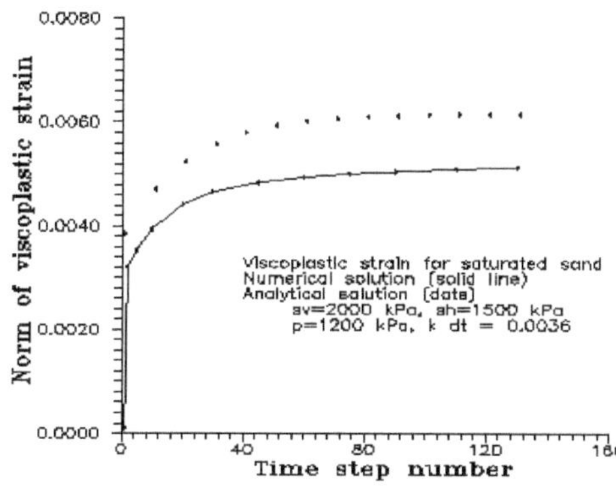


**Figure 14.** The time evolution of viscoplastic strain of tunnel walls for numerical and analytical solutions

## 7.5. Final remarks

Concerning the analytical solution for the problem under consideration the main remark is that it is difficult to obtain the analytical solution and it can be obtained using some (apparently) severe assumptions such as the same stress distribution is used for linear elastic and elasto-viscoplastic condition and has to be invariant in time. On contrary, the numerical solution via Finite Element does not impose such restriction and the results are exhibiting a much slower variation with time during creep. Thus this could attenuate the severity of the assumption of the analytical solution, assumption that could not be considered totally unrealistic.

Since the analytical solution is the exact solution of the governing equations under some restrictions, it can be used to benchmark the FE analysis before putting it in general applications.

Some remarks can also be drawn taking into account the features exhibited by the numerical solution.

Comparison between the numerical solutions obtained in the case of hydrostatic and nonhydrostatic primary stress respectively, shows that a much greater increase in time for the viscoplastic strain in the nonhydrostatic case, and a larger zone in the rock mass is exhibiting viscoplastic behaviour too.

A very slower decrease in time of the radial stress is observed, while the hoop stress has a tendency to increase in time.

In a more realistic case of the lining being simulated as a distinct zone of material with specific mechanical characteristics (concrete in the cases under consideration), instead of being applied as an internal pressure on the tunnel walls, some remarks could be made too. With the concrete lining modelled as elements, the displacement is greatly reduced and the viscoplastic strain is much less than the case of the internal pressure being applied on the tunnel walls. The amount accumulated in time is 5 times smaller than the other ones.

All these features are in good accordance with the practical observation, so, more involved boundary problems could be envisaged to be solved with this code.

## 8. Numerical solution of the three phases tunnel excavation and lining mounting problem

A finite element solution for the problem of a circular tunnel excavated in a homogeneous isotropic elasto-viscoplastic rock mass is presented. The numerical model consists of the successive phases of the excavation and support mounting, emphasizing the role of two important factors of the analysis, namely the time and the tunnel face influence and taking into account the 3D aspect of the problem.

The behavior of rock mass is considered viscoplastic, while the concrete lining is elastic. A possible behavior of sliding at the rock-support interface, which requests some additional contact elements of the mesh, is neglected. The rock mass is homogeneous, for the simplicity of data input, though introducing another rock/soil layer, with different mechanical characteristics represents no difficulty.

It is convenient to lead calculation to a 2D, plane strain or axisymmetrical, if it is possible, since it is less costly in data input and running time than a 3D analysis.

### 8.1. Formulation of the problem

Let us consider the following boundary problem: the rock mass is an infinite body in which a circular opening is made, assuming then that the underground opening is at a certain depth characterized by a hydrostatic primary (initial) stress, $\boldsymbol{\sigma}^P - PI$ where $P = \gamma h$, $h$ is the depth at which the tunnel is dug, $\gamma$ is the specific gravity of the rock and $I$ is the unity tensor.

As the tunnel possesses a circular geometry, the rock-mass and the lining mechanical properties do not depend on the angular coordinate $\theta$ and the far stress field in situ is hydrostatic (primary stress components $\sigma_v$, $\sigma_h$ are assumed equal), the problem is an *axisymmetrical* one in $Orz$ plane (Figure 15).

Consequently, the primary stress components $\sigma_v$, $\sigma_h$ are assumed equal. The boundary conditions are:

On $AB$ *i.e.* $z \in [z_A, z_B]$ and $r = a$ : $\sigma_{rr} = p$ and $\sigma_{rz} = 0$.

On $BC$ *i.e.* $z \in [z_B, z_C]$ and $r = a$ : $\sigma_{rr} = 0$ and $\sigma_{rz} = 0$.

On $CD$ *i.e.* $z = z_C = z_D$ and $r \in [0, a]$: $\sigma_{rr} = 0$ and $\sigma_{rz} = 0$.

On $DE$ *i.e.* $z \in [z_D, z_E]$ and $r = 0$ : $u_r = 0$ and $\tau_{rz} = 0$.

On $EF$ *i.e.* $z = z_E = z_F$ and $r \in [0, r_F]$: $\sigma_{zz} = \sigma_v$ and $\sigma_{rz} = 0$.

On $FG$ *i.e.* $z = z_F = z_G$ and $r = r_F$ : $\sigma_{rr} = \sigma_v$ and $\sigma_{rz} = 0$.

On $GA$ *i.e.* $z = z_A$ and $r \in [r_A, r_G]$: $\sigma_{zz} = \sigma_v$ and $\sigma_{rz} = 0$.

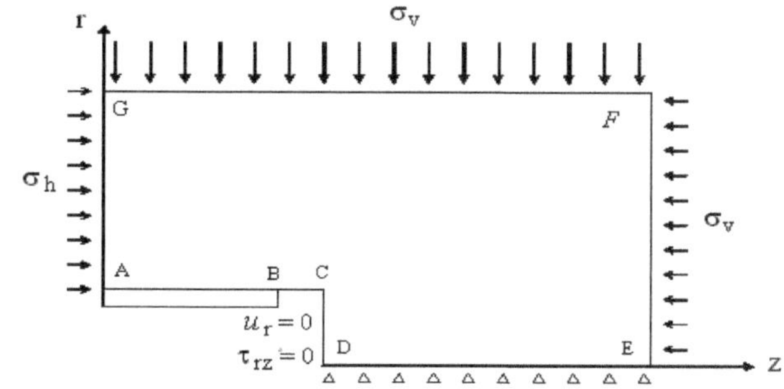


**Figure 15.** Domain and boundary conditions for the problem study in *Orz* plane along the tunnel axis.

Cristescu's elasto-viscoplastic constitutive law is used for the rock-mass and elastic behavior for the concrete lining.

An important factor of the analysis is the time effect and it is involved by two different aspects: the rheological behavior of the rock mass on and the excavation history. Moreover, the tunnel support mounting determines a *problem of interaction*, getting thus a more involved calculus. Another important factor of the analysis of ground-support interaction during the tunnel excavation is the face tunnel. For instance, since the behavior of the rock-mass is viscplastic, rock pressure on the lining increases in time. On the other hand, closer the lining is installed to the tunnel face, more the pressure at the rock-lining interface increases with the advancing of the tunnel face.

The state of stress and strain around a lined tunnel depends explicitly on:

- The mechanical and geometrical characteristic of the rock-mass and the support;
- The excavation conditions, such as excavation rate, generally the excavation phases;
- The support mounting conditions, namely the support mounting time after the excavation and the distance between the lining and the tunnel face.

Concerning the geometry and the loading, the successive phases of the tunnel excavation and support mounting is a *three-dimensional problem*. However, there are certain cases when the problem may be simplified assuming the hypothesis that close to the tunnel face, on the tunnel walls, $r=a$, the decompression of the primary stress component is occurring gradually.

The calculation of tunnel excavation and lining mounting is a complex problem. On one hand, the excavation is a three-dimensional problem that imposes taking into consideration the tunnel face influence that means a gradual decompression of the primary stress $\sigma_h$ of the rock mass on the opening walls. On the other hand, the support mounting determines the problem to be a massive-lining interaction one, mainly based on the behavior of the rock-lining interaction.

The lining is often installed quickly enough after the excavation and at a relatively small distance from the tunnel face that induces a complex combination of the effects mentioned above.

Consequently, it is very important to take into consideration both time and tunnel face effect in the excavation and lining mounting problem. Essentially in this study is the calculation of lining pressure and the convergence of the tunnel walls.

## 8.2. The numerical solution of the successive phases tunnel excavation

The domain discretization of the boundary value problem is presented in Figure 16. As usual, in the tunnel surface region the mesh must be quite refine, while elsewhere a minimum possible number of elements is considered. It is used an 8 nodded-quadrilateral mesh, at least 2 layers of quadrilateral elements in the concrete lining.

We consider the tunnel radius $a$ = 1.2 m and the lining thickness 0.2 m. The depth at which the tunnel is excavated is 273.5 m. One phase duration is 12 hours and respectively 1 day.

For the rock mass the *Borod coal* is used, whose material constants were presented previously. For the concrete the following material constants are used: Young modulus $E$ = 20000 MPa, Poisson coefficient $\nu$ = 0.3 and the volumetric weight $\gamma$ = 0.02 MN/m$^2$.

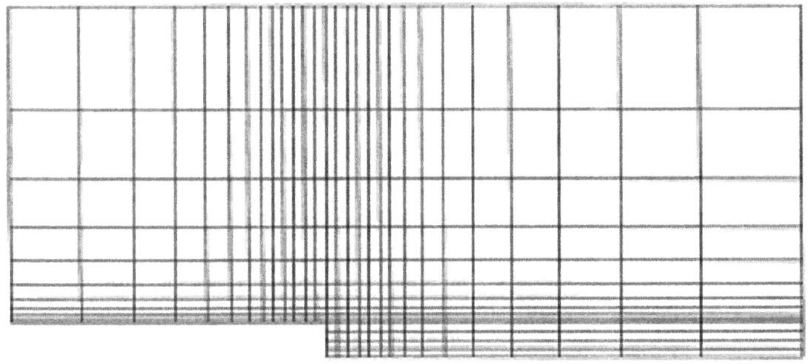

**Figure 16.** The domain discretization of the boundary value problem.

It is used a numbering of elements group as they are activated/deactivated in the excavation and lining installing processes, as follows:

- *1st group* is the rock-mass considered infinite
- *2nd group* corresponds to the already mounted lining
- *3rd group* is in the first phase the rock-mass that is going to be excavated and in the second phase is replaced by the concrete
- *4th group* is in the first phase the rock-mass that is going to be excavated
- *5th group* is in the first two phases the rock-mass that is going to be excavated in the third phase and eventually replaced by concrete in a possible fourth phase
- *6th group* is in the first two phases the rock-mass that is going to be excavated in the third phase

Numerical model concerns three successive phases' tunnel excavation and lining mounting. Let us detail the progression of phases of the example:

- **Phase 1**: tunnel excavation and calculation of strain, stress, damage parameter, without lining mounting on new excavated zone (Excavation of element group 3 and 4). Structure elements corresponding to the support are inactive (null mechanical characteristics). Output storage for the following phase (phase 2) is performed as well.
- **Phase 2**: lining installing at a certain given time $T_0$, on the unexcavated zone in phase 1. Displacement and stress initialization starting from the previous phase output storage and output storage for the following phase (phase 3) are performed as well.
- **Phase 3**: tunnel face advancing on a distance of unit radius, namely 1.2 m. Realization of a new excavation is simulated by inactivation of element group 5 and 6, considering null mechanical characteristic. Displacement and stress initialization with the 2nd phase state is performed as well.

In the following, we present some important results of the calculation concerning, for instance, the normal stress, the equivalent stress or the damage parameter $d_f$.

In Figures 17a, b, c isovalues zones for the damage parameter corresponding to the three phases are presented, respectively. It is observed that in tunnel face zone the damage is maximum (white area).

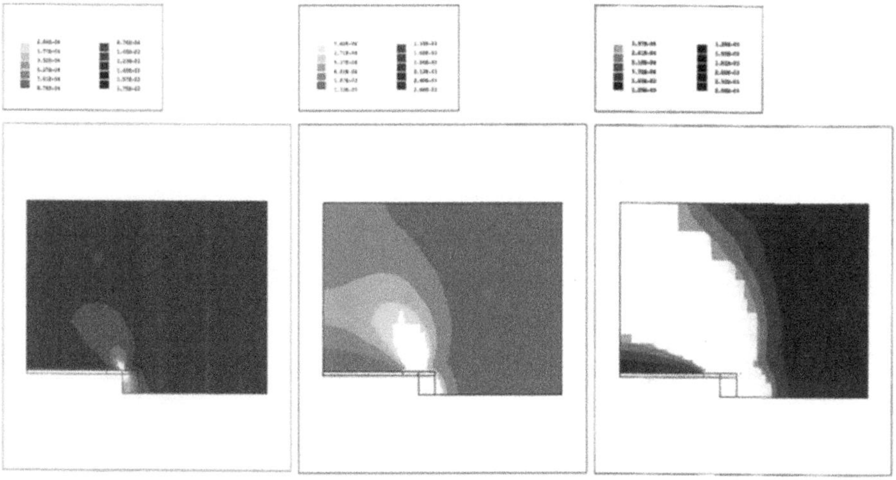

(a) (b) (c)

**Figure 17.** a), b), c) Isovalues zones for damage parameter for the first, second, third phase, respectively

Isovalues zones for normal stress corresponding to the three phases are presented in Figures 18a, 18b, 18c. Great concentrations are observed in tunnel face zone (white area).

Figures 19a, b, c present isovalues zones for equivalent stress corresponding to the three phases, observing small tractions in the second, respectively the third phase. That signifies the possibility of fracture by exceeding the traction resistance, since it is known that it is very low for the rocks.

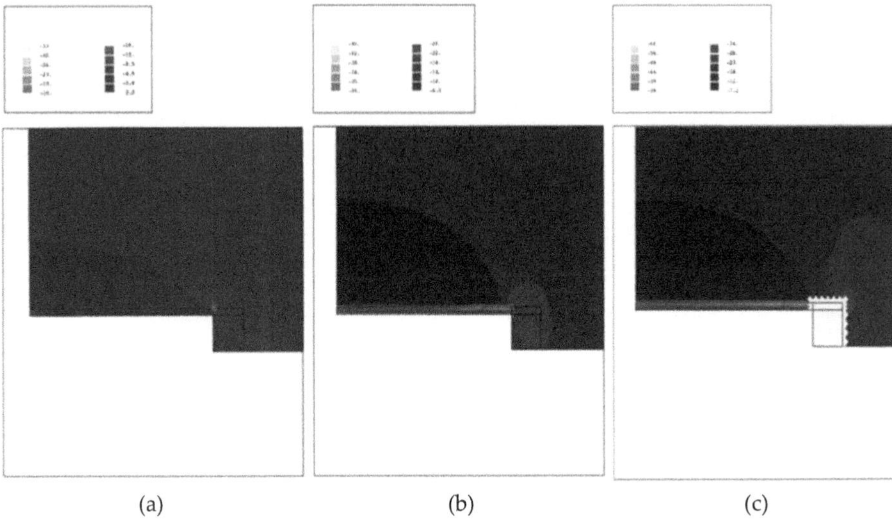

**Figure 18.** a), b), c) Isovalues zones for normal stress for the first, second, respectively third phase.

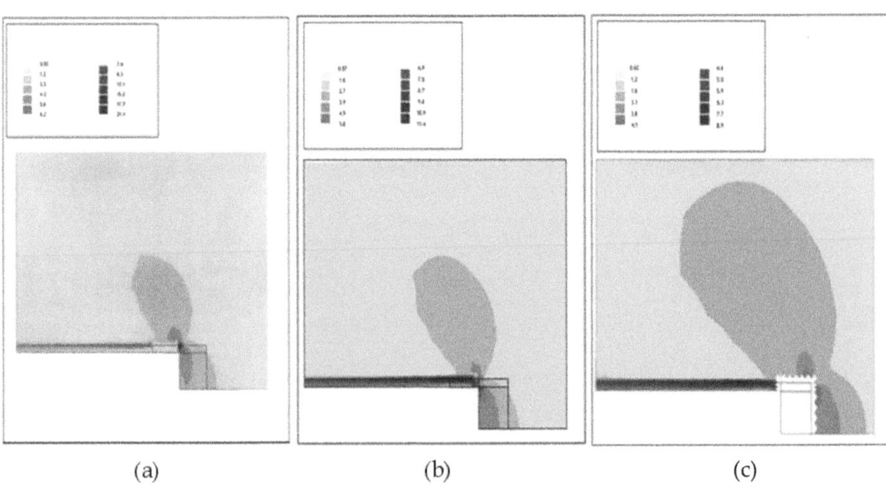

**Figure 19.** a), b), c). Isovalues zones for equivalent stress the first, second, respectively third phase.

## 8.3. Parametric analysis: Influence of tunnel depth, lining stiffness and lining mounting time

### *8.3.1. Support rigidity influence*

The numerical solution is used to analyze the influence of different parameters that intervene in the excavation process. In this chapter we present the influence of lining rigidity and depth at which the tunnel is excavated.

The conclusion is that the displacement and the damage through dilatancy, as it was incorporated in the constitutive law, are decreasing with the increasing of the lining rigidity and increasing with depth increasing. In this paragraph we exhibit these features of the solution by several Figures and observations.

The support rigidity can be calculated by the following formula [1], [20]:

$$q = \frac{E(b^2 - c^2)}{(1+\upsilon)\left[(1-2\upsilon)b^2 + c^2\right]}$$

The previous calculation was performed for a value of Young modulus $E$=20000 Mpa, Poisson coefficient $\nu$ = 0.3, external radius $b$ = 1.2 m and internal radius $c$ = 1 m. We perform the calculation for $E$=2500 Mpa, too. Figures 20a, b present the isovalues zones of damage parameter $d$ defined in relation (2) and show that the dilatancy, as it was incorporated in the constitutive law, is decreasing with the increasing of the lining rigidity. The same results are obtained by author by analytical means in [20], [18].

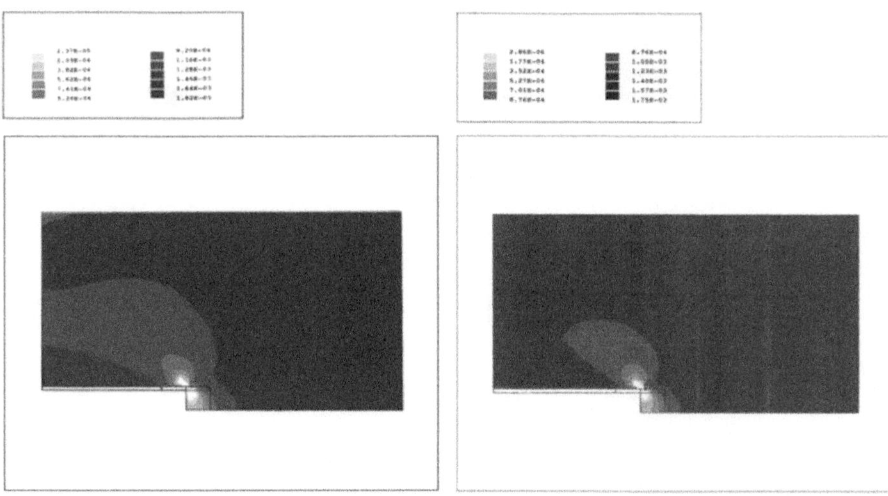

(a) (b)

**Figure 20.** a), b). Isovalues zones for damage parameter for E=2500 Mpa, E=20000 Mpa, respectively

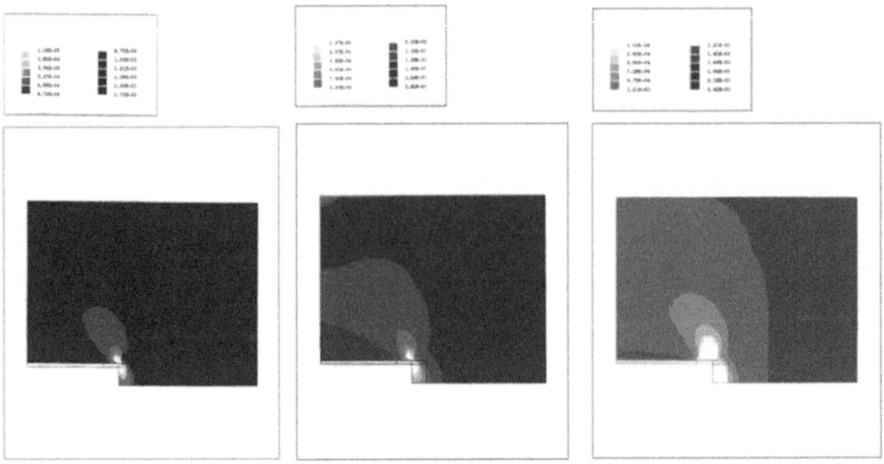

(a) (b) (c)

**Figure 21.** a), b), c). Isovalues zones of dilatancy for tunnel depth of 150 m, 273.5 m, 850 m, respectively

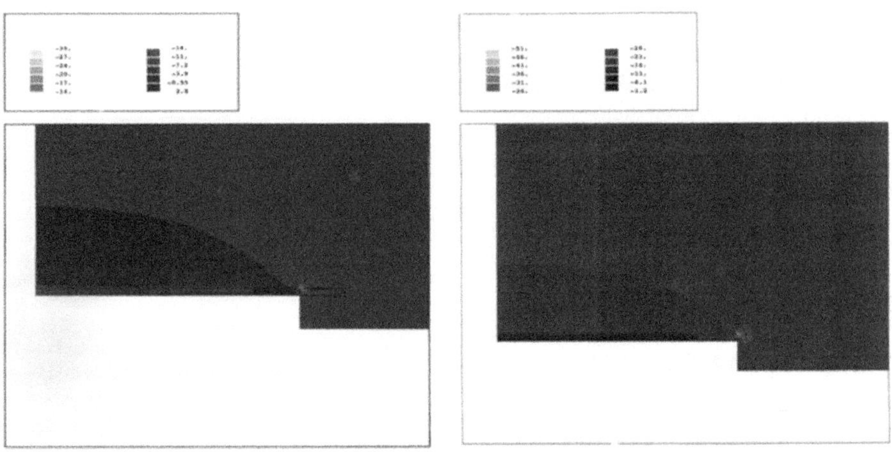

(a) (b)

**Figure 22.** a), b). Isovalues zones for the normal stress for tunnel depth of 150 m, 850 m, respectively

## *8.3.2 Tunnel depth influence*

To analyze the depth influence on the processes we perform the calculation for different values of the depth at which the tunnel is excavated. At the previous calculation performed initially for a depth of 273.5 m, we add another two calculations for 150 m and 850 m depth respectively. The conclusion is that both displacement and damage increase with depth such that: the maximum radial displacement is 6.65 cm at a depth of 150 m, 6.83 cm at a depth of 273.5 m and 7.31 cm at a depth of 850 m. Concerning the damage, for instance, let us observe the Figures 21a, b, c presenting the isovalues zones of the damage parameter for a depth of 150 m, 273.5 and 850 m, respectively. The damage is more pronounced at a depth of 850 m (the white zone is more extended).

Figures 22a, b present isovalues zones of the normal stress $\sigma_{rr}$ for instance, for the same depths of 150 m and 850 m, respectively. It is observed that the normal stress decreases with the depth increase in compression and consequently, it appears the risk of traction at smaller depth, which may induces rock or lining fracture.

## 9. Conclusions

The complex problem of a lined tunnel excavation in a viscoplastic rock mass is approached in this chapter both numerically by a FE code proposed by the author and by an analytical approach too. A good agreement between the numerical and the analytical solutions is obtained.

Both solutions are studied then for further specific features. All these features are in good agrrement with the practical observation, so, more involved boundary problems could be developed with this code and further improvements for the analytical solution.

A special study is devoted to the finite element solution for the simulation of a tunnel excavation with successive tunnel face advancing and the lining mounting. Due to the symmetry of the geometry and loadings, the problem is treated as an axisymmetrical one with an additional emphasis of the three-dimensional aspect of the problem, namely the tunnel face advancing and its proximity influence. So, the approach of a tunnel calculation in two-dimensional analysis along the tunnel axes, simulating thus the three-dimensional aspect of the problem, is more realistic than the classical cross section analysis and obviously less costly than an actual three-dimensional analysis. The parametric analysis performed in this study by the numerical solution is in good accordance with the results obtained by the author by analytical means [19], [20].

## Author details

Simona Roatesi
*Military Technical Academy, Bucharest, Romania*

## 10. References

[1] Cristescu N (1989) *Rock rheology*, Kluwer Academic Publishers, Dordrecht, Holland.
[2] Simionescu O (1999) *Mathematical methods in underground structure termomechanics*, Romanian Academy Publ., Bucharest.
[3] Bernaud D (1991) *Tunnels profonds dans les milieux viscoplastique: approches expérimentale et numérique*. Thése. École National des Ponts et Chaussées, France.

[4] Panet M (1974) *Stabilité et soutenement des tunnels. La méc.des roches appl. aux ouvrages de génie civil*,Ch.IX,ENPC.

[5] Sulem J, Panet M, Guenot A (1987) *An analytical solution for time-dependent displacements in a circular tunnel*, Int.J. Rock Mech. Min. Sci.& Geomech. Abstr., Vol.24, No.3, pp.155–164.

[6] Zienkiewicz O C, Cormeau I C (1974) *Visco-plasticity - plasticity and creep in elastic solids - A unified numerical solution approach*, Int.J.Num.Meth.Engng.,8,pp. 821-845.

[7] Cormeau, I.C., Numerical stability in quasi-static elasto/visco-plasticity (1975) *Int. J. Num. Meth.Engng.*, Vol.9, pp.109–127.

[8] Ionescu I R, Sofonea M (1993) *Functional and numerical methods in viscoplasticity*, Oxford Univ. Press Oxford.

[9] Chan A H, Ou J (2008) *Three-Dimensional Numerical Analysis of a Dynamic Structure, Saturated Soil and Pore Fluid Interaction Problem*, Trends in Eng. Comput. Techn., pp. 335-353.

[10] Zienkiewicz O C, Chan A H C, Pastor M, Schrefler B A, Shiomi T (1999) *Computational Geomechanics with Special Refernces to Earthquake Engineering*, J. Wiley & Sons, Chicester, 1999.

[11] Gioda G (1993) *Finite element analysis of time dependent effects in tunnels*, Advanced school on visco-plastic behaviour of geomaterials, CISM, Udine, Oct.

[12] Lauro F, Bennani B, Drazetic P, Oudin J, Ni X (1997) *Ductile damage and fracture finite element modelling of elasto-viscoplastic voided materials*, Computational Materials Science, Vol. 7, Issue 3, pp. 295-307.

[13] Lin R C, Brocks W (2006) *On a finite-strain viscoplastic law coupled with anisotropic damage: theoretical formulations and numerical applications*, Archive of Appl.Mech.,Vol.75, 6-7, 315-325.

[14] Augarde C E, Burd H J (2001) *3D FE analysis of lined tunnels*, Int. J. Num. Analyt. Meth. in Geomechanics, 25: 243–262.

[15] Komiya K, Soga K, Akagi H, Hagiwara T, Bolton M D (1999) *FE, modeling of excavation and advancement process of a shield tunnelling machine*, Soil &Foundations, Vol. 39, No.3, 37-52.

[16] Leem J, Kemeny J M (1999) *Finite element micromechanical-based model for hydro-mechanical coupling*, The 37th U.S. Symp. on Rock Mech. (USRMS), June 7-9, Balkema, Vail, CO.

[17] Wang J, Wang W (2009) *Multi-field Coupled Model&Num.Sim. During Excav.Tunnel* Modeling, WMSO08,pp.334–337.

[18] Roateşi S, Ariciuc M (2005), *Study on depth dependency in tunnel calculation*, Scientifical Bulletin, UTCB, 1, pp.16-30.

[19] Roateşi S (1996) *The tunnel face influence in the analysis of a circular tunnel with a time-dependent behaviour*, Rev.Roum.Sci.Tech. Mec.Appl., no 3-4, pp. 247-263.

[20] Roateşi S (2005) *Mathematical modeling and FEM problems in tunnel calculation*, Romanian Academy Publisher, Bucharest.

[21] CESAR – LCPC *Documentation*, Version 3.0.x, Publication LCPC, 1992.

[22] Lade P V, Nelson R B, Ito Y M (1987) *Nonassociated flow and stability of granular materials*, J.Engng.Mech., 113, pp. 1302-1318.

# Chapter 14

# Finite Element Modelling of the Dynamic Behaviour of Tubular Footbridges

José Guilherme Santos da Silva, Ana Cristina Castro Fontenla Sieira, Gilvan Lunz Debona, Pedro Colmar Gonçalves da Silva Vellasco and Luciano Rodrigues Ornelas de Lima

Additional information is available at the end of the chapter

## 1. Introduction

Tubular hollow sections are increasingly used in off-shore structures, highway bridges, pedestrian foot-bridges, large-span roofs and multi-storey buildings due to their excellent properties and the associated advances in fabrication technology. The intensive use of tubular structural elements in Brazil, such as the example depicted in Figure 1, mainly due to its associated aesthetical and structural advantages, led designers to be focused on their technologic and design issues.

Nowadays in Brazil, there is still a lack of code that deals specifically with tubular design. This fact induces designers to use other international tubular design codes. Consequently, their design methods accuracy plays a fundamental role when economical and safety points of view are considered. Additionally, recent tubular joint studies indicate that further research is needed, especially for particular geometries. This is even more significant for some failure modes where the collapse load predictions lead to unsafe or uneconomical solutions.

Steel and composite tubular footbridges are currently subjected to dynamic actions with variable magnitudes due to the pedestrian crossing on the concrete deck [1-4]. These dynamic actions can generate the initiation of fractures or even their propagation in the structure. Depending on the magnitude and intensity, these adverse effects can compromise the structural system response and the reliability which may also lead to a reduction of the expected footbridge service life.

Generally, fatigue assessment procedures are usually based on S-N curves which relate a nominal or geometric stress range S to the corresponding number N of load cycles to fatigue failure. In this situation fatigue assessment refers to the nominal stress range $\Delta\sigma$ in a tubular structural member.

**Figure 1.** Example of a tubular steel pedestrian footbridge in Rio de Janeiro/RJ, Brazil.

The fatigue resistance is given according to a classification catalogue in the form of standardized S-N curves. Structural details classified in this catalogue, see e.g. Eurocode 3 Part 1.9 [5], correspond specifically to a situation of stress range, direction, crack position, detail dimension and weld quality which had been characteristic for the tests on which the classification is based [6-7].

The use of circular hollow section members as part of the structure of pedestrian footbridges is a relatively new constructional concept. During the last couple years several steel-concrete composite footbridges had been constructed in Brazil, as illustrated in Figure 1.

The typical cross-section of this type of pedestrian footbridge generally consists of a tubular spatial truss girder carrying the concrete deck slab, as presented in Figure 1. The deck slab is connected directly to the steel structure by either shear studs, concrete dowels or in some cases where no top chord exists. At the bottom chord of the tubular space truss four brace members have to be connected to the continuous bottom chord. This type of joint is usually named K-joint, as depicted in Figure 2 [6].

Steel and composite tubular footbridges can be subjected to the material imperfections of its structural elements, such as mechanical and metallurgic discontinuities. Such defects lead to cracking in the-se structural elements. When these elements are subjected to dynamic actions, the fatigue phenomenon occurs and produces stress concentrations and possible fractures. These fractures are directly responsible for reducing the local or global footbridge stabilities or even its life service [7].

On the other hand, the structural engineers experience and knowledge allied by the use newly developed materials and technologies have produced tubular steel and composite (steel-concrete) footbridges with daring structures. This fact have generated very slender tubular steel and composite pedestrian footbridges and consequently changed the serviceability and ultimate limit states associated to their design. A direct consequence of this design trend is a considerable increase of structural vibrations [1-4, 8-11].

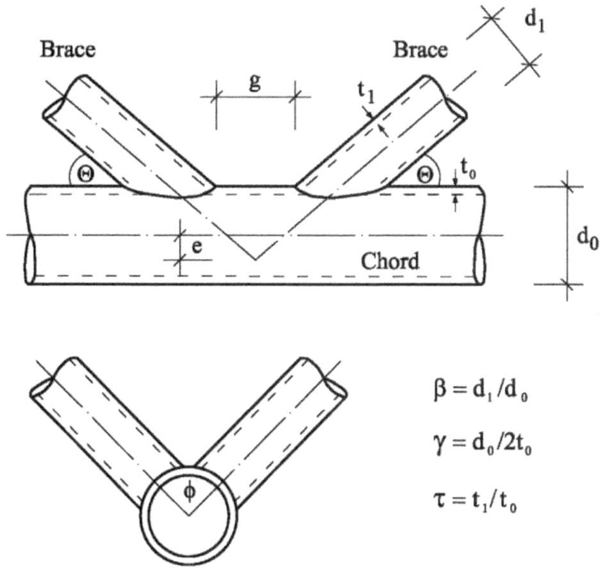


**Figure 2.** Typical multiplanar K-joint with notations.

Considering all aspects mentioned before, the main objective of this investigation is to present the finite element modelling of the dynamic behaviour of tubular composite (steel-concrete) footbridges submitted to human walking vibration. Based on the results obtained in this study, a fatigue assessment will be performed, in order to evaluate the tubular footbridges service life. Further research in this area is currently being carried out.

The investigated structural model was based on a tubular composite (steel-concrete) footbridge, spanning 82.5 m. The structure is composed by three spans (32.5 m, 17.5 m and 20.0 m, respectively) and two overhangs (7.50 m and 5.0 m, respectively). The structural system consists of tubular steel sections and a concrete slab and is currently used for pedestrian crossing [1-2].

The proposed computational model adopted the usual mesh refinement techniques present in finite element method simulations, based on the ANSYS program [13]. The finite element model has been developed and validated with the experimental results. This numerical model enabled a complete dynamic evaluation of the investigated tubular footbridge especially in terms of human comfort and its associated vibration serviceability limit states.

This investigation is carried out based on correlations between the experimental results related to the footbridge dynamic response and those obtained with finite element models [1-2]. The structural system dynamic response, in terms of peak accelerations, was obtained and compared to the limiting values proposed by several authors and design standards [9,14].

The peak acceleration values found in the present investigation indicated that the analysed tubular footbridge presented problems related with human comfort. Hence it was detected that this type of structure can reach high vibration levels that can compromise the footbridge user's comfort and especially its safety.

## 2. Human walking modelling

Human loads comprise a large portion of the acting live loads in offices and residential building floors. In general, the human live loads are classified into two broad categories: in situ and moving. Periodic jumping due to music, sudden standing of a crowd, and random in-place movements are examples of in situ activities. Walking, marching, and running are examples of moving activities. As the main purpose of footbridges is the pedestrian's crossing, they must be safe and do not cause discomfort to users [1-4].

On the other hand, human activities produce dynamic forces and their associate vibration levels should not disturb or alarm their users. Therefore, this investigation describes four different load models developed to incorporate the dynamic effects induced by people walking on the footbridges dynamic response. It must be emphasized that the geometry of the human body walking is an organized leg motion that cause an ascent and descending movement of the effective body mass at each step [1-4].

The human body mass accelerations are associated to floor reactions, and are approximately periodic to the step frequency. The two feet produce this type of load, as function of the static parcel associated to the individual weight and three or four harmonic load components. These harmonics appear due to the interaction between the increasing loads, represented by one foot, and the simultaneous unload of the other foot [1-4].

However, it is also necessary to incorporate several other parameters in the human walking representation, like step distance and speed. These variables are related to the step frequency and are depicted in Table 1 [12]. Table 1 presents a detailed description of the excitation frequency values, dynamic coefficients, as well as the phase angles to be employed in the mathematical representation of the four dynamic loading models implemented and used in the present investigation.

| Activity | Velocity (m/s) | Step Distance (m) | Step Frequency (Hz) |
|---|---|---|---|
| Slow Walking | 1.1 | 0.60 | 1.7 |
| Normal Walking | 1.5 | 0.75 | 2.0 |
| Fast Walking | 2.2 | 1.00 | 2.3 |

**Table 1.** Characteristics of the human walking [12].

## 2.1. Load model I (LM-I)

This walking load model can be represented by the load static parcel, related to the individual weight, and a combination of harmonic forces whose frequencies are multiples or harmonics of the basic frequency of the force repetition, e.g. step frequency, $f_s$, for human activities. This load model considers a space and temporal variation of the dynamic action over the structure and the time-dependent repeated force can be represented by the Fourier series, see Equation (1).

$$F(t) = P[1 + \sum \alpha_i \cos(2\pi i f_s t + \phi_i)] \quad (1)$$

Where:

F(t) : dynamic load;
P : person's weight (800 N [1-4]);
$\alpha_i$ : dynamic coefficient for the harmonic force;
i : harmonic multiple (i = 1,2,3…,n);
$f_s$ : walking step frequency;
$\phi$ : harmonic phase angle;
t : time.

In this load model, five harmonics were considered to represent the dynamic load associated to human walking [12]. Table 2 shows the dynamic coefficients and phase angles used in this load model. Figure 3 presents a dynamic loading function for one person walking with step frequency equal to 2 Hz.

| Harmonic i | Dynamic Coefficients $\alpha_i$ | Phase Angles $\phi_i$ |
|---|---|---|
| 1 | 0.37 | 0 |
| 2 | 0.10 | $\pi/2$ |
| 3 | 0.12 | $\pi/2$ |
| 4 | 0.04 | $\pi/2$ |
| 5 | 0.08 | $\pi/2$ |

**Table 2.** Dynamic coefficients and phase angles [12].

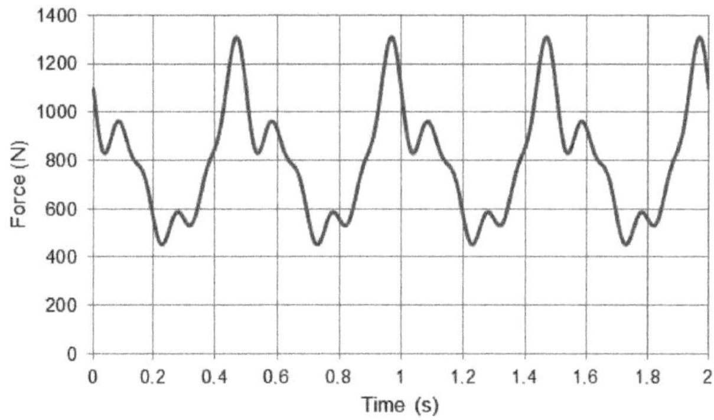


**Figure 3.** LM-I: dynamic load function for one person walking ($f_s$ = 2.0 Hz).

## 2.2. Load model II (LM-II)

In this load model, the time-dependent repeated force also can be represented by the Fourier series, as expressed in Equation (1) and four harmonics were considered to represent the dynamic action associated to human walking [9]. This model also considers a space and temporal variation of the dynamic action over the structural system. Table 3 shows the dynamic coefficients and phase angles used in this model. Figure 4 presents a dynamic loading function for one person walking with step frequency equal to 2 Hz.

| Harmonic i | Dynamic Coefficients $\alpha_i$ | Phase Angles $\phi_i$ |
|---|---|---|
| 1 | 0.50 | 0 |
| 2 | 0.20 | $\pi/2$ |
| 3 | 0.10 | $\pi$ |
| 4 | 0.05 | $3\pi/2$ |

**Table 3.** Dynamic coefficients and phase angles [9].

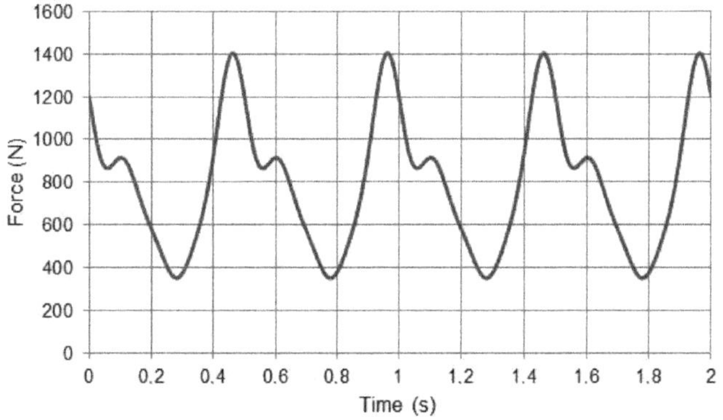


**Figure 4.** LM-II: dynamic load function for one person walking ($f_s$ = 2.0 Hz).

## 2.3. Load model III (LM-III)

In this case a general expression is used to represent the excitation produced by an individual walking throughout time. These loads are produced with both feet, as function of a static part associated to the individual weight and three harmonics were considered to represent the dynamic action related to human walking [15], as illustrated in Equation (2). This dynamic loading model considers a space and temporal variation of the dynamic action over the footbridge.

$$F(t) = P + \Delta P_1 \sin(2\pi f_s t - \phi_1) + \Delta P_2 \sin(4\pi f_s t - \phi_2) + \Delta P_3 \sin(6\pi f_s t - \phi_3) \quad (2)$$

Where:

F(t) : dynamic load;

P : person's weight (800 [1-4]);

$f_s$ : walking step frequency;

$\phi_i$ : harmonic phase angle;
t : time.

In Equation (2), the magnitudes ΔP1, ΔP2 and ΔP3 are associated with harmonic amplitudes. The first harmonic amplitude, ΔP1, is equal to 0.4P for $f_s$ equal to 2.0 Hz and 0.5P for $f_s$ equal to 2.4Hz. A simple interpolation between these two values can be used in intermediate cases. The second and third harmonic amplitudes, ΔP2 e ΔP3, were assumed to be equal to 0.1P for $f_s$ equal to 2.0Hz [15].

The phase angles $\phi_2$ and $\phi_3$ depend on various other factors and should represent the most favourable used load combinations. In the present study the phase angles $\phi_2$ and $\phi_3$ were assumed to be equal to $\pi/2$ and phase angle $\phi_1$ was assumed to be equal to zero [15]. Figure 5 presents a dynamic loading function for one person walking with step frequency equal to 2 Hz.

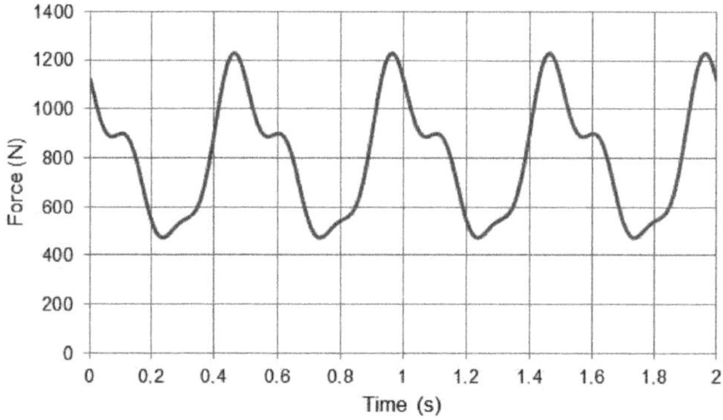


**Figure 5.** LM-III: dynamic load function for one person walking ($f_s$ = 2.0 Hz).

## 2.4. Load model IV (LM-IV)

The fourth walking load model considered the same idea of the previous models. The main difference was the incorporation of the human heel effect in this particular load representation with the aid of Equations (3) to (6). The mathematical model behind this strategy was proposed by Varela [8] as well as a numerical approach to evaluate the floor structure reaction, as presented in Figure 6.

The proposed mathematical model, see Equations (3) to (6), used to represent the dynamic actions produced by people walking on floor slabs is not simply a Fourier series. This is due to the fact that the mentioned equations also incorporate the heel impact effect [8]. This loading model also considers a space and temporal variation of the dynamic action over the structure and is evaluated considering four harmonics.

Additionally, Load Model IV (LM-IV) also incorporates the transient effect due to the human heel impact [8]. The present investigation used a heel impact factor equal to 1.12 ($f_{mi}$

= 1.12). However, it must be emphasized that this value can vary substantially from person-to-person. Figure 6 illustrates the dynamical load function for an individual walking at 2 Hz, based on Equations (3) to (6) and Tables 1 and 3 [9,12].

## 3. Investigated structural model

The structural model consists of tubular steel sections and a 100 mm concrete slab and is currently submitted to human walking loads [1-2]. The structure was based on a tubular composite (steel-concrete) footbridge, spanning 82.5 m. The structure is com-posed by three spans (32.5 m, 17.5 m and 20.0 m, respectively) and two overhangs (7.50 m and 5.0 m, respectively), as illustrated in Figures 7 and 8.

The steel sections used were welded wide flanges (WWF) made with a 300 MPa yield stress steel grade. A Young's modulus equal to 2.05 x $10^5$ MPa was adopted for the tubular footbridge steel beams and columns. The concrete slab has a 20 MPa specified compression strength and a 2.13 x $10^4$ MPa Young's Modulus.

$$F(t)=\begin{cases}\left(\dfrac{f_{mi}F_m-P}{0.04T_p}\right)t+P & \text{if } 0\le t<0.04T_p\\[2ex] f_{mi}F_m\left[\dfrac{C_1\left(t-0.04T_p\right)}{0.02T_p}+1\right] & \text{if } 0.04T_p\le t<0.06T_p\\[2ex] F_{\mathrm{m}} & \text{if } 0.06T_p\le t<0.15T_p\\[2ex] P+\sum_{i=1}^{nh}P\,\alpha\,sen\left[2\pi i f_p\left(t+0.1T_p\right)+\phi_i\right] & \text{if } 0.15T_p\le t<0.90T_p\\[2ex] 10\left(P-C_2\right)\left(\dfrac{t}{T_p}-1\right)+P & \text{if } 0.90T_p\le t<T_p\end{cases} \tag{3}$$

$$F_m=P\left(1+\sum_{i=1}^{nh}\alpha_i\right) \tag{4}$$

$$C_1=\left(\frac{1}{f_{mi}}-1\right) \tag{5}$$

$$C_2=\begin{cases}P\left(1-\alpha_2\right) & \text{if } nh=3\\ P\left(1-\alpha_2+\alpha_4\right) & \text{if } nh=4\end{cases} \tag{6}$$

Where:

$F_m$ : maximum Fourier series value, given by Equation (4);
$f_{mi}$ : human heel impact factor;
$T_p$ : step period;
$f_s$ : step frequency;
$\phi_i$ : harmonic phase angle;
P : person's weight;
$\alpha_i$ : dynamic coefficient for the harmonic force;
i : harmonic multiple (i = 1,2,3…,n);
t : time;
C1, C2 : coefficients given by Equations (5) and (6).

## 4. Finite element model

The developed computational model adopted the usual mesh refinement techniques present in finite element method simulations, based on the ANSYS program [13]. The finite element model has been developed and validated with the experimental results [1-2]. This numerical model enabled a complete dynamic evaluation of the investigated tubular footbridge especially in terms of human comfort and its associated vibration serviceability limit states, see Figure 9. In this model, all steel tubular sections were represented by three-dimensional beam elements (PIPE16 and BEAM44) with tension, compression, torsion and bending capabilities. These elements have six degrees of freedom at each node: translations in the nodal x, y, and z directions and rotations about x, y, and z axes, see Figure 9.

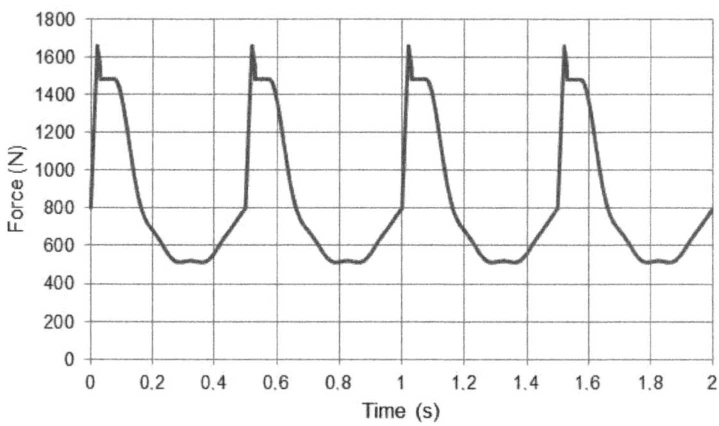


**Figure 6.** LM-IV: dynamic load function for one person walking ($f_p$ = 2.0 Hz).

On the other hand, the reinforced concrete slab was represented by shell finite elements (SHELL63), as presented in Figure 9. This finite element has both bending and membrane capabilities. Both in-plane and normal loads are permitted. The element has six degrees of freedom at each node: translations in the nodal x, y, and z directions and rotations about the nodal x, y, and z axes.

**Figure 7.** Investigated steel-concrete composite tubular footbridge.

**Figure 8.** Internal section of the investigated structural model.

The footbridge pier bearings were represented by a non-linear rotational spring element (COMBIN39). This element is a unidirectional element with non-linear generalized force-deflection capability that can be used in any analysis.

The finite element model presented 71540 degrees of freedom, 11938 nodes and 15280 finite elements (BEAM44: 1056; PIPE16: 5642; SHELL63: 8580 and COMBIN39: 8), as presented in Figure 9. It was considered that both structural elements (steel tubular sections and concrete slab) have total interaction with an elastic behaviour.

## 5. Dynamic analysis

Initially, the steel-concrete composite tubular footbridge natural frequencies, vibration modes and peak accelerations were determined based on experimental tests [2]. The peak acceleration values were obtained considering three types of human walking: slow walking, regular walking and fast walking.

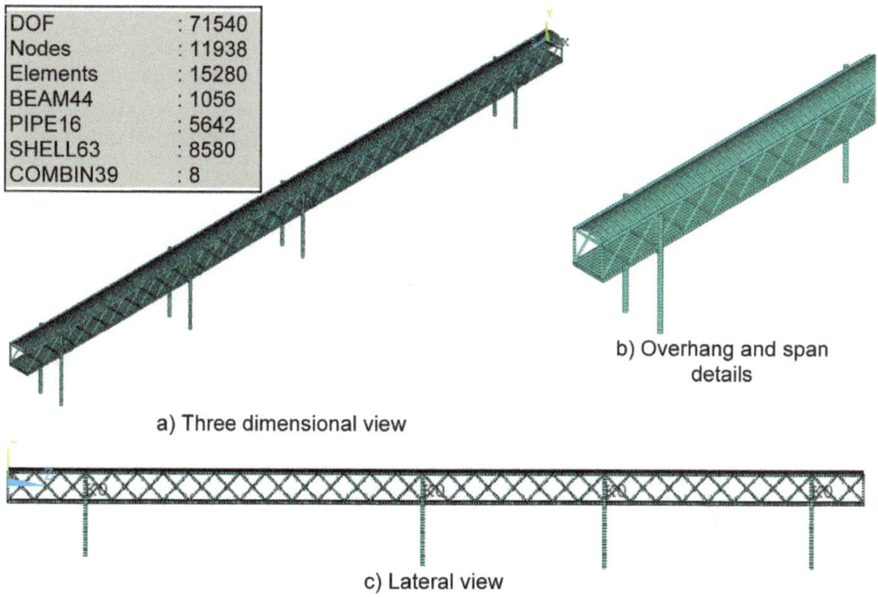


**Figure 9.** Tubular footbridge finite element model

In a second phase, the steel-concrete composite tubular footbridge natural frequencies vibration modes and peak accelerations were determined with the aid of the numerical simulations [1], based on the finite element method using the ANSYS program [13].

## 5.1. Natural frequencies and vibration modes

It can be clearly noticed that there is a very good agreement between the structural model natural frequency values calculated using finite element simulations [1] and the experimental results [2], see Table 4. Such fact validates the finite element model here presented, as well as the results and conclusions obtained throughout this investigation. The vibration modes of the tubular footbridge are depicted in Figures 10 to 12.

| Tubular Footbridge Natural Frequencies (Hz) | $f_{01}$ | $f_{02}$ | $f_{03}$ |
|---|---|---|---|
| Finite Element Model (see Figure 9) | 1.61 | 2.12 | 5.39 |
| Experimental Results | 1.56 | 2.34 | 5.08 |
| Error (%) | 3.20 | 9.40 | 6.10 |

**Table 4.** Tubular footbridge natural frequencies.

When the tubular footbridge freely vibrates in a particular mode, it moves up and down with a certain configuration or mode shape. Each footbridge natural frequency has an associated mode shape. It was verified that longitudinal amplitudes were predominant in the fundamental vibration mode ($f_{01}$ = 1.61 Hz), related with Z axis direction, see Figure 10. In the second mode shape lateral displacements were predominant ($f_{02}$ = 2.12 Hz), associated with X axis direction, as presented in Figure 11. On the other hand, in the third vibration mode ($f_{03}$ = 5.39 Hz), the flexural effects were predominant, related to vertical amplitudes in the Y axis direction, as illustrated in Figure 12.

**Figure 10.** Vibration mode associated with the 1st footbridge natural frequency ($f_{01}$=1.61 Hz).

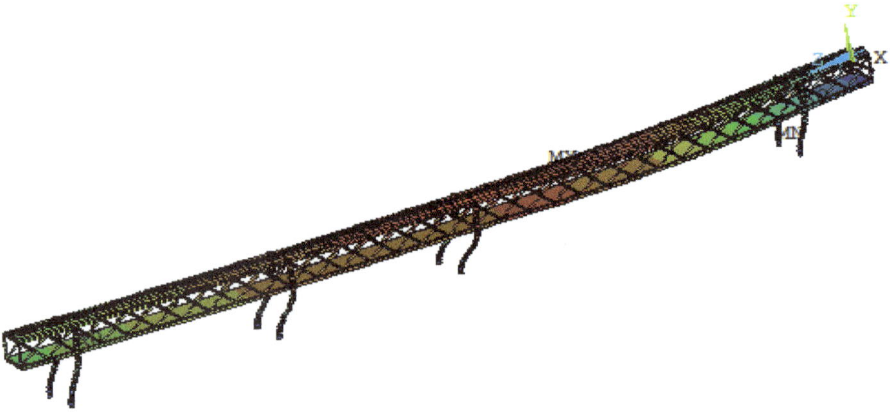


**Figure 11.** Vibration mode associated with the 2nd footbridge natural frequency ($f_{02}$=2.12 Hz).

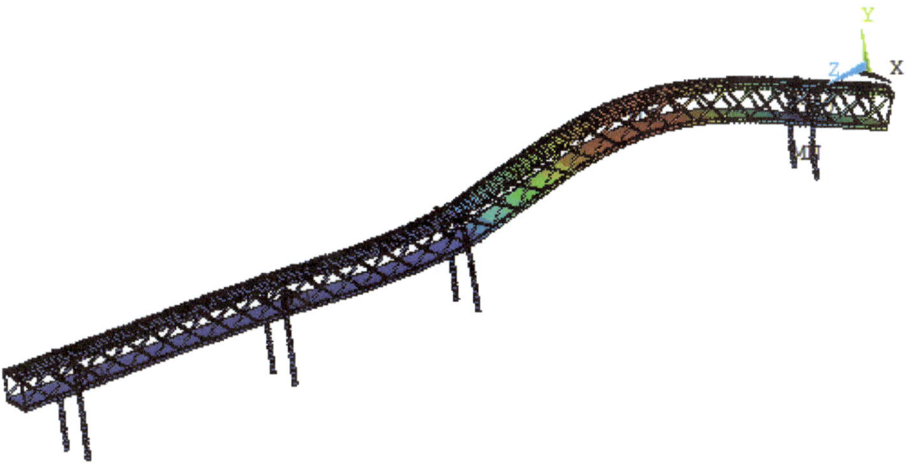


**Figure 12.** Vibration mode associated with the 3rd footbridge natural frequency ($f_{03}$=5.39Hz).

## 5.2. Determination of the tubular footbridge peak accelerations

The finite element modelling follows with the evaluation of the footbridge performance in terms of vibration serviceability due to dynamic forces induced by people walking. The first step of this investigation concerned in the determination of the tubular footbridge peak accelerations, based on a linear time-domain dynamic analysis.

The dynamic loading models (see Equations (1) to (6) and Figures 3 to 6), related to one, two and three people crossing the tubular footbridge on the concrete slab centre, in normal walking, see Figures 13 to 15, were applied on the investigated footbridge over 55.0 s.

The maximum accelerations (peak accelerations) were obtained utilizing an integration time step of $2x10^{-3}$ s ($\Delta t = 2x10^{-3}$ s). In this investigation, seven sections of the structural model were analysed, see Figure 16. These maximum accelerations were compared to the limits recommended by design codes [9,14]. The structural damping coefficient adopted in this investigation was equal to 0.01 ($\zeta$=1%), in accordance with the measured experimental damping [2].

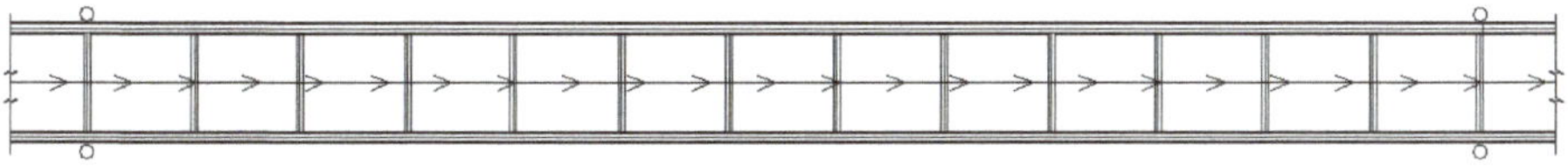

**Figure 13.** One person walking on the footbridge (regular walking).

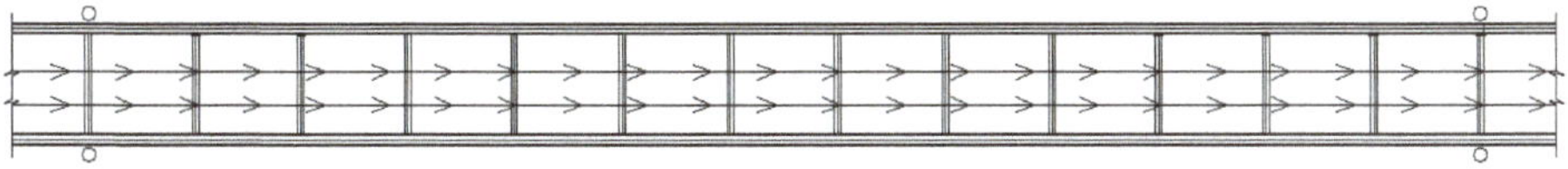

**Figure 14.** Two people walking on the footbridge (regular walking).

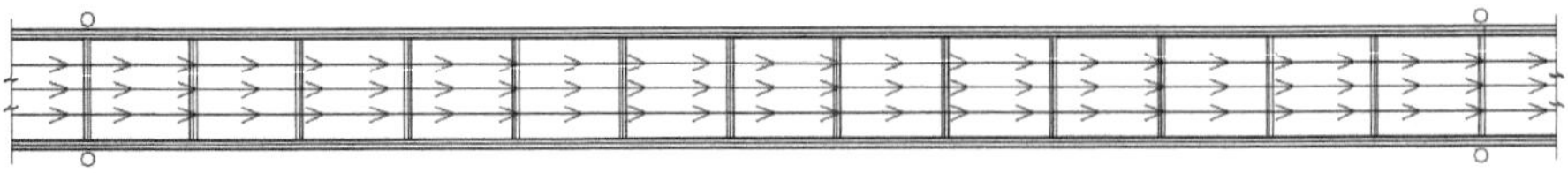

**Figure 15.** Three people walking on the footbridge (regular walking).

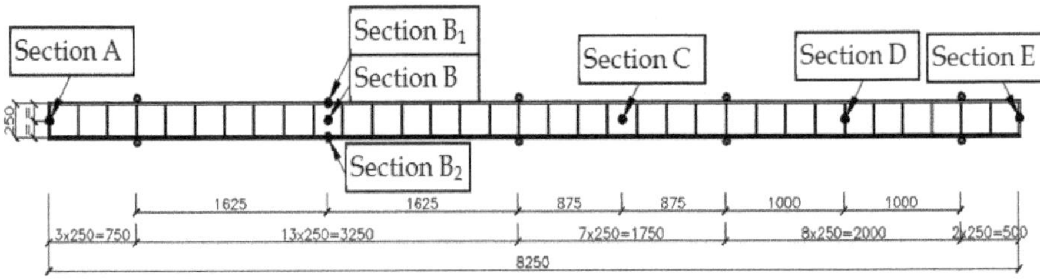


**Figure 16.** Tubular composite (steel-concrete) footbridge investigated sections.

In sequence, Figures 17 to 20 illustrate the tubular footbridge dynamic response, along the time, related to the section B (see Figure 16), when one pedestrian crosses the footbridge in regular walking (resonance condition).

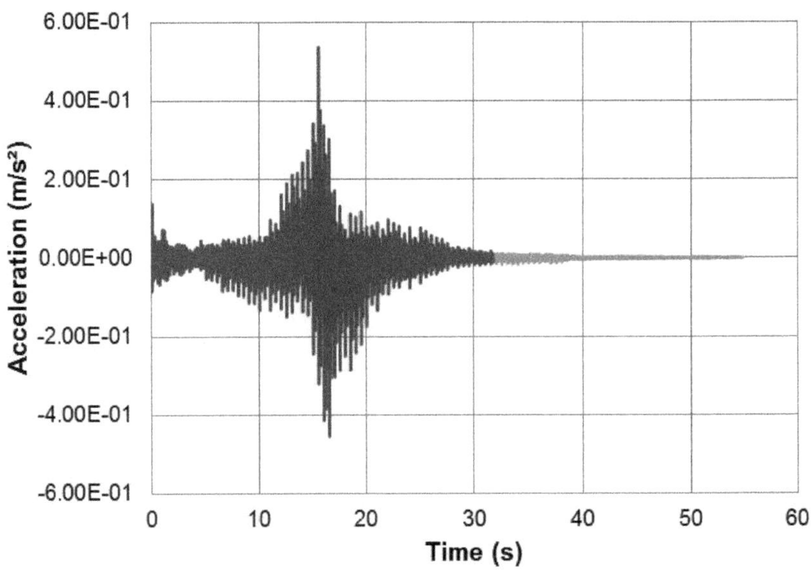


**Figure 17.** LM-I: tubular footbridge acceleration response at section B. One pedestrian crossing the concrete slab centre at resonance condition. Normal walking.

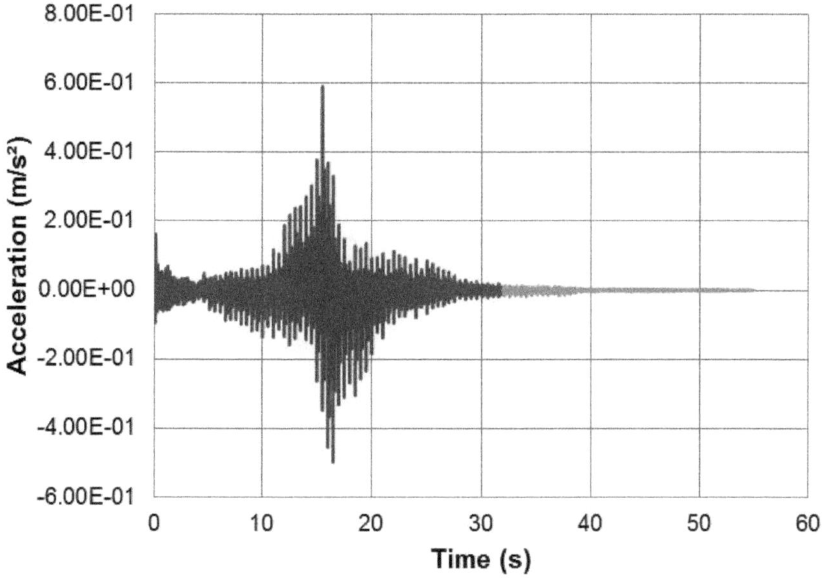


**Figure 18.** LM-II: tubular footbridge acceleration response at section B. One pedestrian crossing the concrete slab centre at resonance condition. Normal walking.

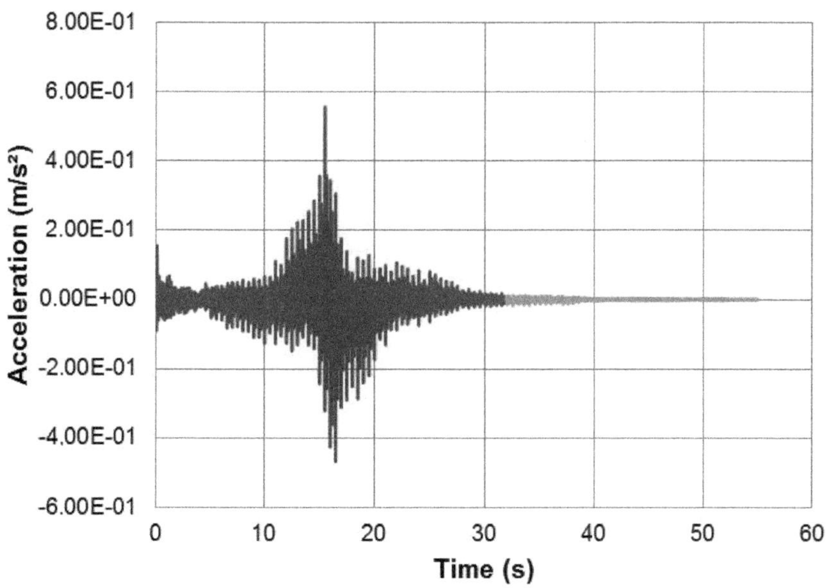


**Figure 19.** LM-III: tubular footbridge acceleration response at section B. One pedestrian crossing the concrete slab centre at resonance condition. Normal walking.

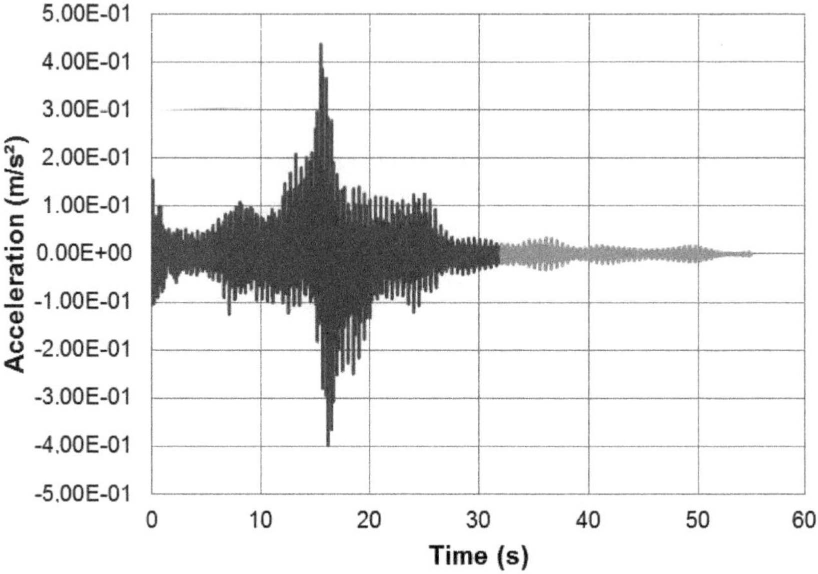


**Figure 20.** LM-IV: tubular footbridge acceleration response at section B. One pedestrian crossing the concrete slab centre at resonance condition. Normal walking.

Figures 17 to 20 present the vertical acceleration versus time graph for the tubular footbridge at section B (see Figure 16). These figures show that the vertical acceleration of the structure gradually increase along the time. In this particular case, the third harmonic with a 2.0 Hz step frequency ($f_s$ = 2.0 Hz), was the walking load resonant harmonic.

The maximum acceleration value found at section B (see Figure 16) was equal to 0.53 m/s$^2$ (LM-I), 0.58 m/s$^2$ (LM-II), 0.55 m/s$^2$ (LM-III) and 0.44 m/s$^2$ (LM-IV), as illustrated in Figures 17 to 20. These figures also indicate that from the moment that the pedestrian leaves the footbridge span (Section B, see Figure 16), when the time is approximately equal to 26 s, the structural damping minimises the dynamic structural model response, as presented in Figures 17 to 20. This assertive occurs only in dynamic loading models that consider the load spatial variation.

The peak acceleration analysis was focused on the steel-concrete composite tubular footbridge dynamic behaviour when the pedestrian normal walking was considered in this work. In sequence, Tables 5 to 7 present the maximum accelerations (peak accelerations: $a_p$ in m/s$^2$), related to seven structural sections of the investigated footbridge (A, B, B1, B2, C, D and E), as illustrated in Figure 16.

The maximum acceleration values (peak accelerations) found in this investigation are respectively equal to 1.50 m/s$^2$ (Section A), 0.18 m/s$^2$ (Section B1), 0.58 m/s$^2$ (Section B), 0.18 m/s$^2$ (Section B2), 0.58 m/s$^2$ (Section C), 0.43 m/s$^2$ (Section D) and 0.79 m/s$^2$ (Section C), corresponding to one individual crossing the composite footbridge in normal walking (resonance condition), as illustrated in Figure 13.

On the other hand, these maximum acceleration values increases when the normal walking associated to two and three people (see Figures 14 and 15) is considered in the analysis, as presented in Tables 5 to 7.

| Dynamic Loading Models | Tubular Footbridge Peak Accelerations ($a_p$ in m/s$^2$) Investigated Sections | | | | | | | Limit Accelerations ($a_{lim}$ in m/s$^2$)* |
|---|---|---|---|---|---|---|---|---|
| | A | $B_1$ | B | $B_2$ | C | D | E | |
| Load Model I (LM-I) | 1.38 | 0.17 | 0.53 | 0.17 | 0.53 | 0.39 | 0.72 | 0.49 |
| Load Model I (LM-II) | 1.50 | 0.18 | 0.58 | 0.18 | 0.58 | 0.43 | 0.79 | |
| Load Model I (LM-III) | 1.38 | 0.17 | 0.55 | 0.17 | 0.55 | 0.40 | 0.74 | |
| Load Model I (LM-IV) | 1.00 | 0.16 | 0.44 | 0.16 | 0.38 | 0.33 | 0.78 | |

*$a_{lim}$ = 1.5%g = 0.15 m/s$^2$: indoor footbridges [9,14]
*$a_{lim}$ = 5.0%g = 0.49 m/s$^2$: outdoor footbridges [9,14]

**Table 5.** Structural model peak accelerations corresponding to one individual walking.

| Dynamic Loading Models | Tubular Footbridge Peak Accelerations ($a_p$ in m/s$^2$) Investigated Sections | | | | | | | Limit Accelerations ($a_{lim}$ in m/s$^2$)* |
|---|---|---|---|---|---|---|---|---|
| | A | $B_1$ | B | $B_2$ | C | D | E | |
| Load Model I (LM-I) | 2.14 | 0.19 | 0.74 | 0.19 | 0.72 | 0.61 | 1.23 | 0.49 |
| Load Model I (LM-II) | 2.32 | 0.21 | 0.81 | 0.21 | 0.80 | 0.67 | 1.35 | |
| Load Model I (LM-III) | 2.14 | 0.20 | 0.76 | 0.19 | 0.75 | 0.63 | 1.27 | |
| Load Model I (LM-IV) | 1.87 | 0.21 | 0.57 | 0.22 | 0.58 | 0.55 | 1.36 | |

*$a_{lim}$ = 1.5%g = 0.15 m/s$^2$: indoor footbridges [9,14]
*$a_{lim}$ = 5.0%g = 0.49 m/s$^2$: outdoor footbridges [9,14]

**Table 6.** Structural model peak accelerations corresponding to two people walking.

| Dynamic Loading Models | Tubular Footbridge Peak Accelerations ($a_p$ in m/s$^2$) Investigated Sections | | | | | | | Limit Accelerations ($a_{lim}$ in m/s$^2$)* |
|---|---|---|---|---|---|---|---|---|
| | A | $B_1$ | B | $B_2$ | C | D | E | |
| Load Model I (LM-I) | 2.71 | 0.25 | 0.97 | 0.25 | 0.94 | 0.81 | 1.68 | 0.49 |
| Load Model I (LM-II) | 2.93 | 0.27 | 1.06 | 0.27 | 1.04 | 0.89 | 1.84 | |
| Load Model I (LM-III) | 2.70 | 0.26 | 1.00 | 0.25 | 0.97 | 0.84 | 1.73 | |
| Load Model I (LM-IV) | 2.61 | 0.31 | 0.76 | 0.32 | 0.75 | 0.73 | 1.86 | |

*$a_{lim}$ = 1.5%g = 0.15 m/s$^2$: indoor footbridges [9,14]
*$a_{lim}$ = 5.0%g = 0.49 m/s$^2$: outdoor footbridges [9,14]

**Table 7.** Structural model peak accelerations corresponding to three people walking.

It must be emphasized that the footbridge overhang sections (Sections A and E, see Figure 16) have presented very high peak accelerations, in all investigated situations, as presented in Tables 5 to 7. This is explained due to the fact that the transient impact produced by the pedestrians on the entrance and exit of the investigated structure have generated high acceleration values.

Based on a quantitative analysis of the maximum accelerations, it was verified that the loading model I (LM-I) has produced the highest peck acceleration values practically in all investigated cases, as illustrated in Tables 5 to 7.

These peak accelerations presented in Tables 5 to 7 are related to a pedestrian normal walking situation. It must be emphasized that the limit acceleration value is equal to 0.49 $m/s^2$ [9,14], when the outdoor footbridges are considered in the analysis.

Based on the finite element modelling of the steel-concrete composite tubular footbridge dynamic behaviour, the numerical results presented in Tables 5 to 7 indicated that the dynamic actions produced by human walking led to peak accelerations higher than the limiting values present in design code recommendations (Outdoor footbridges: $a_{lim} = 5\%g = 0.49\ m/s^2$ [9,14]), as depicted in Tables 5 to 7.

## 6. Final remarks

This contribution covers the application of tubular structural elements in pedestrian footbridge design and tries to give an overview about the evaluation of tubular footbridges dynamic behaviour, objectifying to help practical structural engineers to deal with this kind of problem and to allow for a further application of tubular structural elements in pedestrian footbridge design.

The present investigation was carried out based on four dynamic loading models (LM-I to LM-IV) implemented objectifying to incorporate the dynamic effects induced by people walking on the footbridges dynamic response. In these models, the position of the human walking load was changed according to the individual position. However, a more realistic loading model (LM-IV) considered the ascent and descending movement of the human body effective mass at each step load (human walking load) and additionally also incorporates the transient effect due to the human heel impact.

The proposed analysis methodology considered the investigation of the dynamic behaviour, in terms of serviceability limit states, of a composite tubular footbridge, spanning 82.5 m. The structure is composed by three spans (32.5 m, 17.5 m and 20.0 m, respectively) and two overhangs (7.50 m and 5.0 m, respectively). The structural system is constituted by tubular steel sections and a concrete slab and is currently used for pedestrian crossing.

A computational model, based on the finite element method, was developed using the ANSYS program. This model enabled a complete dynamic evaluation of the investigated tubular footbridge especially in terms of human comfort and its associated vibration serviceability limit states.

The results found throughout this investigation have indicated that the dynamic actions produced by human walking could generate peak accelerations that surpass design criteria limits developed for ensuring human comfort. Hence it was detected that this type of structure can reach high vibration levels that can compromise the footbridge user's comfort and especially its safety.

The analysis methodology presented in this paper is completely general and is the author's intention to use this solution strategy on other pedestrian foot-bridge types and to investigate the fatigue problem. The fatigue problem is a relevant issue and certainly much more complicated and is influenced by several design parameters and footbridge types. Further research in this area is currently being carried out.

## Author details

José Guilherme Santos da Silva, Ana Cristina Castro Fontenla Sieira, Gilvan Lunz Debona, Pedro Colmar Gonçalves da Silva Vellasco and Luciano Rodrigues Ornelas de Lima
*State University of Rio de Janeiro (UERJ), Rio de Janeiro/RJ, Brazil*

## Acknowledgement

The authors gratefully acknowledge the support for this work provided by the Brazilian Science Foundation CAPES, CNPq and FAPERJ.

## 7. References

[1] Debona GL. Modelagem do comportamento dinâmico de passarelas tubulares em aço e mistas (aço-concreto) (Modelling of the dynamic behaviour of steel-concrete composite tubular footbridges), MSc Dissertation (in Portuguese), Civil Engineering Post-Graduate Programme, PGECIV, State University of Rio de Janeiro, UERJ, Rio de Janeiro, Brazil, pp. 1-154; 2011.

[2] Zúñiga JEV. Análise da resposta dinâmica experimental de uma passarela tubular mista, aço-concreto, submetida ao caminhar humano (Dynamic experimental analysis of a steel-concrete composite tubular footbridge submitted to human walking), MSc Dissertation (in Portuguese), Civil Engineering Post-Graduate Programme, PGECIV, State University of Rio de Janeiro, UERJ, Rio de Janeiro, Brazil, pp. 1-135; 2011.

[3] Figueiredo FP, Silva JGS da, Vellasco PCG da S, Andrade SAL de, Andrade, SAL de. A parametric study of composite footbridges under pedestrian walking loads. Engineering Structures, 2008; 30:605-615.

[4] Silva JGS da, Vellasco PCG da S, Andrade SAL de, Lima LRO de, Figueiredo FP. Vibration analysis of footbridges due to vertical human loads. Computers & Structures, 2007; 85:1693-1703.

[5] Eurocode 3: Design of steel structures. Part 1.9: General rule - Fatigue. European Committee for Standardisation; 2005.

[6] Kuhlmann U, H-P Günther, Saul R, Häderle, M.-U. 2003. Welded circular hollow section (CHS) joints in bridges. ISTS 2003: Proceedings of the 10th International Symposium on Tubular Structures, Madrid, Spain.

[7] Leitão FN, Silva JGS da, Vellasco PCG da S, Lima LRO de, Andrade SAL de. Composite (steel-concrete) highway fatigue assessment. Journal of Constructional Steel Research, 2011; 67(1):14-24.

[8] Varela WD. Modelo teórico-experimental para análises de vibrações induzidas por pessoas caminhando sobre lajes de edifícios (Theoretical-experimental model to analyse vibrations induced by people walking on floor slabs of buildings), PhD Thesis (in Portuguese), Federal University of Rio de Janeiro, Civil Engineering Department, COPPE/UFRJ, Rio de Janeiro, Brazil, pp. 1-309; 2004.

[9] Murray TM, Allen DE, Ungar EE. Floor vibrations due to human activity, Steel Design Guide Series, American Institute of Steel Construction, AISC, Chicago, USA; 2003.

[10] Pimentel RL, Pavic A, Waldron, P. Evaluation of design requirements for footbridges excited by vertical forces from walking. Canadian Journal of Civil Engineering, 2001; 28(5),769-776.

[11] Chen, Y. Finite element analysis for walking vibration problems for composite precast building floors using ADINA: modelling, simulation and comparison. Computer & Structures, 1999; 72:109-126.

[12] Bachmann, H, Ammann, W. Vibrations in structures induced by man and machines, Structural Engineering Document 3e. International Association for Bridges and Structural Engineering; 1987.

[13] ANSYS. Swanson Analysis Systems, Inc. P.O. Box 65, Johnson Road, Houston, PA, 15342-0065, Version 10.0, Basic analysis procedures, Second edition; 2003.

[14] International Standard Organization / ISO 2631-2. Evaluation of human exposure to whole-body vibration, Part 2: Human Exposure to Continuous and Shock-Induced Vibrations in Buildings (1 to 80Hz), International Standard; 1989.

[15] Comité Euro-international du Béton. CEB-FIP: Bulletin d'information, N⁰ 209, England, London, August; 1993.

# Chapter 15

# Damage-Tolerant Design of Stiffener Run-Outs: A Finite Element Approach

S. Psarras, S.T. Pinho and B.G. Falzon

Additional information is available at the end of the chapter

## 1. Introduction

Modern aerostructures are predominantly of semi-monocoque construction characterized by a thin skin and stiffeners. The latest generation of large passenger aircraft also use mostly carbon-fibre composite material in their primary structure and there is a trend towards the utilization of bonding of subcomponents in preference of mechanical fastening. Current design philosophy requires that certain stiffeners are terminated, for example due to an intersecting structural feature or an inspection cut-out. In these circumstances, the loading in the stiffener must be diffused into the skin, leading to complex three-dimensional stress-states. The development and utilization of reliable virtual component testing, in the design of composite aerostructures, can potentially yield significant cost reductions. Such reliability requires a thorough understanding of the damage mechanisms and failure processes in realistic aerostructures, particularly in critical regions such as stiffener run-outs.

When a stiffener is terminated, the loads which it carries must be transferred to the skin, making the design of the run-out region vital, hence improved design methodologies are required. Several studies of skin-stiffener failure [1-8] have been carried out. Falzon et al. [7, 8] investigated the failure of realistic stiffener run-outs loaded in uniaxial compression. Different stacking sequences and skin thicknesses at the run-out region were tested and a wealth of complexity in the response and subsequent failure was reported. For all tests, failure initiated at the edge of the run-out and propagated across the skin–stiffener interface. It was found that the failure load of each specimen was greatly influenced by changes in the geometric features of these specimens. Falzon and Hitchings [8] used a Virtual Crack Closure Technique (VCCT) [9] to predict the crack growth characteristics of the modelled specimens and reported shortcomings in the quantitative correlation between the predicted and observed failure loads and modes.

Bisagni et al. [10] investigated, experimentally and numerically, the postbuckling response of hat-stiffened composite panels using cohesive zone models and predicted collapse loads which were in agreement with experiments. Krueger [6] used VCCT and a numerically effective shell/3D modelling approach to predict the delamination failure of skin-stiffener run-outs. Camanho et al. [11] implemented a cohesive element to numerically investigate the debond strength of skin-stiffener composite specimens using cohesive zone models and compared with experimental results.

In a recent study by the authors [12], a parametric numerical analysis was conducted to optimize the design of the run-out section to increase the crack growth stability under axial compression. Improved damage tolerance (stable crack propagation) was reported in the modified stiffener run-out design as compared to the baseline configuration. The modified design eventually failed catastrophically by interlaminar delamination, not bondline failure (debonding), which had not been considered in the numerical study. A more detailed analysis of different configurations, which accounts for delamination, was therefore undertaken in the work presented in this chapter. Building on the previous findings, the merits of a *Compliant* termination scheme are presented.

The research in this paper focuses on stiffener run-outs loaded in compression with a selection of stiffener termination schemes. These schemes are analysed numerically (Section 2) in order to compare the influence of the design on the energy release rate for debonding and delamination. The *Baseline*, *Tapered* and *Compliant* schemes were manufactured (Section 3) and tested to failure (Section 4). The experimental results are presented in Section 5, and compared to the predictions in Section 6.

The failure mode of the *Tapered* specimen configuration was found to be a combination of interlaminar and intralaminar failure [12]. The *Tapered* stiffeners had an unexpected delamination in the flange between the bottom-most $0^{\circ}$ ply and above the $45^{\circ}$ ply. Furthermore, the delamination led to an intralaminar failure in the form of a matrix crack across the $0^{\circ}$ ply near the filler ends and continued delaminating between the filler and $0^{\circ}$ ply, Figure 19a. For this reason, a new finite element (FE) model was created in order to investigate these modes of damage, Figure 19b and c.

## 2. Numerical analysis

### 2.1. Skin-stiffener configurations

The main focus of this study was a new configuration to improve the crack growth stability of a stiffener run-out. The structural performance of a baseline skin-stiffener configuration under longitudinal compression, with geometry and dimensions shown in Figure 1a, was compared to that of a modified parametric configuration shown in b. The modified configuration has a widening flange towards the termination end of the stiffener but this added material is offset by the taper of the stiffener web. This results in a stiffener design with a similar overall weight to the baseline design. For the parametric configuration, various values of *b*, *c* and *d* were analysed. The materials used in this study were IM7/8552 carbon/epoxy pre-preg, with ply thickness 0.25 mm, for the skin and the stiffener, and FM300 adhesive film (0.15 mm thick) for the bondline.

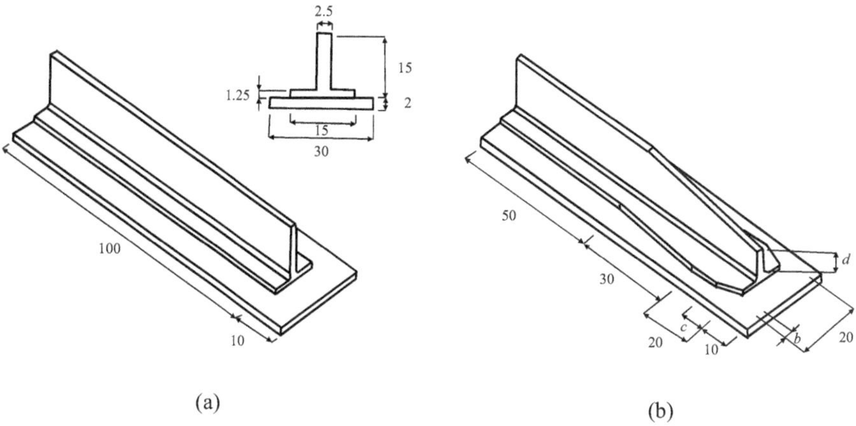


**Figure 1.** Designs and dimensions in mm of a) the baseline stiffener and b) the parametric stiffener.

| Material | $E_{xx}$ [GPa] | $E_{yy}$ [GPa] | $G_{xy}$ [GPa] | $\upsilon_{xy}$ | $X$ [MPa] | $Y$ [MPa] | $S$ [MPa] | $G_{Ic}$ [kJ/m²] | $G_{IIc}$ [kJ/m²] | $\eta$ |
|---|---|---|---|---|---|---|---|---|---|---|
| **IM7/8552** | 154.1 | 9.8 | 4.48 | 0.34 | 1572.9 | 254.6 | 101.2 | 0.21 | 0.61 | - |
| **FM300** | 2.38 | - | 0.68 | - | 61 | - | 49.8 | 0.9 | 2.5 | 8.0 |

**Table 1.** Material properties for IM7/8552, measured in-house, and FM300

In previous test results with geometry similar to that in Figure 1a, the specimens failed by unstable debonding of the stiffener from the skin. Therefore, the different configurations in this study were assessed by comparing the energy release rates of the run-outs for a given displacement and for several debond lengths.

## 2.2. The FE model

All the FE simulations of the parametric study were carried out in ABAQUS [13] and the parameterized models were created in Python [14]. The main model has five different parts, the skin, the adhesive between the skin and the stiffener, the parts of the stiffener and the filler. The material properties for IM7/8552 measured using standard tests and are shown in Table 1. A mesh sensitivity study was carried out to ensure that all results presented are mesh converged.

### *2.2.1. Element Idealisation*

All parts were modelled with three dimensional hexahedral solid elements C3D8 to accurately capture stresses in the through-thickness direction. Also, solid elements are capable of modelling several layers of different materials for the analysis of laminated composites which is ideal for our numerical study. In consideration of computational cost, linear order elements were chosen and to ensure higher fidelity results whew required, the models executed using a mesh density based on a ability to define a detailed mesh convergence study. One of the major advantages of solid elements is the feature of defining several ply stack ups within a single element along the stacking direction with an improved element formulation, illustrated in the Figure 2 below.

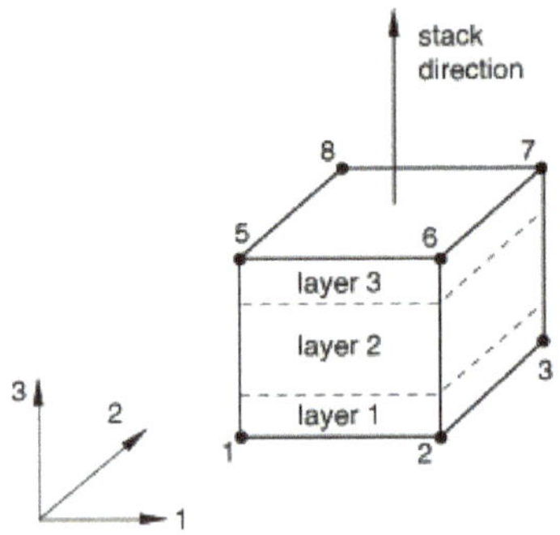


**Figure 2.** Solid element with composite layups

All the models were meshed with fully integrated solid elements (C3D8) rather using reduced integration elements (C3D8R), since the reduced integration elements were not able to capture damage variable calculations at all the nodes while delaminating, Figure 3. On further investigating this issue, it was observed that the problems arose because of

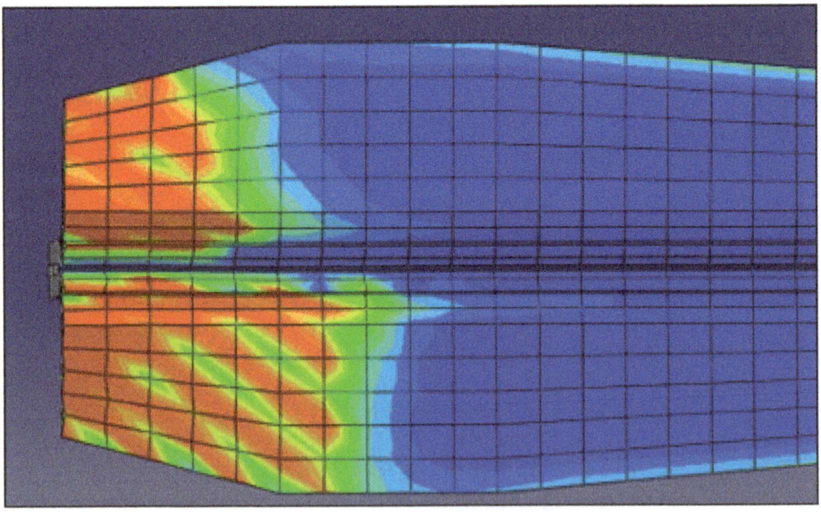

**Figure 3.** Contour plot of cohesive damage variable with Hourglass modes

hourglassing. Even though this particular problem could be partially mitigated by using the hourglass stiffness enhanced element formulation, the use of C3D8R elements in the numerical models was avoided for greater accuracy. Consequently, C3D8 elements were used in all the models.

### 2.2.2. *Stacking Sequence*

The skin consists of eight plies, while the stiffener consists of five plies. The stacking sequences used in [1] for the skin and the stiffener, are shown in Table 2. Figure 4 shows the ply orientations for the skin. The thickness of each ply is 0.25mm (double thickness) and the number of integration points is set to three within each ply/element through the thickness.

| Part | Stacking Sequence |
|---|---|
| Skin | $[45/-45/0/90]_s$ |
| Stiffener (per half section) | [0/90/-45/45/0] |

**Table 2.** Stacking sequence for the skin and the stiffener

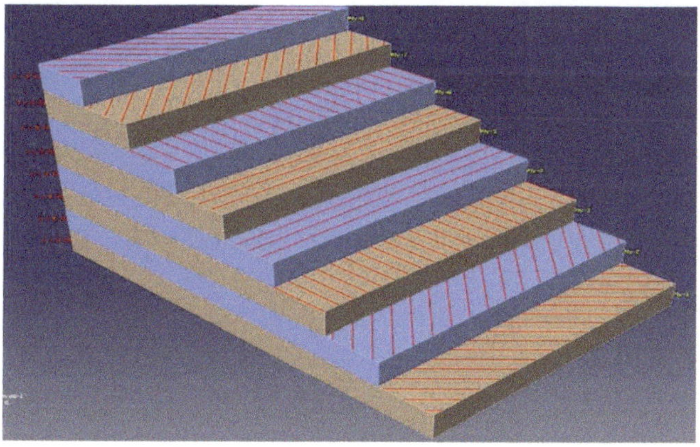

**Figure 4.** Stacking sequence for the skin

### 2.2.3. *Mesh*

The number and distribution of elements is a crucial step when using the FE method to solve structural problems. Figure 5 shows the element families that are most commonly used in stress analysis. The main difference among the element families is the geometry type that each of the family represents. As mentioned earlier, an 8-node linear brick element was used and a course mesh of the modified design is shown in Figure 6.

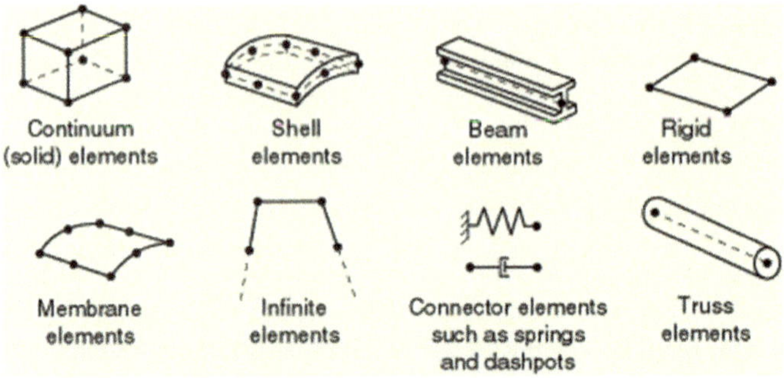


**Figure 5.** Commonly used element families in ABAQUS [13]

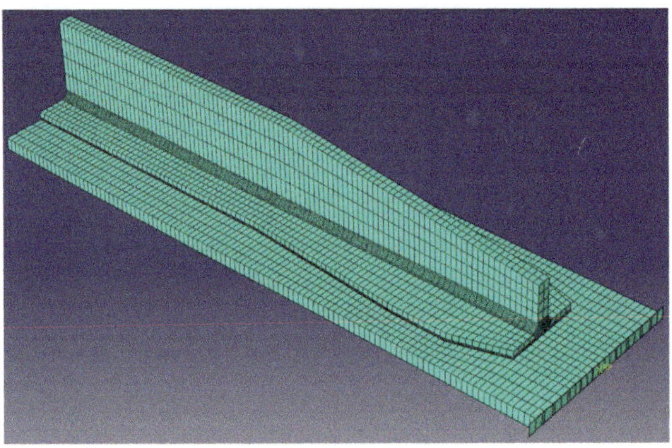

**Figure 6.** Course mesh for the modified design with 3D continuum elements

### *2.2.4. Constraints*

Constraints were defined in the *Assembly module* [13] for the initial positions of *instances*. The type of constraint created for this model was a "tie constraint", which allows the user to fuse together two regions even though the meshes created for sectors may be dissimilar [8]. For the current model, sixteen constraints were created between all the parts being in contact. In Figure 7 it can be seen how the master and the slave surfaces are displayed in the model.

### *2.2.5. Boundary conditions*

In the current study, two boundary conditions were defined. The first condition, named BC-1, was a displacement/rotation type boundary condition. All displacements and rotations were set to zero. BC-1 was applied to one edge of the stiffener as shown in Figure 8. The second BC, named BC-2, was a displacement/rotation type BC as well, where all displacements and rotations were set to zero apart from U3, which was set to U3=1. U3 is the displacement in the axial direction and represents the testing machine's displacement.

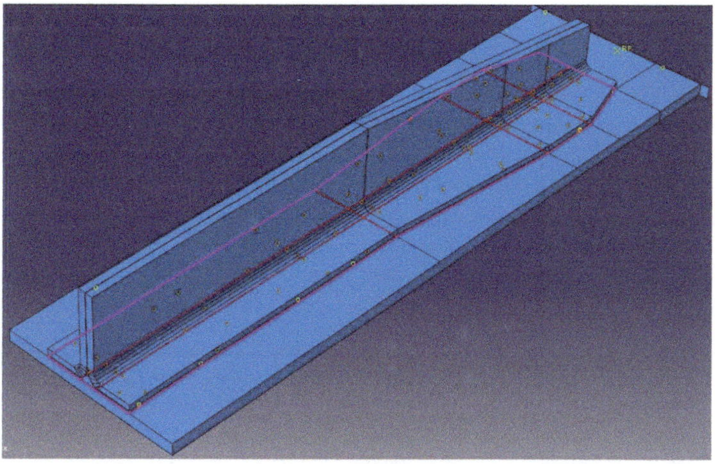

**Figure 7.** Surfaces constraints between the skin (master surface-red) and the adhesive (slave surface-pink)

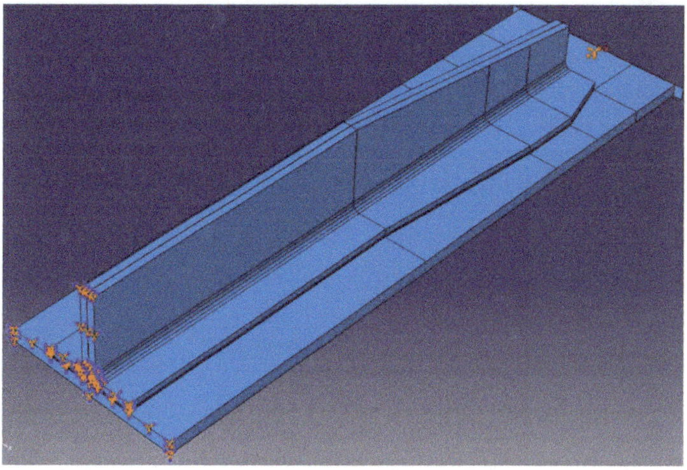

**Figure 8.** Modified model with BC-1 on the left edge (clamped) and BC-1 on the right edge (U3=1)

### *2.2.6. Mesh sensitivity study*

Finite element analysis provides an approximate solution and it can only guarantee that equilibrium is satisfied on average over an element. This does not mean that it will satisfy equilibrium over any smaller volume compared to a complete element. As a consequence, equilibrium is enhanced when the size of the element is decreased. Moreover, in regions of stress concentrations, it is necessary to increase the accuracy of the FE solution by either using elements with higher-order shape functions (p-refinement) or by using a finer mesh of elements (h-refinement). The goal that a designer needs to achieve is to select the best mesh density which is not prohibitively expensive to run and at the same will provide accurate and acceptable results [15].

In this study, three different meshes were used (a coarse mesh, an intermediate mesh and a fine mesh). 13100 elements were used for the coarse mesh, 18700 elements for the intermediate mesh and 51000 elements for the fine mesh. Table 3 illustrates the three different meshes used, the strain energy and the running time of each model. According to the values of Table 3, the solution is converged. Since the running time for the intermediate mesh was 35 minutes and the results appeared accurate and acceptable, this specific density of elements was selected for the rest of this study.

| Mesh | Strain Energy (KJ) | Number of Elements | Running Time (minutes) | Deviation (%) |
|---|---|---|---|---|
| **Coarse** | 30150 | 13100 | 25 | - |
| **Intermediate** | 30138 | 18700 | 35 | 0.039 |
| **Fine** | 30109 | 51000 | 90 | 0.096 |

**Table 3.** Mesh Sensitivity Study results.

Moreover, investigation was undertaken to assess the accuracy of using multiple ply orientations within a solid finite element. Abaqus® provides the option to make a partition through the ply thickness and define the material orientation. As a result, eight partitions were created for the skin and five for each stiffener using material orientations given in Table 2. The results obtained from both approaches were similar, as illustrated in Figure 9.

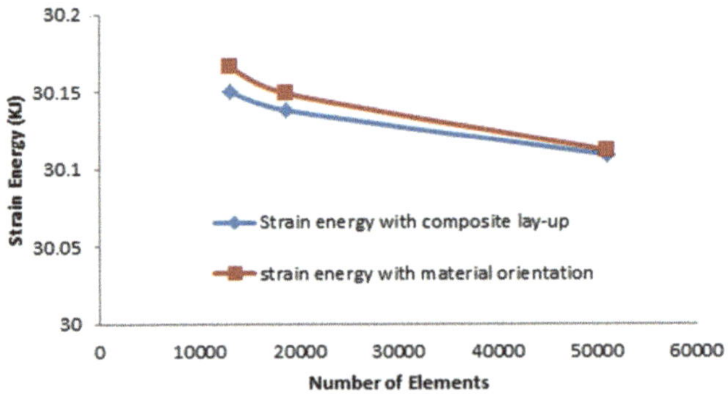


**Figure 9.** Strain energy with composite lay-up (blue line) and with material orientations (red line)

### 2.2.7. *The Python script*

In order to perform the parametric study, a large number of models were created and a script that generates these models was needed. This script was written in Python [14] because of the advantage of using the Abaqus scripting interface to generate automate repetitive models and execute.

The parametric script generates the stiffener run-out models with different crack lengths. The script uses a basic stiffener run-out configuration and changes the design each time according to the parameter values. When a new design is generated, the script propagates the crack and the energy release rate is calculated for every step.

## 2.3. Results from modelling

### 2.3.1. *Energy Release rate along crack*

After converging to an optimum mesh density, failure in the stiffeners was captured by consecutive models, propagating a delamination crack or a debonding crack, 1 mm at a time. The results of the parametric study are shown in Fig. 10, where the values of

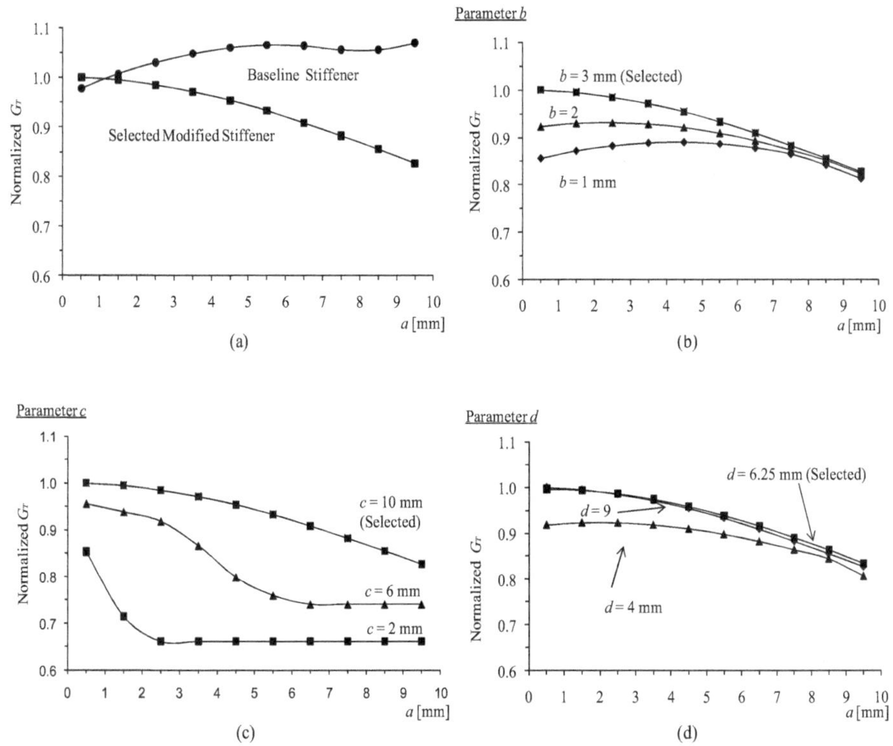


**Figure 10.** Normalized energy release rates as a function of crack length (a) comparison between baseline stiffener design (Fig. 1a) and selected modified stiffener (Fig 1b with $b$ = 3 mm, $c$ = 10 mm and $d$ =6.25 mm), (b) Influence of parameter $b$ on G, (c) Influence of parameter $c$ on G and (d) Influence of parameter $d$ on G.

$G_T = G_I + G_{II} + G_{III}$, the total energy release rate, have been normalized to the $G_T$ of the reference parametric stiffener for 0.5 mm of crack length. Fig. 10a compares the normalised energy release rate for the baseline and selected parametric stiffeners. The negative slope of the *G(a)* curve for the latter indicates stability of crack growth (assuming constant fracture toughness). The influence of parameters *b*, *c*, and *d* is presented in Fig. 10b, c and d. Given the objective of optimising for stability of crack growth, the configuration with $b = 3$ mm, $c = 10$ mm and $d = 6.25$ mm was selected to be carried out for the consequent stages of this study.

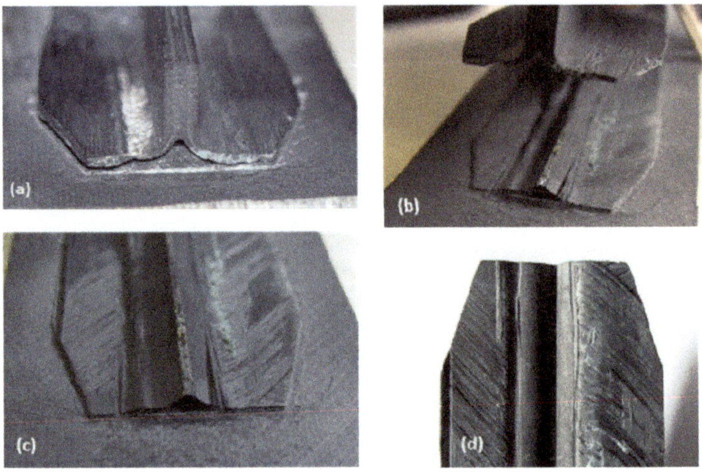


**Figure 11.** (a) Front view of failed specimen; (b) Exploded view showing the failed area; (c) Front view of bottom part showing $0^0$ plies ; (d) Bottom view of the upper part showing delaminated $45^0$ plies

Interlaminar and intralaminar failures were observed in the modified run-out stiffener. The stiffener had an unexpected delamination between the $0^0$ ply and the $45^0$ ply. After examination of the specimens it was assumed that the delamination led to an intralaminar failure in the form of a matrix crack across the $0^0$ ply near the filler ends and continued delaminating between the filler and $0^0$ ply. In Figure 11, the failed specimens clearly shows that delaminations and matrix cracks occurred in the $0^0$ ply on both sides of the specimen.

## 3. Modeling the debonding of the stiffener

The finite element model of the stiffener run-out had the following features:

1. Skin - single part of $(45^0/-45^0/0^0/\ 90^0)_s$ laminate
2. Filler - made-up of $0^0$ fibres
3. Left - $0^0$ lamina of the stringer
4. Left - single part of $(0^0/\ 90^0/-45^0/45^0)$ laminate of the stringer
5. Right - $0^0$ lamina of the stringer
6. Right - single part of $(0^0/\ 90^0/-45^0/45^0)$ laminate of the stringer

The part descriptions are clearly illustrated in a front view of the skin-stiffener assembly, Figure 18.

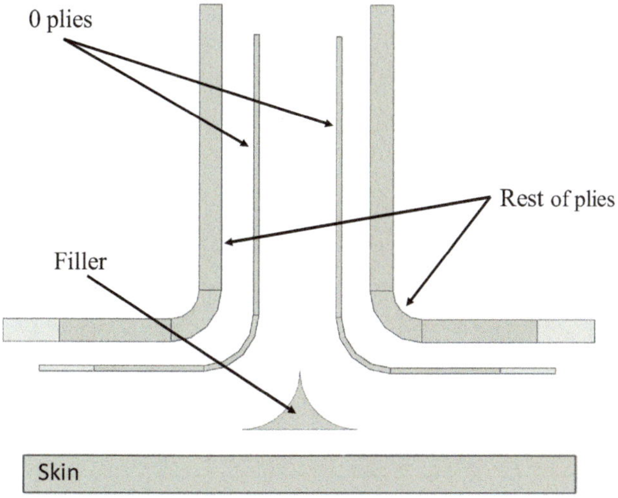


**Figure 12.** The parts of the FE model

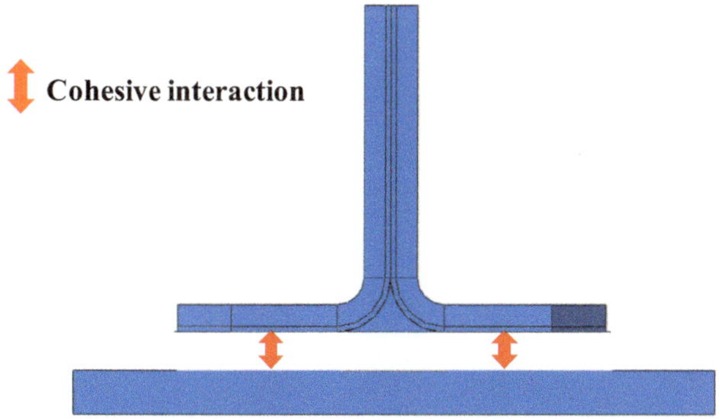


**Figure 13.** Illustration of imposed cohesive properties for debonding mode.

The interface between the skin and the stiffener was connected by using a surface-based cohesive layer, as seen in Figure 12. Cohesive zone modelling is generally used for the numerical simulation of interlaminar failure. Damage initiation is driven by a traction seperation law and the value of the maximum traction, $t^o$, Figure 14(a). New crack surfaces are formed when the fracture toughness $G_c$ is equal to the area surface under the traction-separation curve. Considering the nature of predictions made by the parametric study, the model was analysed with cohesive interaction properties of adhesive FM 300 and listed in Table 4.

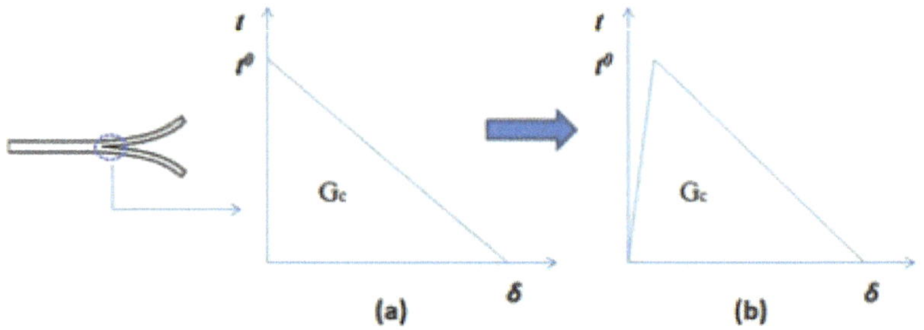


**Figure 14.** Illustration of (a) Traction-separation law for cohesive zone models (b) Modified law to implement in FEM

| FM300 (adhesive) | Normal direction | First shear direction | Second shear direction | BK law $\eta$ |
|---|---|---|---|---|
| Initial linear elastic behaviour | $K_{nn}$ *in* $Nmm^{-3}$ | $K_{ss}$ *in* $Nmm^{-3}$ | $K_{tt}$ *in* $Nmm^{-3}$ | |
| | $10^6$ | $10^6$ | $10^6$ | |
| Damage initiation | $N$ *in MPa* | $S_1$ *in MPa* | $S_2$ *in MPa* | |
| | 50 | 100 | 100 | |
| Damage evolution | $G_n$ *in N/mm* | $G_s$ *in N/mm* | $G_t$ *in N/mm* | 8 |
| | 0.9 | 2.5 | 2.5 | |

**Table 4.** Cohesive interaction properties

### 3.1. Initial linear elastic behaviour

In order to accurately predict damage initiation, the interaction of traction components was taken into account and the quadratic stress criterion used. This criterion is included in Abaqus and was formulated based on Ye's criterion [13] including the interaction between traction components. According to this criterion, the damage initiates when a quadratic interaction function reaches a value of one. The quadratic interaction function is shown in the following equation.

$$\left\{\frac{t_n}{t_n^0}\right\}^2 + \left\{\frac{t_s}{t_s^0}\right\}^2 + \left\{\frac{t_t}{t_t^0}\right\}^2 = 1 \qquad (1)$$

where, $t^0{}_n$ , $t^0{}_s$ and $t^0{}_t$ are the peak values of the stress when the separation is either purely normal to the interface or purely in the first or the second shear direction respectively. In the present numerical study, these values for IM7/8552 are 50 MPa in the normal direction and 100 MPa in the two shear directions[10], see Table 4.

When the initiation criterion is met, the cohesive stiffness degrades at a rate defined by the damage evolution model. The overall damage of the damage zone is represented by a scalar damage variable, D, and is implemented in the damage evolution model. After the damage initiation, the damage variable monotonically evolves from 0 to 1 on increasing the loading.

For the mix-mode definition of fracture energies, Benzeggagh-Kenane's criterion [16] (BK law) was used which defines the energy dissipated as in equation below,

$$G_n^c + \left(G_s^c - G_n^c\right)\left\{\frac{G_S}{G_T}\right\}^{\eta} = G^c \tag{2}$$

with,

$$G_T = G_n + G_s \tag{3}$$

$$G_S = G_t + G_s \tag{4}$$

where $G_n$, $G_s$ and $G_t$ are the fracture toughness values in the normal and two shear directions respectively which were measured in-house for IM7/8552 and given in Table 4. In this study the value of the BK mode-mixity power parameter, $\eta=1.6$, was obtained from Maimi [17, 18].

### 3.2. Response of the Numerical Model

The debonding failure initiation load predicted by the model was compared with the debonding loads predicted by using the energy release rate and the experimental results, Table 5.

| | Experiment | $G_{avg}$ Prediction | $G_{max}$ Prediction | Cohesive layer Prediction |
|---|---|---|---|---|
| Failure load [kN] | 17.7 | 19.2 | 17.5 | 18.7 |
| Difference from experiments | - | 8.5% | 1.1% | 5.6% |

**Table 5.** Debonding loads

Recalling the results from the energy release rate analysis, when the maximum energy release rate across the width of the stiffener was used, the predictions were closer to the experimental results. In addition, when the average energy release rate was used, the difference from the experimental results had an over-prediction of 8.5%. By using the cohesive zone model, the difference with the experimental results was 5.6%.

Also, the debonding growth along the width of the specimen correlates well with the predictions using strain energy release rate across the width of the stiffener as can be seen in Figure 15. Comparing the FE model pattern with the *Tapered* slop on the right, it is clearly observed that there is a correlation between the extent of damaged and the value of normalized strain energy release rate across the width of the specimen.

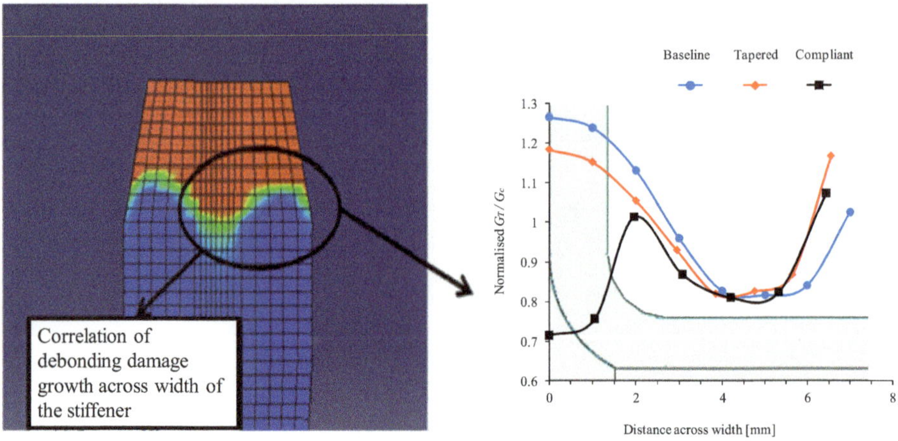


**Figure 15.** Damage growth pattern compared with energy release rate predictions for tapered specimen

## 3.3. Modeling the debonding failure using VCCT

The implemented VCCT method in Abaqus standard, developed by Boeing and Simulia, suggested promising results, especially when the mismatched, tie-constrained, meshes had only 3% error comparing with pairing meshes. As can be seen in Figure 16(a) there was good correlation between the VCCT and the parametric study. On the other hand, the VCCT method could not capture the increase in normalised GT/GC arising from the geometric discontinuity at the edge of the flange. In order to capture the detail at the edge of the flange, the mesh resolution was increased and was biased towards these edges, Figure 17, and the results can be seen in Figure 16(b). Despite the good results, the size of the model and the time needed for execution made the use of VCCT impractical and was not used in further investigations.

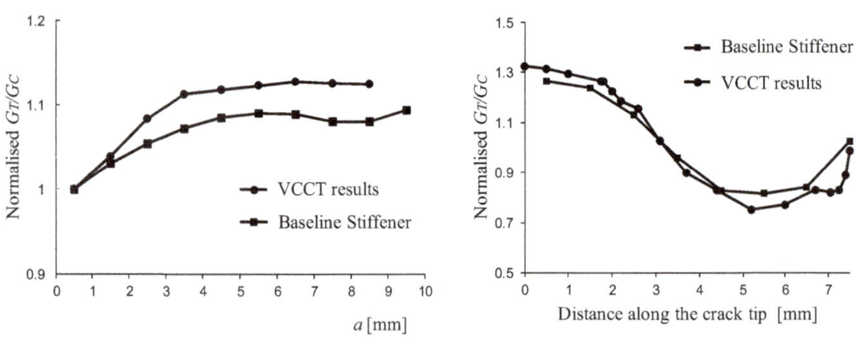


(a) (b)

**Figure 16.** Comparing the results of the parametric study with the VCCT method (a) along the crack and (b) along the width of the stiffener.

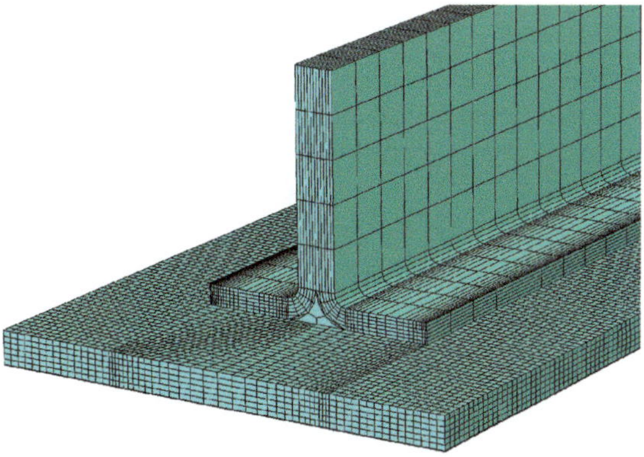

**Figure 17.** The refined model that used in the VCCT method

## 4. 2nd iteration

The objectives defined in the first part of this study were to create two FE models for the baseline and the modified configuration. The baseline stiffeners failed due to debonding of the skin/stiffener interface, while the modified stiffeners failed by delamination between the $0^0$ and $45^0$ ply interface. Improved damage tolerance (stable crack propagation) was reported in the modified stiffener run-out design as compared to the baseline configuration. The modified design eventually failed catastrophically by interlaminar delamination, not bondline failure, which had not been considered in the numerical study. A more detailed analysis of different configurations, which accounted for delamination, was therefore undertaken. Building on the previous findings, the merits of a compliant termination scheme are presented.

### 4.1. Skin stiffener runout configurations

The structural performance of three different skin-stiffener configurations – *Baseline* (B), *Tapered* (T) and *Compliant* (C) – under longitudinal compression, with geometry and dimensions shown in Figure 18, was assessed. Compared to the *Baseline* stiffener (Figure 18a), the other two configurations have a widening flange towards the termination end of the stiffener but this added material is offset by the taper of the stiffener web (*Tapered* configuration, Figure 18b). The third configuration includes a curved tape (*Compliant*, Figure 18c).

The proposed *Compliant* design was developed by considering potentially beneficial local stiffness variations. This resulted in a stiffener design with a similar overall weight to the *Baseline* design. For the modified configurations, various values of *b*, *c* and *d* (see Figure 18b) were analysed [12].

The failure mode of the *Tapered* specimen configuration was found to be a combination of interlaminar and intralaminar failure [12]. The *Tapered* stiffeners had an unexpected delamination in the flange between the bottom-most $0^\circ$ ply and above the $45^\circ$ ply.

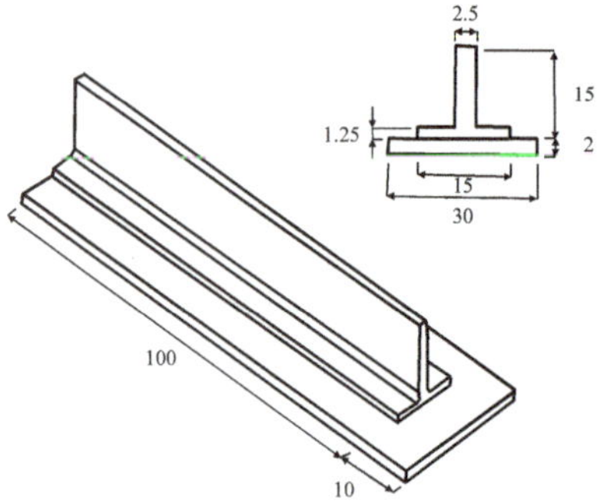


(a) *Baseline* stiffener configuration

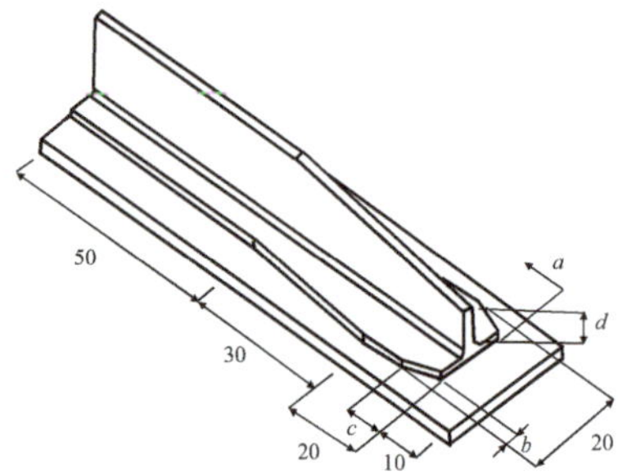


(b) *Tapered* stiffener configuration

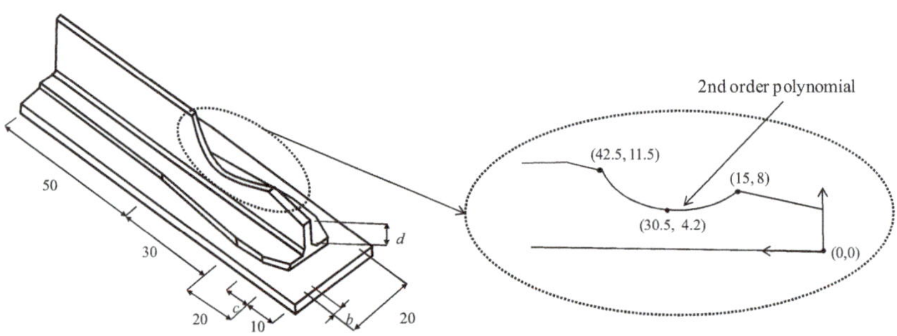


(c) *Compliant* stiffener configuration

**Figure 18.** Stiffener design configurations (dimensions in mm).

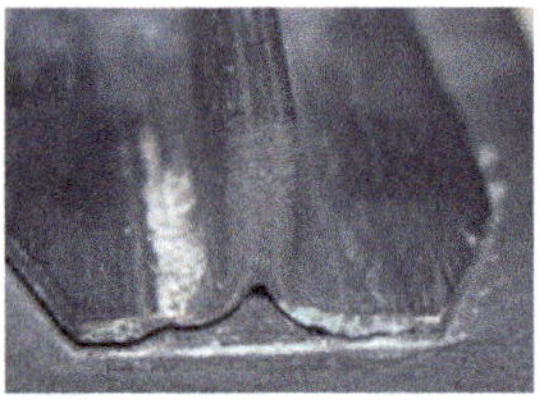

(a)

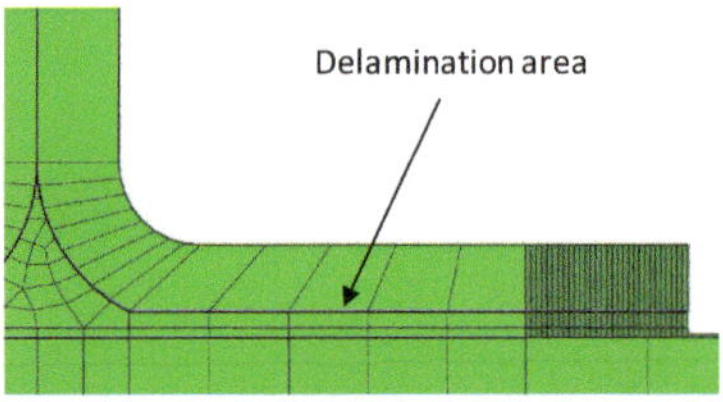


(b)

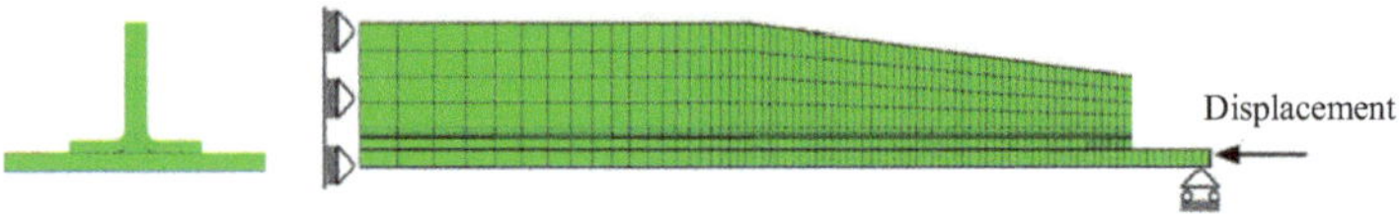


(c)

**Figure 19.** a) *Tapered* stiffener after testing, b) FE model showing delamination path, and c) FE model of a specimen with boundary conditions.

Furthermore, the delamination led to an intralaminar failure in the form of a matrix crack across the 0° ply near the filler ends and continued delaminating between the filler and 0° ply, Figure 19a. For this reason, a new FE model was created in order to investigate these modes of damage, Figure 19b and Figure 19c.

### 4.2. Energy release rate along crack

The three different configurations, *Baseline*, *Tapered* and *Compliant* (see Figure 18), were analysed for debonding and delamination growth stability. The results of this analysis are presented in Figure 20, where the values of $G_T = G_I + G_{II} + G_{III}$, the total strain energy release rate, have been normalized to the $G_T$ of the reference parametric stiffener for 0.5 mm of crack length. Figure 20 compares the normalised strain energy release rate for the *Baseline*, the *Tapered* and the *Compliant* stiffeners. The negative slope of the *G(a)* curve indicates crack growth stability, while a positive slope indicates instability (assuming constant fracture toughness). Recalling the failure modes obtained experimentally [12] the *Baseline* stiffener failed by debonding and the *Tapered* stiffener initially experienced debonding until it finally failed by delamination. This is in agreement with predictions, Figure 20. Consequently, both models were able to correctly describe these experimental results [12]. In addition, the stability analysis for the *Compliant* stiffener predicts that this design will fail stably by debonding, Figure 20.

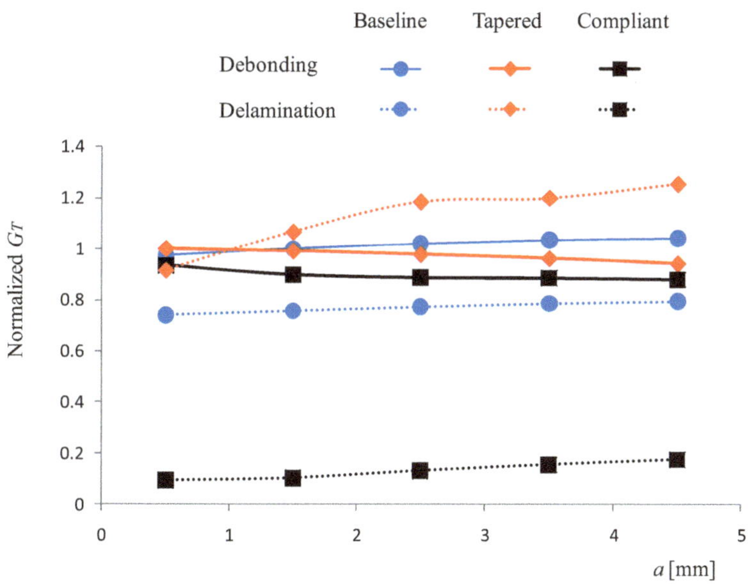


**Figure 20.** Normalized strain energy release rates as a function of crack length; comparison between *Baseline* stiffener design, *Tapered* stiffener and *Compliant* stiffener with $b$=3 mm, $c$=10 mm and $d$=6.25 mm.

### 4.3. Energy release rate along the width of the crack tip

The strain energy release rate along the width of the crack tip was calculated for the *Baseline*, *Tapered* and *Compliant* configurations (Figure 18). Fracture initiation is expected when the $G_T$ exceeds the fracture toughness $G_c$ for a given mixed-mode ratio $G_{II}$ / $G_T$ at each point along the crack tip. In other words, propagation at each point occurs when $G_T$ / $G_c$ >1 [19, 20]. The interlaminar fracture toughness $G_c$ can be calculated by using the equation 2 [20]

The value of $G_c$ is normalised to the width-average value for the *Tapered* specimen. It can be observed that the trend is similar for the *Baseline* and *Tapered* specimen types but is different in the centre of the *Compliant* stiffener. This is due to the difference in the web of the stiffeners. The curved taper has reduced the normalized strain energy release rate in the centre without affecting the trend in the flange. The maximum value of the energy release rate can be used to predict the load corresponding to the initiation of fracture using

$$\frac{P}{P_{FE}} = \sqrt{\frac{G_c}{G_T}} \tag{5}$$

where $P$ is the load at initiation of fracture, $P_{FE}$ is the load from the FE model, $G_c$ the critical strain energy release rate (Equation 1.26), and $G_T$ is the strain energy release rate predicted by the FE model as defined previously. Two different predictions for $P$ can be made: one using the maximum value of $G_T$ along the width, and another using the average, Table 6.

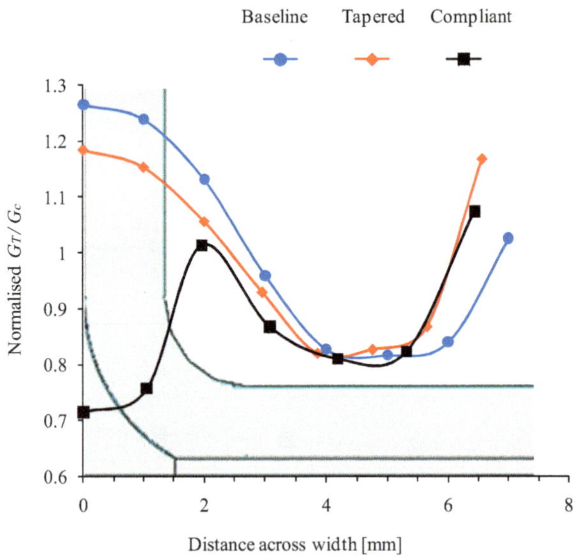


**Figure 21.** Normalized $G_T/G_c$ across the crack tip for crack $a$ = 1 mm for the *Baseline*, *Tapered* and *Compliant* specimens.

## 5. Experiments

The tests were carried out in an Instron testing machine, equipped with a 100 KN load cell, at a loading rate of 0.5 mm/min. Load and crosshead displacement were recorded continuously by a PC data logger connected to the load cell and the Instron machine at a sampling rate of 2 Hz. The specimens were aligned by careful measurement in the loading direction to avoid bending. The Imperial Data Acquisition (IDA) program was used to record load and displacement during the tests.

| | Predicted failure load [kN] (% difference with respect to experimental) | | Experimental failure load [kN] |
|---|---|---|---|
| | Based on $G_{avg}$ | Based on $G_{max}$ | |
| ***Baseline* Stiffener** | 19.00 (+15.2%) | 16.56 (+0.4%) | $16.49^{+0.34}_{-0.39}$ |
| ***Tapered* Stiffener** | 19.17 (+8.2%) | 17.45 (-1.5%) | $17.72^{+0.16}_{-0.22}$ |
| ***Compliant* Stiffener** | 19.93 (+10.6%) | 18.17 (+0.8%) | $18.02^{+0.16}_{-0.29}$ |

**Table 6.** Failure loads for the different specimen types, as well as the predicted failure loads using Eq. 2.

AE sensors were used to identify and investigate failure, within the specimens, during testing. The AE equipment was manufactured by Physical Acoustic Corporation (PAC) and failure was monitored by AEwin software. Broadband (WD) sensors with an operating frequency range of 100 Hz to 1000 kHz were used and positioned in order to obtain the best results without affecting the specimens behaviour [21].

The *Baseline* stiffeners had an average failure load of 16.5 kN while the *Tapered* stiffeners had an average failure load of 17.7 kN and the *Compliant* an average failure load of 18 kN, Table 6. The fracture surfaces for selected specimens are shown in Figure 23, and the load versus displacement curves for selected specimens of the three stiffener designs are shown in Figure 22. The predicted loads (using Eq. 2) match well with the experimental values when the maximum $G$ across the width is used, Table 6.

The acoustic emission signals (Figure 22) show that there was an increase in AE activity 0.1 mm before catastrophic failure for the *Baseline* specimen. For the *Tapered* specimen type, the increase in AE emission started about 0.05 mm before catastrophic failure and for the *Compliant* specimen, 0.02 mm. Figure 22 also shows the peak frequency during the tests for all specimen types. It can be observed that there is some very low-energy micro-cracking from the start of the test and this is possibly at the resin pots. The *Tapered* specimen configuration promoted a combination of failure modes including delamination and fibre bridging which preceded catastrophic failure. In addition, the *Compliant* stiffener, according to AE data and as visually observed (Figure 23c), suffered only from debonding.

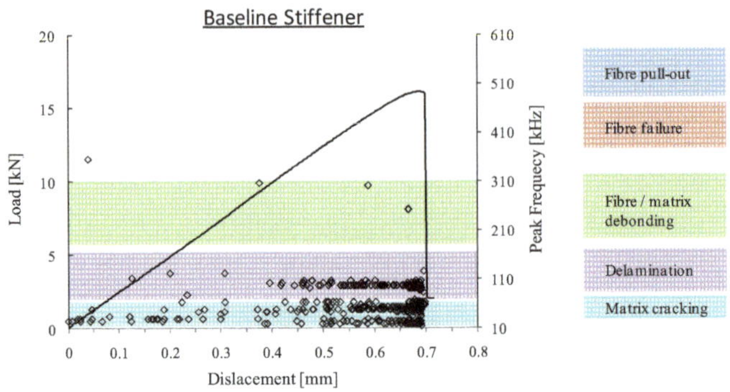


(a)

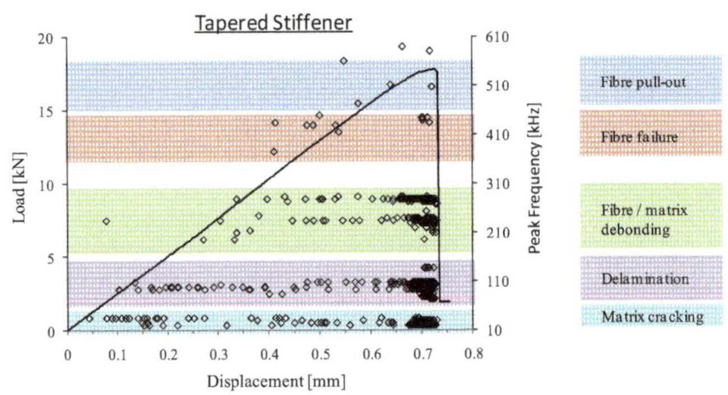


(b)

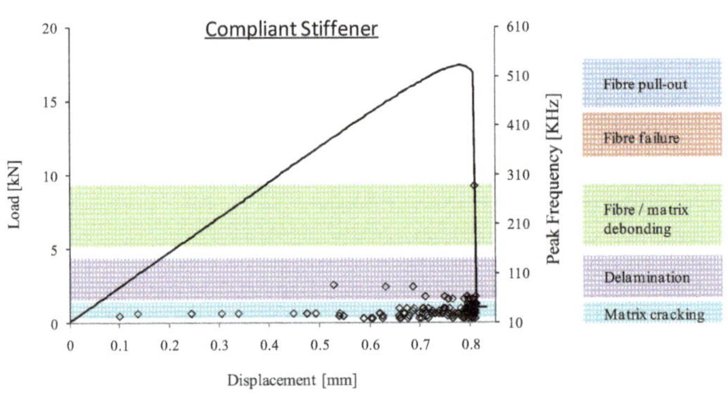


(c)

**Figure 22.** Loads and Peak frequencies versus displacement for a) the *Baseline* b) the *Tapered* and c) the *Compliant* stiffeners. A scale on the right hand side indicates the mode of failure typically associated with these peak frequencies [21].

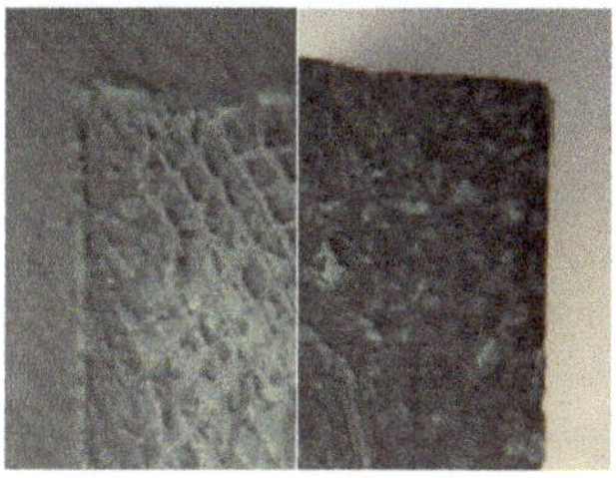

(a)

(b)

(c)

**Figure 23.** (a) *Baseline* stiffener, (b) *Tapered* Stiffener and (c) *Compliant* Stiffener after failure respectively.

## 6. Remarks

The strain energy release rate analysis yielded good results in the investigation of the run-out design influence in debonding/delamination for stiffener terminations. The FE models accurately predicted the failure loads and failure modes for the specimens tested and the predictions were improved when the distributions of the strain energy release rate across the width was considered. The differences in the predictions using the average and the maximum energy release rates are shown in Table 6 and can be compared to the experimental failure loads. The load-displacement, as well as the peak frequency-displacement plots (Figure 22), show that the *Tapered* design is slightly more damage tolerant than the *Baseline* one and this improved further with the *Compliant* design. The AE monitoring proved to be valuable in detecting and analysing the failure modes experienced by the specimens.

## 7. Conclusions

This study was based on the strain energy release rates for debonding and delamination and successfully predicted the failure loads for the three different specimen types. The predictions were more accurate when the maximum strain energy release rate across the width was used. It can be concluded that the variation of the energy release rate across the width should be considered when stiffener run-outs are designed. AE data recorded during skin-stiffener run-out compression tests proved useful to analyse the failure processes which take place in these specimens. The results show that in the design of skin-stiffener run-outs it is important to consider the possibility of failure modes other than debonding, and that compliant termination schemes offer the possibility of improved damage tolerance.

## Author details

S. Psarras* and S.T. Pinho
*Dept. Aeronautics, Imperial College London, SW7 2AZ, United Kingdom*

B.G. Falzon
*Dept. Mechanical and Aerospace Engineering, Monash University, Victoria, Australia*

## 8. References

[1] A. Faggiani and B. G. Falzon, "Numerical analysis of stiffener runout sections," *Applied Composite Materials,* vol. 14, pp. 145-58, 2007.

[2] B. G. Falzon, *et al.*, "Postbuckling behaviour of a blade-stiffened composite panel loaded in uniaxial compression," *Composites Part A (Applied Science and Manufacturing),* vol. 31A, pp. 459-68, 2000.

[3] E. Greenhalgh and M. H. Garcia, "Fracture mechanisms and failure processes at stiffener run-outs in polymer matrix composite stiffened elements," *Composites Part A (Applied Science and Manufacturing),* vol. 35A, pp. 1447-58, 2004.

---

* Corresponding Author

[4] H. Hosseini-Toudeshky, *et al.*, "Analysis of composite skin/stiffener debounding and failure under uniaxial loading," *Composite Structures,* vol. 75, pp. 428-436, 2006.

[5] S. Mahdi, *et al.*, "The mechanical performance of repaired stiffened panels. Part II. Finite element modelling," *Composites Part B:Engineering,* vol. 33, pp. 355-366, 2002.

[6] R. Krueger and P. J. Minguet, "Analysis of composite skin-stiffener debond specimens using a shell/3D modeling technique," *Composite Structures,* vol. 81, pp. 41-59, 2007.

[7] B. G. Falzon and G. A. O. Davies, "The Behavior of Compressively Loaded Stiffener Runout Specimens – Part I:Experiments," *Journal of composite materials,* vol. 37, 2002.

[8] B. G. Falzon and D. Hitchings, "The Behavior of Compressively Loaded Stiffener Runout Specimens – Part II: Finite Element Analysis," *Journal of composite materials,* vol. 37, 2002.

[9] R. Krueger, "Virtual crack closure technique: history, approach and applications," *Appl Mech Rev,* vol. 57:109–43, 2004.

[10] C. Bisagni, *et al.*, "Assessment of the damage tolerance of postbuckled hat-stiffened panels using single-stringer specimens," *AIAA-2010-2696, 51st AIAA/ASME/ASCE/AHS/ASC Structures, Structural Dynamics, and Materials Conference,* 12-15 April , Orlando, Florida 2010.

[11] P. P. Camanho, *et al.*, ". Fracture analysis of composite co-cured structural joints using decohesion elements," *Fatigue Fract Engng Mater struct 27,* pp. 745- 757, 2004.

[12] S. Psarras, *et al.*, "Design of composite stiffener run-outs for damage tolerance," *Finite Elements in Analysis and Design,* vol. 47, pp. 949-954, 2011.

[13] Simulia, *Rising Sun Mills, 166 Valley Street, Providence, RI 02909-2499, USA, ABAQUS 6.10,* 2011.

[14] Python, "Python Software Foundation (PSF)," *Wolfeboro Falls, NH 03896-0037, PO Box 37, USA, Python 2.6.2,* 2010.

[15] D. Hitchings, "Finite element modelling of composite materials and structures," *Cambridge (UK): WOODHEAD PUBLISHING LIMITED,* 2009.

[16] M. K. Benzeggagh and M. Kenane, "Measurement of mixed-mode delamination fracture toughness of unidirectional class/epoxy with mixed-mode bending apparatus," *Composites Science and Technology,* vol. 56, pp. 439-449, 1996.

[17] P. Maimi, *et al.*, "A continuum damage model for composite laminates: Part I-Constitutive model," *Mechanics of Materials,* vol. To appear, 2007.

[18] P. Maimi, *et al.*, "A continuum damage model for composite laminates: Part II-Computational implementation and validation" *Mechanics of Materials,* vol. To appear, 2007.

[19] R. Krueger, *et al.*, "Panel stiffener debonding analysis using a shell/3D modeling technique," *Composites Science and Technology,* vol. 69, pp. 2352-2362, 2009.

[20] M. L. Benzeggagh and M. Kenane, "Measurement of mixed-mode delamination fracture toughness of unidirectional glass/epoxy composites with mixed-mode bending apparatus," *Composites Science and Technology,* vol. 56, pp. 439-49, 1996.

[21] R. Gutkin, *et al.*, "On acoustic emission for failure investigation in CFRP," *Mechanical Systems and Signal Processing 25,* vol. 1393–1407, 2011.

# Chapter 16

# Finite Element Analysis of Loading Area Effect on Sandwich Panel Behaviour Beyond the Yield Limit

Salih Akour and Hussein Maaitah

Additional information is available at the end of the chapter

## 1. Introduction

Research efforts continuously are looking for new, better and efficient construction materials. The main goal of these researches is to improve the structural efficiency, performance and durability. New materials typically bring new challenges to designer who utilizes these new materials. In the past decades various sandwich panels have been implemented in aerospace, marine, architectural and transportation industry. Light-weight, excellent corrosion characteristics and rapid installation capabilities created tremendous opportunities for these sandwich panels in industry. Sandwich panel normally consists of a low-density core material sandwiched between two high modulus face skins to produce a lightweight panel with exceptional stiffness as shown in Figure 1. Face skins act like flanges of an I-beam. These faces are typically bonded to a core to achieve the composite action and to transfer the forces between sandwich panel components.

### 1.1. Main principles of sandwich structures

Typical sandwich composite construction consists of three main components as illustrated in Figure 1. The sandwich consists of two thin, stiff and strong faces are separated by thick, light and weaker core. Faces and core materials are bonded together with an adhesive to facilitate the load transfer mechanism between the components, therefore effectively utilize all the materials used. The two faces are placed at a distance from each other to increase the moment of inertia, and consequently the flexural rigidity, about the neutral axis of the structure.

In sandwich structure, typically the core material is not rigid compared to face sheets; therefore, the shear deflection within the core is insignificant in most cases. The shear deflection in the faces can be also neglected. The effect of shear rigidity in the core is shown in Figure 2. Figure 2a shows an ideal sandwich beam using relatively stiff core, therefore the two faces cooperate without sliding relative to each other. Figure 2b shows a sandwich beam using weak core, therefore the faces are no longer coupled together effectively and each face works independently as plates in bending. The use of weak core in shear results in significant loss of the efficiency of the sandwich structures. In a typical sandwich panel the faces carry the tensile and compressive stresses. The local flexural rigidity of each face is typically small and can be ignored. Materials such as steel, stainless steel, aluminum and fiber reinforced polymer materials are often used as materials for the face. The core has several important functions. It has to be stiff enough to maintain the distance between the two faces constant. It should be also rigid to resist the shear forces and to prevent sliding the faces relative to each other. Rigidity of the core forces the two faces to cooperate with each other in composite action. If these conditions are not fulfilled, the faces behave as two independent beams or panels, and the sandwich effect will be totally lost. Furthermore, rigidity of the core should be sufficient to maintain the faces nearly flat, therefore prevent possibility of buckling of the faces under the influence of compressive stress in their plane. The adhesive between the faces and the core must be able to transfer the shear forces between the face and the core.

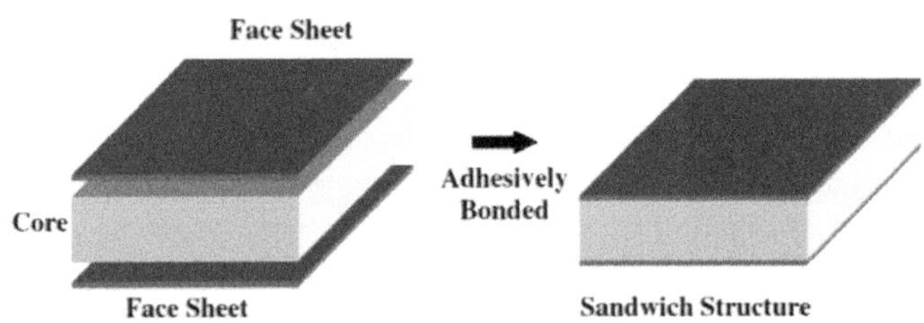


**Figure 1.** Schematic of sandwich construction

## 1.2. Applications

Sandwich construction provides efficient utilization of the materials used for each component to its ultimate limit (Zenkert, 1997). The sandwich structure offers also a very high stiffness-to-weight ratio. It enhances structure flexural rigidity without adding substantial weight and makes it more advantageous as compared to composite materials. Sandwich constructions have superior fatigue strength and exhibit superior acoustical and thermal insulation. Sandwich composites could be used in a wide variety of applications such as:

*Aerospace Industry:* Sandwich composites are increasingly being used in the aerospace industry because of their bending stiffness-to-weight ratio. Floorboards, composite wing, horizontal stabilizer, composite rudder, landing gear door, speed brake, flap segments, aircraft interior and wingspans are typically made of sandwich composites.

*Marine Industry:* Sandwich composites are ideally suited for the marine industries most advanced designs. The foam cores meet the critical requirements of strength, buoyancy and low water absorption. Applications include the construction of bulkheads, hulls, decks, transoms and furniture.

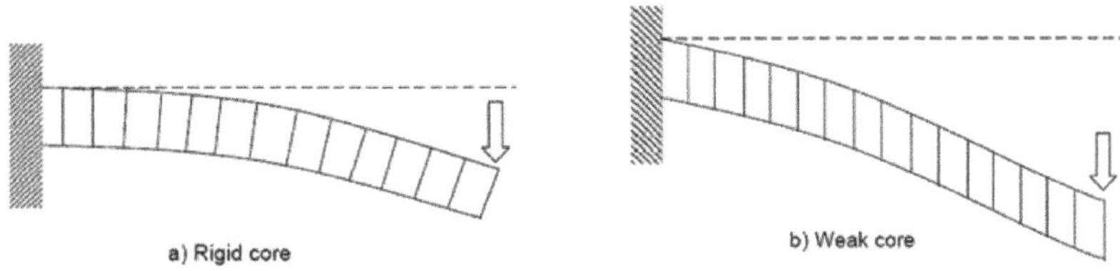


**Figure 2.** Presentation of the effect of a) rigid and b) week core.

*Transportation Industry:* High strength-to-weight ratios of sandwich composites offer great advantages to the transportation industry. The insulating, sound damping properties and low cost properties make them the choice materials for the constructions of walls, floors, doors, panels and roofs for vans, trucks, trailers and trains.

*Architectural Industry:* The foam offers an excellent thermal and acoustical insulation which makes it ideal choice for the architectural industry. Typical applications include structural columns, portable buildings, office partitions, countertops and building facades.

## 1.3. Literature review

Work on the theoretical description of sandwich structure behaviour began after World War Two. (Plantema, 1966) published the first book about sandwich structures, followed by books by (Allen, 1969), and more recently by (Zenkert, 1995). Although (Triantafillou and Gibson, 1987) developed a method to design for minimum weight, and reported the failure mode map of sandwich construction, without considering the post yield state of the sandwich structure.

The basic sandwich structure theory presented in all these texts is generally called the classical sandwich theory. This theory assumes that:

- The core carries the entire shear load in sandwich beams and plates.
- The face sheets carry the entire bending load.
- Core compression is negligible.
- This theory states that the above–mentioned assumptions are true if:
    1. The core and face sheets are elastic.
    2. The overall length to thickness ratio is high.
    3. The face sheet thickness is small compared to the overall thickness.
    4. The ratio of mechanical properties between the face sheet and the core is high.

With these assumptions, a sandwich structure is considered to be incapable of acquiring additional load carrying capacity once the core yields.

(Mercado and Sikarskie, 2000) reported that the load carried by sandwich structures continue to increase after core yielding. Knowing that the core could not carry additional load after yield, this increasing load carrying capacity of post yield sandwich structure initiates the postulation that the additional shear load was transferred to the face sheets. To account for the above-mentioned phenomenon, (Mercado et al, 1999) developed a higher order theory by including a bilinear core material module. This theory yields a fairly accurate prediction on the deflection of a foam cored sandwich structure in four point bending (Mercado et al, 2000). In addition, this theory does not take into account the core compression under localized load, or any geometric non-linearity. The classical sandwich beam theory also assumes that in-plane displacements of the core through its depth are linear. In other words, it was assumed that the core thickness remains constant and cross-sections perpendicular to the neutral axis remain plane after deformation. This assumption is generally true for traditional core material such as metallic honeycomb. However, this assumption is not suitable for soft, foam-based cores, especially when the sandwich structure is subjected to a concentrated load (Thomsen, 1995). With a much lower rigidity compared to metallic honeycomb, foam-based cored sandwich structures are susceptible to localized failure. Insufficient support to the face sheets due to core compression near the application points of concentrated loads can lead to failures such as face sheet/ core delamination, face sheet buckling, and face sheet yielding. This localized non-linearity is reported by many researchers such as (Thomsen, 1995), (Thomsen, 1993), (Rothschild 1994), (Caprino, 2000), and (Gdoutos et al, 2001). The shear distribution at localized failure points has not been well defined. (Miers, 2001) investigated the effect of localized strengthening inserts on the overall stiffness of a sandwich structure. This localized strengthening increases the rigidity of the sandwich structure, but the addition of high stiffness inserts complicates the manufacturing process of sandwich structure.

To design an efficient sandwich structure, it is vital to understand the behavior of each layer in the structure. Classical sandwich theory (Zenkert 1995, Plantema 1966, Allen 1969), higher order theory by Mercado (2000) and high order theory developed by Frostig et al. (1992) could predict the sandwich panel behavior fairly accurate in the linear range. However, these theories could not give an accurate prediction of the sandwich structure behavior after core yielding. Large deflection of sandwich structures due to core yielding could vary the direction of the applied load on the structure.

### 1.4. Research objective

To design an efficient sandwich structure, it is vital to understand the load distribution pattern in each layer of the structure. Most of the previous efforts are made by using classical sandwich theory, and higher order theory, where high order theory predicted the sandwich panel behavior fairly well in the linear range. However, these theories could not give an accurate prediction of the shear distribution in each layer after core yielding. Large deflection of sandwich structures due to core yielding could vary the direction of the applied load on the structure. Change in loading direction would obviously change the shear distribution in the sandwich structure. In order to investigate the exact change of shear distribution due to distributed loads, as well as geometric nonlinearity and localized core failure, finite element analysis is used in this research effort. The main objective of this research is to investigate the following:

1. Post yield behavior of sandwich panel.
2. Effect of geometric non-linearity under distributed loads.
3. The effect the size of the distributed load area on the behavior of the sandwich panel beyond the core yield limit for different types of materials is investigated. These parameters are the determining factors of the significance of geometric non-linearity and core material nonlinearity

The above investigation is done in view of the following points:

1. Localized core yielding occurs mainly through core compression. Therefore, analysis should be done using material properties determined from compression test.
2. For practical purposes, the assumptions that have been made in developing the sandwich panel theory eliminated part of the problem physics.
3. The Finite Element Model (FEM) is extended to include the relative dominance of core shear failure and face sheet yielding.
4. Localized loads are modeled as load on small partitioned area to better simulate the actual loading condition.
5. Experimental verification is conducted for selected cases.

## 2. Physical model

This section presents the physical model of the sandwich panel, which includes geometry, boundary conditions as well as the materials used in the investigation.

### 2.1. Sandwich panel geometry

The sandwich panel consists of two face sheets made of metal. The thickness of each face is **t**. Soft core of **c** thickness is sandwiched between those face sheets. The core material is made of foam which is soft compared to the face sheets .The panel is square in shape. The side length is designated by **a.** Figure 3 illustrates the sandwich panel geometry whereas the dimensions of the sandwich panel are shown in Table 1.

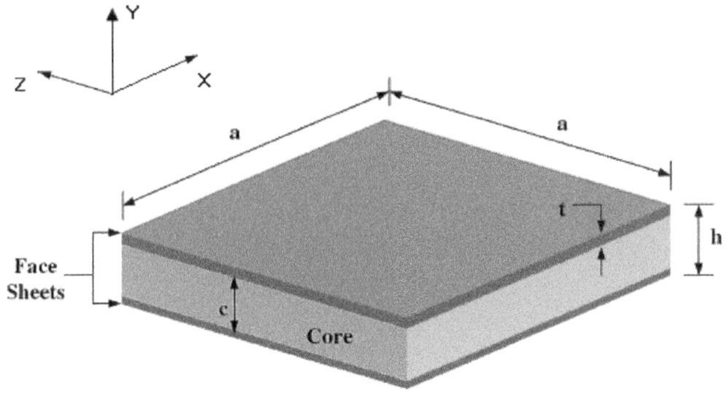


**Figure 3.** Illustration sandwich plate geometry.

| Parameter | Dimension | Note |
|---|---|---|
| a | 600 mm | Constant |
| t | 1.0 mm | Constant |
| c | 30 mm | Constant |

**Table 1.** The value of the parameters shown in Figure 3.

### 2.2. Assumptions

This research takes into consideration the geometric non-linearity as well as the material nonlinearity. The following assumptions are made to simplify the model without losing the problem physics:

1. Face sheets and core are perfectly bonded.
   The FEM model assumes no delamination occur between layers.
2. Face sheets remain elastic at all time.
   Due to the significantly higher yield strength and modulus of elasticity of the face sheets compared to the core, face sheets are assumed to remain elastic throughout the loading for simply supported panel. The analysis stops when the face sheets start to yield.
3. Geometric non-linearity has a significant effect:
   Geometric non-linearity is considered to have significant effect on the load distribution on each layer of the sandwich structure.

### 2.3. Boundary condition

Due to the symmetry of the sandwich panel (symmetric over X-axis and symmetric over Z-axis), only quarter of it is being modeled. Such symmetric boundary conditions are applied of the X-axis and Z-axis. The two planes of symmetry of the panel have symmetric boundary conditions, (see Fig. 4). A simply supported boundary condition is applied to strip area of the quarter panel as shown in Fig. 5. This simulates the simply supported condition of the panel. The loading area is square in shape, its side length varies in steps of 100, 200, 400 and 600mm for full panel dimension. But when dealing with quarter panel, the side length is 50, 100, 200, and 300mm

### 2.4. Study parameters

The main parameters that have influence on the performance of the sandwich plate are, the loading area on which the load is distributed and the core material stiffness.

#### *2.4.1. Loading*

The load is applied to the sandwich top face sheet as a distributed load which is increased gradually (step by step) till the face sheet stress reaches yield stress or the core material reaches fracture limit. The distributed load is applied on the top surface of the sandwich panel. The area on which the distributed load is applied (see Figure 5 and 7) is located at the middle of the top face sheet plate. The loading area at the middle top face of sandwich panel is square in shape. This area has been varied from 100X100 mm$^2$ through 200X200 mm$^2$, 400X400 mm$^2$, and 600X600 mm$^2$ so the ratio of these areas relative to the total area of the sandwich panel is 1/36, 4/36, 16/36 and 36/36 respectively.

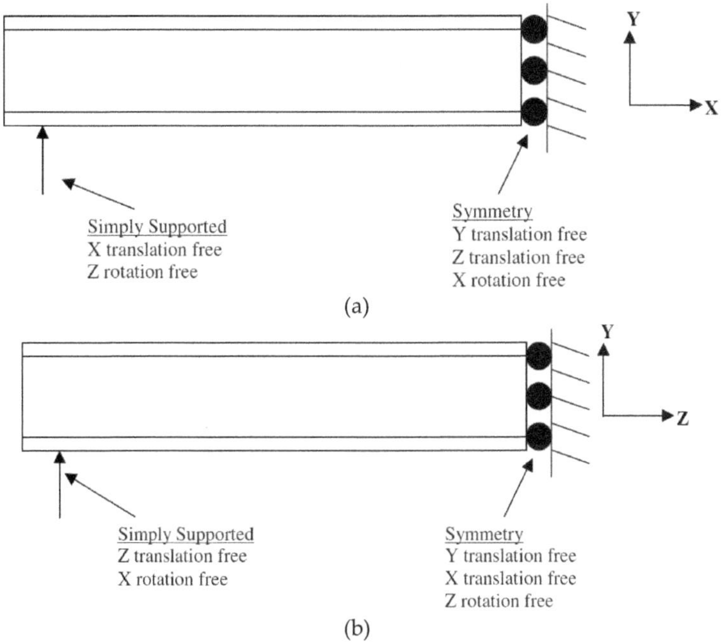


**Figure 4.** Sandwich panel boundary condition for a) X-Y plane and b) Y-Z plane.

### 2.4.2. *Core material*

In the current research, different materials are used. Their modulus of elasticity is varying from 37.5 MPa through 138.6 MPa, 180 MPa, and 402.6 MPa as shown in Table 2. Core thickness is selected to be 30mm as shown in Table 1.

## 2.5. Material properties

The core of sandwich structure is used to separate the two faces, most often identical in material and thickness, which primarily resist the in plane and bending load. The core is mainly subjected to shear so that the core shear strain produces global deformations and core shear stresses. Thus, core must be chosen such that not to fail under applied transverse load. It should have shear modulus that is high enough to give the required stiffness.

Furthermore, its young's modulus normal to the faces should be high enough to prevent contraction of the core thickness and therefore a rapid decrease in flexural rigidity. The core should have low density in order to add as little as possible to the total weight of sandwich structure. Because of low density requirement, core materials are very different from face sheet materials. A detailed characterization of their mechanical behavior is essential for their efficient use in structural application. Four types of foam H100, H250, AirexR63.50 and Herex C70.200 are investigated.

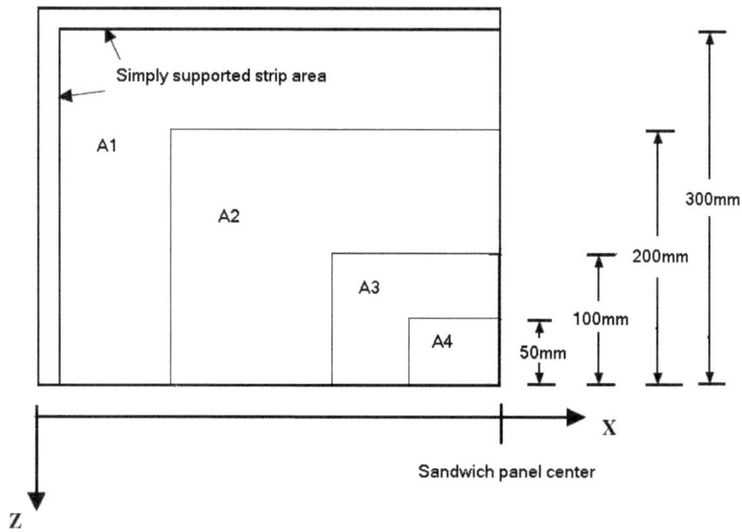


**Figure 5.** Panel span overview of quarter sandwich panel for different loading area

### *2.5.1. Mechanical properties for face sheet*

Material properties for the sandwich plate face sheets are taken from (Boyer and Gall (Eds.), 1991). Aluminum 3003-H14 is a type of aluminum alloy that has high resistance to corrosion and is easy to weld is used in this investigation. The 3003-aluminum family is normally used in the production of cooking utensils, chemical equipment, and pressure vessels. The face sheets are assumed to remain elastic at all times. Therefore only elastic material properties are required for the face sheets and they are presented in Table 2.

### *2.5.2. Mechanical properties for core*

This subsection presents the core material properties used to model the sandwich panel. In all cases, face sheets of the sandwich structures are assumed to remain elastic throughout the analyses. Therefore, only core materials require a good post yield behavior descriptions. The core materials undergo plastic deformation; hence there is a need to obtain a full description of the core materials' behavior upon yield initiation.

Airex R63.50 has high fatigue strength, high three-dimensional formability, and high resistance to dynamic loads. Materials in Airex R63 family are widely used in the production of marine hulls and lightweight cars due to the appreciation of their low density and high strength and stiffness to weight ratio. Airex R63.50 is presented in Table 2.

Material properties of the HerexC70.200 foam core is obtained from (Rao, 2002) work. Herex C70.200 is an isotropic and stiff foam material with high stiffness and strength to weight ratios. The materials in Herex C70 family have excellent chemical resistance and low thermal conductivity and water absorption. The appreciation of these inherent properties of Herex C70 materials makes this material a popular choice for the core materials of structural sandwich structures in marine and railway applications. The stress strain curve of this material is presented in Figure 6.

In this research a first-order idealized core material property module suggested by (Mercado, Sikarskie, 1999) is used. This first-order idealized model, also called the bi-linear model, describes the material properties of the core with the stress strain curve as shown on Figure 6a and 6c.

The other material used in this research is linked PVC close called cellular foam (divinycell). The type of divinycell, H100, H250 with densities of 100 and 250 kg/m$^3$, their mechanical properties are stated in Table 2 and their stress strain curves are shown in Figure 6b and 6d respectively.

| Material | Property source | Young's modulus (MPa) | Poisson's ratio | Shear modulus (Mpa) | Shear strength (Mpa) | 0.2% offset yield strength (Mpa) | Strain at yield popup (mm/mm) |
|---|---|---|---|---|---|---|---|
| Face sheet : Aluminum 3003-H14 | Boyer and Gall 1991 | 69,000 | 0.33 | 25,000 | 120 | 145 | Not available |
| Core A : AirexR63.50 | Rao, 2002 | 37.5 | 0.335 | 14.05 | 0.45 | 0.637 | 0.019 |
| Core B: H100 | Kuang, 2001 | 138.6 | 0.35 | 47.574 | 1.2 | 1.5 | 0.0108225 |
| Core C: Herex C70.200 | Rao, 2002 | 180 | 0.37 | 65.69 | 1.6 | 2.554 | 0.0162 |
| Core D: H250 | Kuang, 2001 | 402.6 | 0.35 | 117.2 | 4.5 | 5 | 0.014 |

**Table 2.** Compression of sandwich panel material properties

## 3. Finite element model

This section presents the development of finite element models for simply supported sandwich panel. Detailed descriptions of the boundary conditions, element types, and the loading are presented in the coming subsections. The finite element software used in the development of the finite models is (I-DEAS Master Series 10 1999). The relatively robust and user-friendly solid modeling and finite element meshing interface are the main advantages of this solid modeling and finite element software.

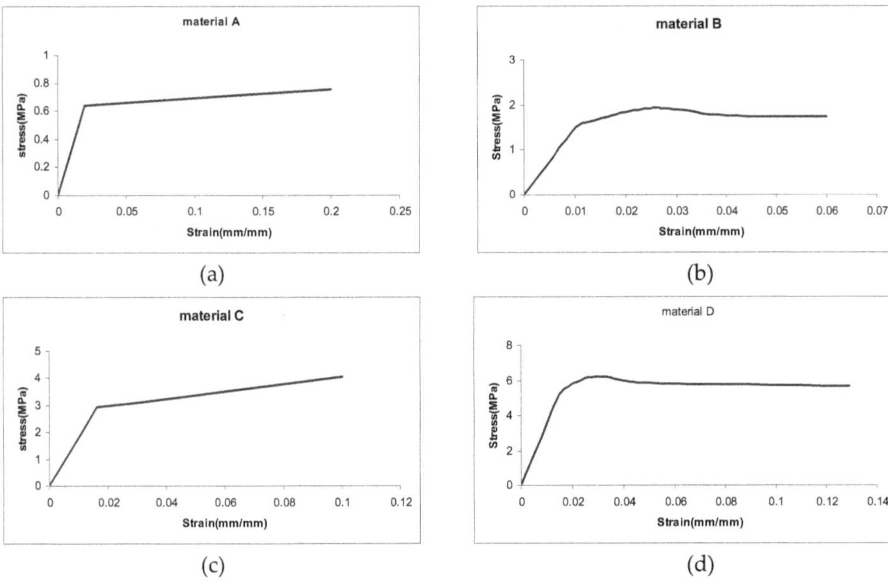


**Figure 6.** Stress strain curve for a)material A: AirexR63.50 (Rao, 2002), b) material B: H100 (Kuang, 2001), c)material C: Herex C70.200 (Rao, 2002), d) material D: H250 (Kuang, 2001)

## 3.1. Model assumptions

All the finite element model analyses done in this research involves the use of non-linear analysis capability of I-DEAS, which includes geometric non-linearity and material nonlinearity. With geometric non-linearity, the software takes the effect of geometry changes into account while calculating the solution. Using material non-linearity option the non-linear behavior of the material response (i.e. post yield material properties) is taken into account.

Below are the assumptions made for the Finite Element Model:

1. Face sheets and core are perfectly bonded:
   The numerical model assumes no delamination occur between layers. This assumption is applied by utilizing the partitioning option in the preprocessing module of the software. This option allows the analyst to deal with the whole volume of the structure as one unit also it allows the analyst to assign different material for each partitioned volume.

2. Face sheets remain elastic all the time:
   Due to the high yield strength and high modulus of elasticity of the sandwich face sheets compared to the core, face sheets are assumed to remain elastic throughout the loading for the simply supported panel.
3. Load scenarios are quasi-static:
   The loading cases considered are modeled quasi-static instead of dynamic. Incremental loadings are applied slowly during the actual experiments (i.e. simulates exactly the real situation). Therefore, the type of analysis done for this research effort is "static, non-linear analysis".
4. Geometric non-linearity has a significant effect:
   Geometric non-linearity is considered to have significant effect on the load distribution on each layer of the sandwich structure. Therefore, all finite element analysis that is done takes into consideration the geometric non-linearity. This is the main difference between the numerical models and the theoretical models. Classical sandwich plate theory and higher order theory do not take shape change of the sandwich structures into account.
5. The panel is simply supported from all sides. It is partitioned into three layers, forming three bonded material layers.

### 3.2. Finite element mesh and boundary conditions

The symmetric nature of the problem allows only quarter of the whole panel to be meshed. The boundary conditions applied are shown on Figures 4 and 5. The two planes of symmetry of the panel have symmetric boundary conditions, where in-plane displacements and rotation about an axis respective normal to the symmetry plane is allowed. A simply supported boundary condition is applied to the two other sides of the quarter panel. A distributed load is applied on the top surface of the sandwich panel. The area in which the distributed load is applied is varying as shown in Figures 5 and 7.

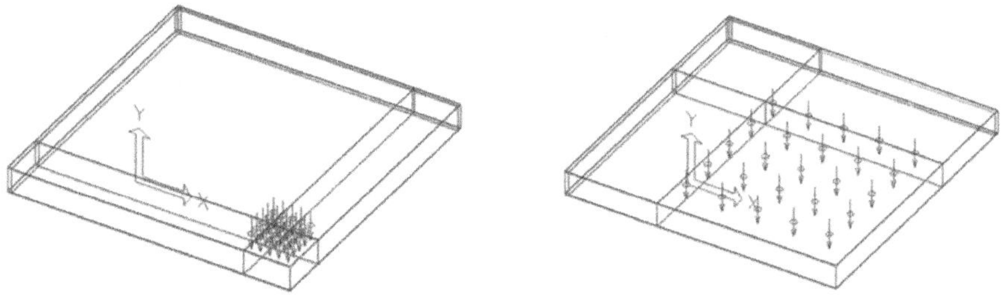

(a) (b)

**Figure 7.** The loading area with side length a) 50 mm and b) 200 mm.

The panel is loaded with a set of loads that are varying slowly with time, and the analysis is carried out at each load step. The finite element software is set in such a way to solve the model at each load step. This allows all the analysis to be done in a single run of the finite element model. As a result of this, the model would consume less memory space because one single solid model and finite element model can be used for all load steps.

The numerical model utilizes the map meshing facility in I-DEAS. By controlling the number of nodes along each edge of the solid model, this function provides full control of the mesh size. The element size is chosen by referring to (Miers, 2001) work in mesh refinement. (Mires, 2001) recommended a core element size of 1.5 mm and face element size of 3 mm in order to achieve convergence in the data obtained. For the current case constant mesh density is ensured with the mapped meshing function. This is important because constant mesh density ensures that the data collected from any region in the panel are of the same degree of resolution. Three-dimensional solid brick elements (20 node brick element) are used in this analysis. Second order (parabolic) brick elements are chosen over the first order (linear) brick elements in order to better interpolate the data between nodes. Figure 8 shows the FEM mesh model of the sandwich panel and the brick element utilized in FEM.

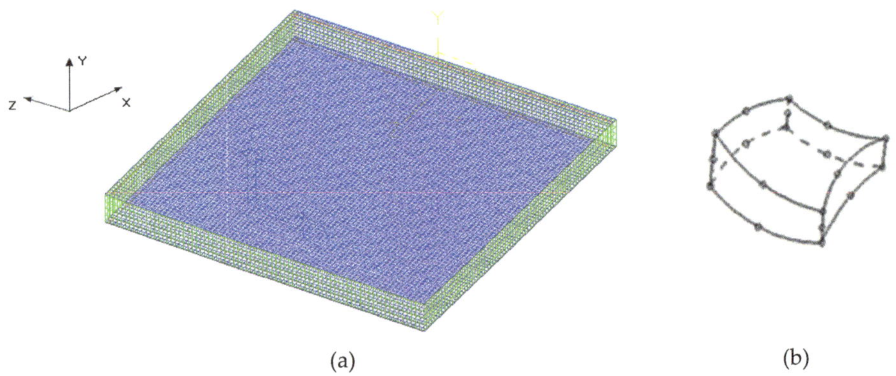


**Figure 8.** Illustration of a) Meshed quarter sandwich panel and b) Solid Brick Element (20 node brick element) used in mesh generation.

Since the analysis involves material non-linearity, a yield function or yield criteria needs to be defined for the model. Von Mises yield criteria and its associated flow rule is used in this analysis. Isotropic hardening is also used to describe the change of the yield criterion as a result of plastic straining. Only the core elements are assigned a yield function due to the assumption that only core yielding occurs throughout the loading process. The face sheets are assumed to remain elastic at all time; hence no yield function needs to be assigned to the face sheet elements. However the yield point of the face sheet material is fed to the software to be used as indicator for stopping the analysis.

### *3.2.1. FEM challenges*

The following challenges are experienced:

- One of these challenges is extracting the element force and storing them in a file. This problem is solved by displaying the data on the screen and then copied and stored in a separate file for further analysis.
- Identifying the nodes for a surface of interest so that they can be extracted from the file in which the elemental forces are stored. Since I-DEAS labels the nodes, the nodes corresponding to the surface of interest are copied and stored in a separate node labels file for further analysis.
- MATLAB program is developed to extract the elemental forces of the surface of interest from the file in which they are stored by matching the node labels of the surface that are stored in node labels file.
- Singularity is a serious problem. The post processing analysis for the quality of the elements is utilized to identify the poor elements. The problem is solved by refining the element size.

#### 3.2.2. *Advantages of FEM*

The following are some advantages of using FEM over other methods:

- FEM is capable of capturing the problem details with little approximations compared to the analytical techniques.
- FEM provides solution for many problems like the current case that they do not have analytical solution.
- FEM method is cheap compared to the experimental models. There is no need to produce a prototype or to have high tech facility to conduct the investigation.
- There is no need for the investigator to be available in a certain place to perform the investigation.

### 3.3. FEM verification

The finite element model is verified analytically and experimentally. The analytical verification is based on the classical sandwich panel theory whereas the experimental investigation is carried out for selected cases.

#### 3.3.1. *Analytical verification*

Classical sandwich theory has been utilized to obtain close form solution (Zenkret, 1995). The comparison between the numerical and theoretical models in the linear rang are presented in Figure 9. The Figures show very good agreement between theoretical and numerical solution. The classical sandwich plate theory is therefore used to compare and validate the FEM predicted shear distribution of the panel in the linear range. Comparison between the FEM determined shear distribution and the classical sandwich plate theory distribution is performed at all load steps. It is assumed that is the core in the linear range carries the entire shear load. Results obtained from the closed form solution are compared with the total resultant shear load in the global Y direction, $R_{TOT}$ $(Yg)$, obtained numerically using MATLAB.

Sample of the total shear resultant comparisons between the numerical and theoretical models in the linear rang are shown Figure 9a and 9b for load of 17.2 kPa and 51.7 kPa respectively.

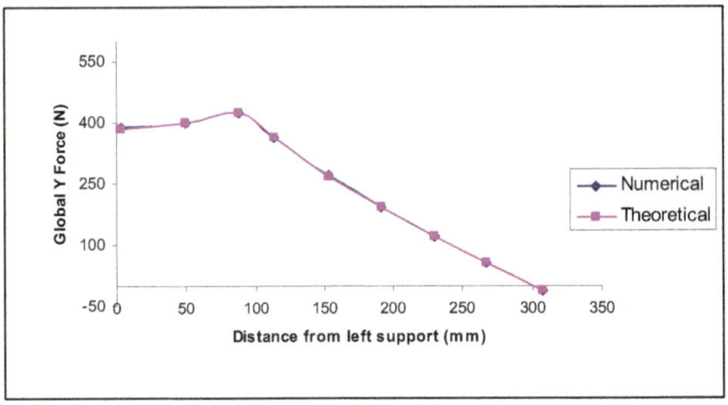


(a)

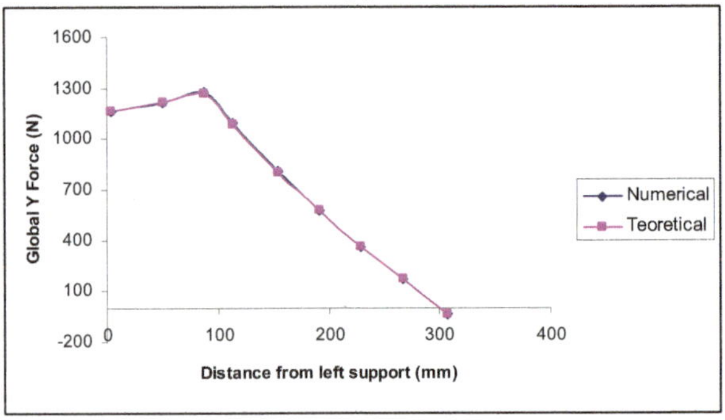


(b)

**Figure 9.** Total plate shear distribution comparison along X-axis at a) 17.2 kPa and b) 51.7 kPa.

### *3.3.2. Experimental verification*

To assure accuracy and validity of the results some selected cases are investigated experimentally. The results obtained from the FEM are compared against those obtained experimentally. Both results show excellent agreement.

#### *3.3.2.1. Test setup*

Here is a description of the experimental setup used in the study and consists of the following:

1. The core of the sandwich panel is made of polyurethane foam. Top and bottom sheets of the sandwich panel are made of steel. The dimension of the panels used in the investigation is 250X250 mm². Mechanical properties of the sheet metal are obtained experimentally.

2. Fixture for applying simply supported boundary condition is produced. Figure 10 shows two different views of the fixture.
3. The test is performed on a uniaxial testing machine that is shown in Figure 11.
4. Distributed load is applied to the specimen by adaptors manufactured for this purpose. Figure 12 illustrates the adapters used in experimental setup.

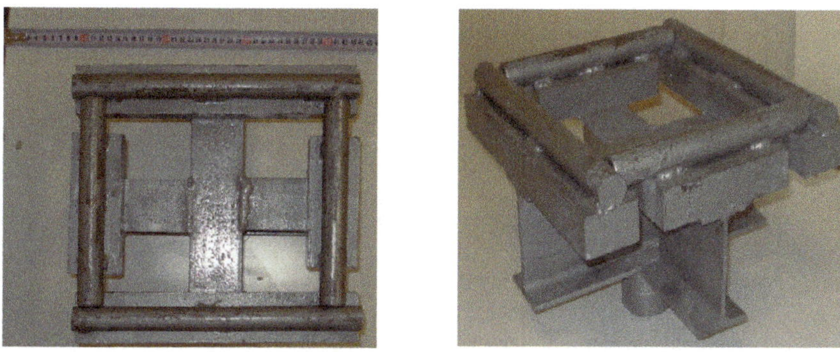

(a) (b)

**Figure 10.** Pictures of the fixture that is produced for applying simply supported boundary condition, a) top view and b) 3D view.

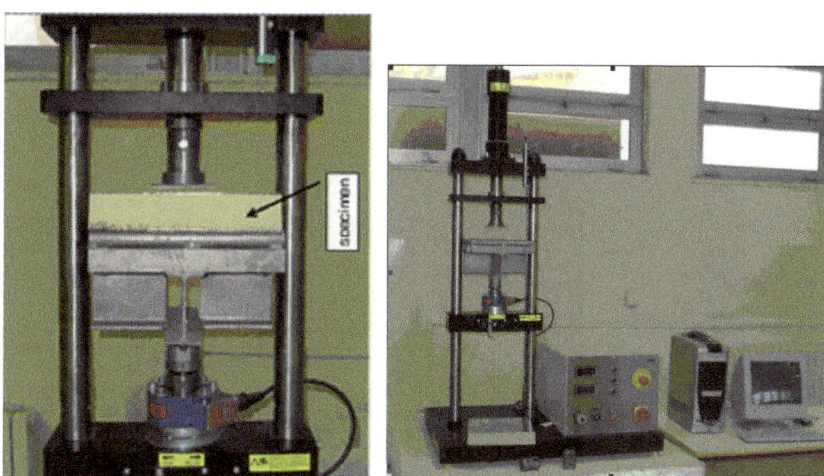


(a) (b)

**Figure 11.** Uniaxial testing machine a) with specimen and b) without specimen

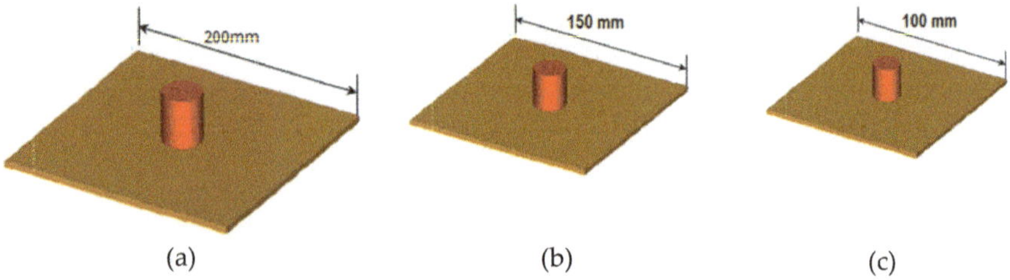


**Figure 12.** The adapters used in the experiments for applying distributed load on specimen of side length a) 200mm, b) 150mm and c) 100mm.

*3.3.2.2. Mechanical properties of the specimen*

The sandwich panel is made of polyurethane foam and steel sheets. The mechanical properties are obtained experimentally for both the sheets and the core. ASTM Designation: C 365 – 00 used for testing the core material whereas ASTM Designation: D 638 – 00 used for testing the sheets.

*3.3.2.3. Analysis*

The relation between the applied load and the deflection of the specimen center point are shown in Figures 13 and Figure 14 that present a comparison between the experimental results and FEM results. It may be seen that the results are in very good agreement.

To assure accuracy of the experimental results, the experiment is performed many times and the average values are plotted. The variation in the experimental results dose not exceeds 7% of the average value.

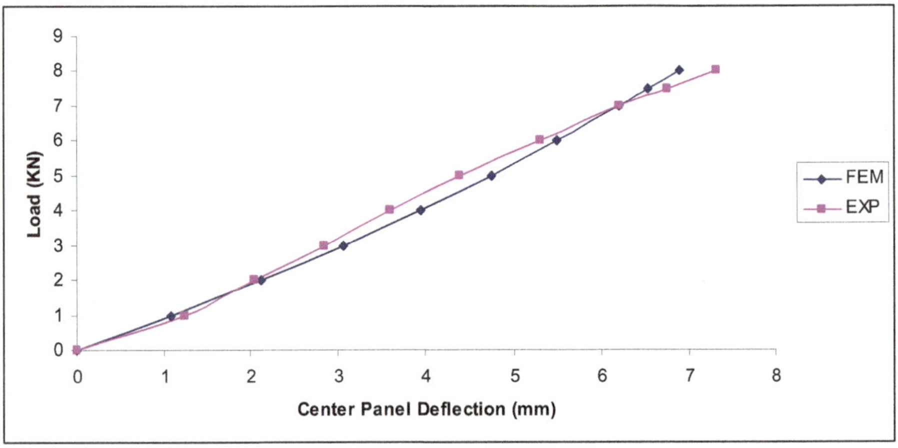


**Figure 13.** Comparison of load versus center deflection for core thickness = 49 mm, Sheet Thickness = 0.5 mm, applied load area = 200 X 200 mm$^2$.

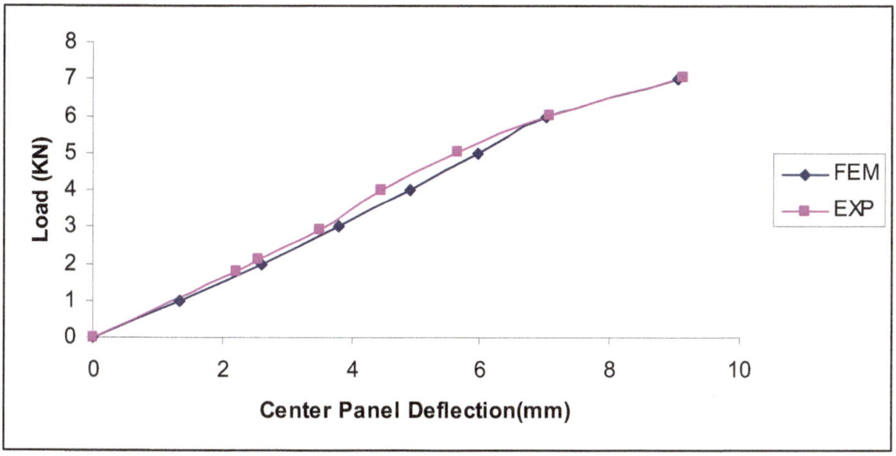


**Figure 14.** Comparison of load versus center deflection for core thickness=71 mm, sheet Thickness = 0.5 mm, applied load area = 150 X 150 $mm^2$.

## 4. Results

The main advantage of these results over the sandwich panel theory is that both geometric and material nonlinearities are considered without approximation. Usually these approximations eliminate part of the problem physics. By utilizing "I-DEAS' post processing module, stress and its all components, strain and it is all components including the plastic strain, and deformations are obtained.

Figures 15a and 15b present Von Mises stress contours for both panel and core respectively whereas Figures 16a and 16b present the plastic strain for both panel and core respectively. It is clear from Figure 16a and 16b that the plastic deformation occurs close to the panel support (close to the area where boundary conditions are applied).

The criterion, which is adopted by this investigation at what load step the FEM should stop the analysis, is when any of face sheets starts to yield or core material reaches fracture limit. This criterion fulfills the need of the designer; in general design engineer tries to avoid panel face sheets permanent distortion. As soon as the face sheet metal starts to yield, this means that permanent deformation is taking place. So all results produced neither exceed the loading that could cause face - sheet yielding nor exceed core fracture limit.

Figure 17a present the effect of loading area (area on which the load is applied) for core material A. It is obvious as the loading area increases the stress decreases for the same amount of loading. Same thing can be said for the bottom face sheet in Figure 17b. The core material (Figure 17a) reaches yield at low loads when the loading area is small.

The effect of loading area at sandwich panels of cores A, B, C and D (see Table 2) are presented 18 through 21. The maximum shear stress of each core in these graphs is normalized by the maximum shear yield strength of its corresponding core material and the loading area is normalized by the total surface area of the panel. Also the shear stress of the face sheets is normalized by the corresponding shear yield strength of the face sheets. It can be seen from Table 2, the core materials are labeled from A to D in ascending order according to their stiffness. It is obvious from Figures 18 through 21 that the load carrying capacity of sandwich panel increases by increasing core stiffness. It is observed through all the results that the lower face sheet reaches yield limit before the top face sheet so in the Figures 18 through 21 the lower face sheet is presented. The results of this work are generated according to the univariate search optimization technique (Chapra and Canal, 2006).

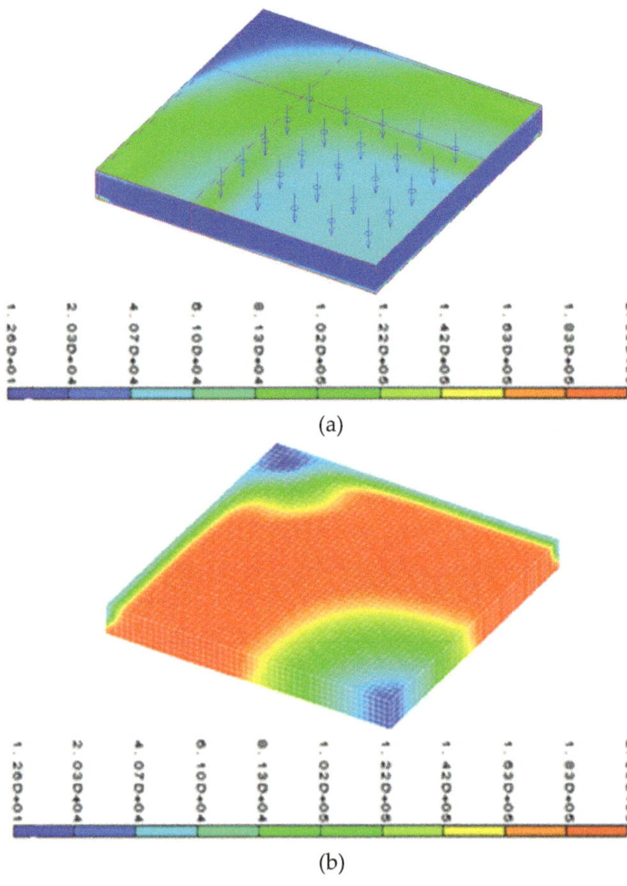


**Figure 15.** Von Mises stress contour (in MPa) for panel A of loading area 4/36 at load step 145kPa for a) the whole panel and b) the core of the panel.

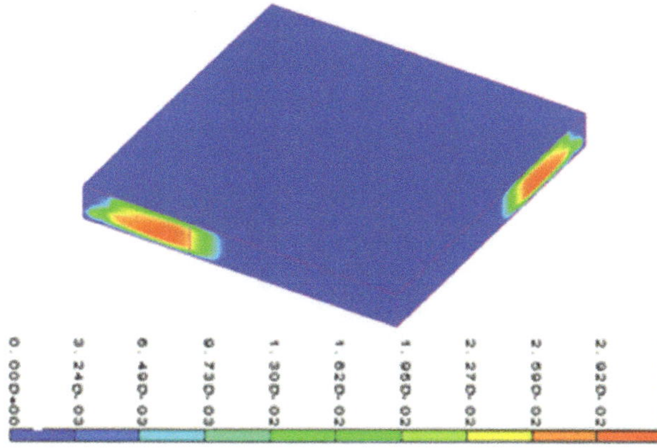

(a)

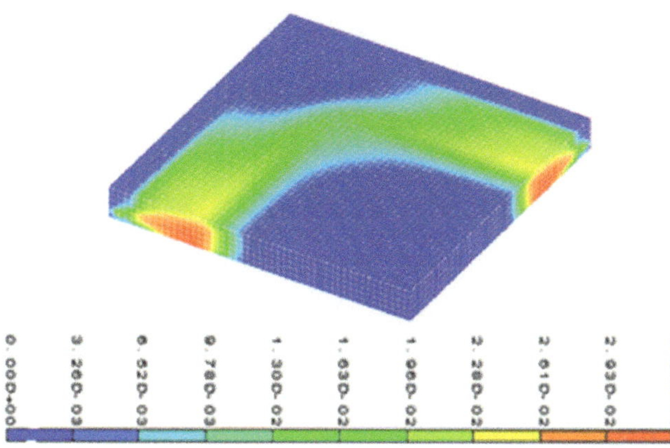

(b)

**Figure 16.** Demonstration of the plastic deformations contour for panel A of loading area 4/36 at load step 145 kPa for a) the whole panel and b) the core of the panel.

## 5. Discussion

As illustrated in Figure 16, the face sheet material starts to yield (entering the plastic range) close to the support (where the boundary conditions are applied). This is physically true, the distributed load over the loading area becomes concentrated reaction force on the strip area on which the boundary conditions (simply supported boundary condition) are applied, i.e., distributed load is converted to concentrated load. So the area where the boundary conditions are applied reaches the yield stress range before any other part of the panel.

As the loading area decreases the load is getting closer to the concentrated load, this is why in Figure 17 panel A of area ratio 1/36 reaches yield (plastic range) at lower load, than the other panels presented in the Figure. Increasing the loading area increases the load carrying capacity of the panel.

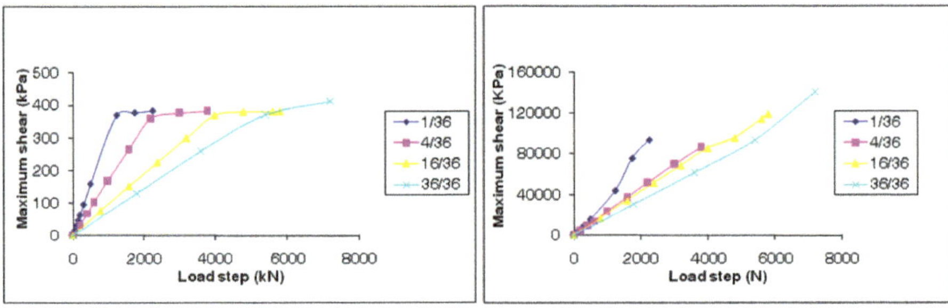


(a) (b)

**Figure 17.** Presentation of panel A maximum shear stress versus loading for different load area ratio for a) Core and b) Lower Sheet.

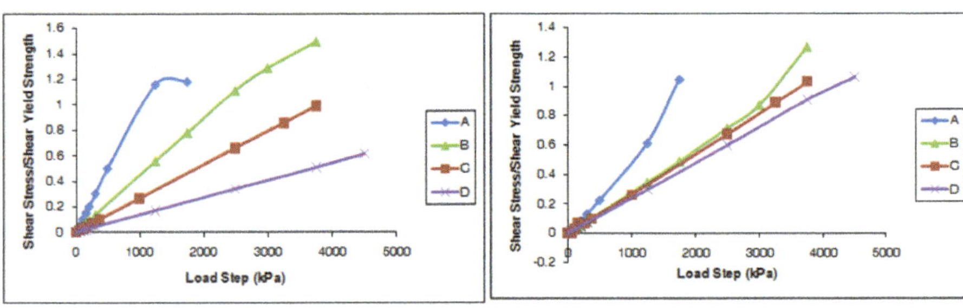


(a) (b)

**Figure 18.** Presentation of maximum shear stress versus loading for A, B, C, and D core material panels of load area ratio 1/36 for a) Core and b) Lower Sheet.

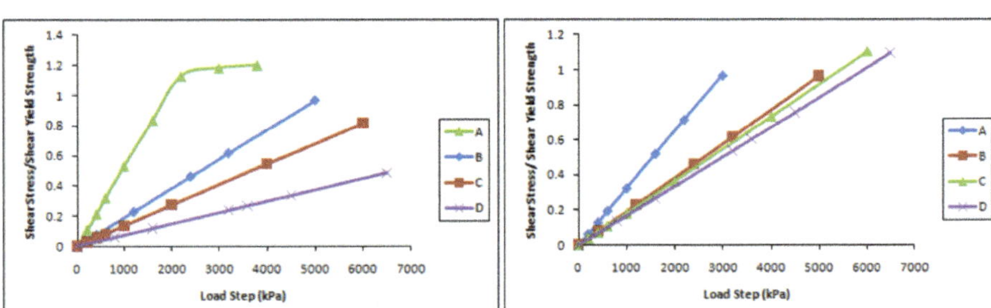


(a) (b)

**Figure 19.** Presentation of maximum shear stress versus loading for A, B, C, and D core material panels of load area ratio 4/36 for a) Core and b) Lower Sheet.

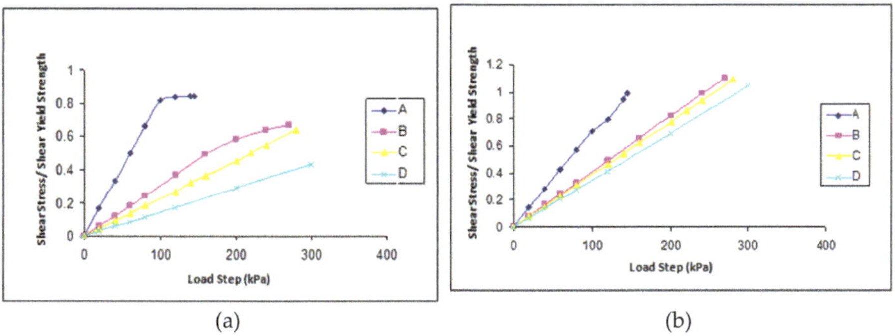


**Figure 20.** Presentation of maximum shear stress versus loading for A, B, C, and D core material panels of load area ratio 16/36 and core thickness 30mm for a) Core and b) Lower Sheet.

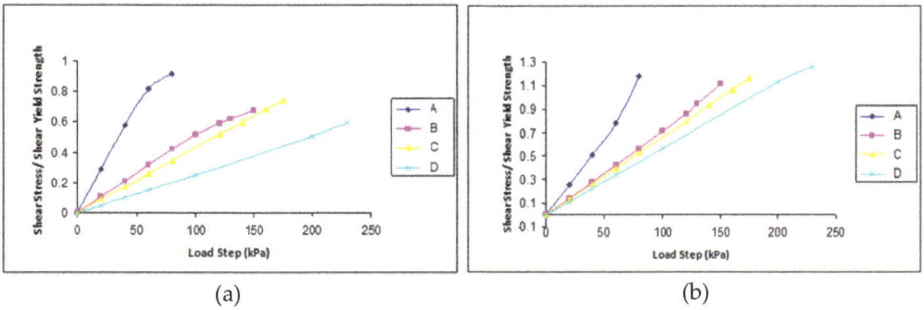


**Figure 21.** Presentation of maximum shear stress versus loading for A, B, C, and D core material panels of load area ratio 36/36 for a) Core and b) Lower Sheet.

Figure 19 through 21 present that the lower face sheet for core material B, C and D reaches yield limit before their corresponding core material. This can be referred to the high stiffness of its core material, i.e., the panel gets closer in its behavior to isotropic plate.

It is obvious from Figure 17 through 21 that panel carrying capacity increases beyond core yield limit. In yield range the core material keeps deforming while the stress is constant (see Figure 22). This deformation works as a mechanism for transferring the excess load to the face sheets. For example in Figure 20, the shear stress of core material A after 100kPa load does not change whereas the shear stress of the corresponding lower face sheet keeps increasing.

To replace the core material with same material of the top and bottom sheets, core's width should be shrunk according to the ratio of the modulus of elasticity of the core to that of the metal. The materials B, C and D are relatively stiff in comparison with A. Equivalent cross-section of core material (see Figure 23) has the same height for all cases and the width is increasing according to the modulus of elasticity ratios. For a rectangle the second moment of area ($wh^3/12$) is varying linearly with the width (equivalent width). The effect of the difference between the materials B, C, and D is relatively small. So the stress curves for these panels are close to each other and the differences are small as it can be seen in Figures 17 through 21.

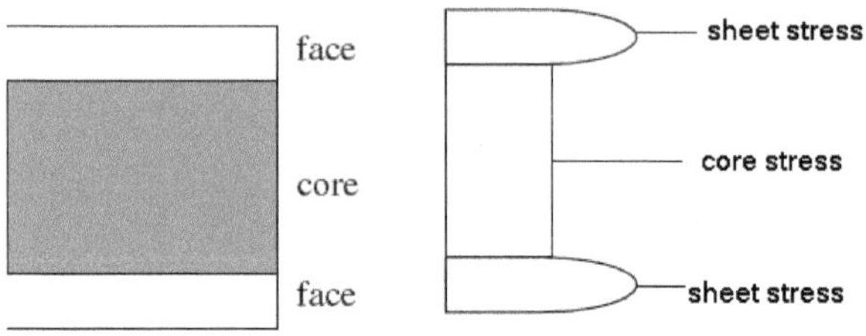


**Figure 22.** Schematic drawing of the shear stress for both face sheets and the core within plastic range.

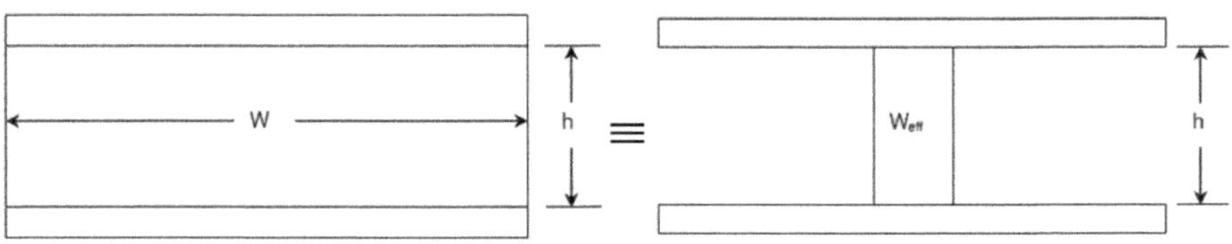


**Figure 23.** Equivalent cross-section of core material with the same height

## 6. Conclusions

- Investigation of sandwich panel behavior beyond core material yield is carried out. The investigation is accomplished in sight of the core material nonlinearity and the geometric nonlinearity of the whole panel. High tech software 'I-DEAS' (Integrated Design Engineer Analysis software) is utilized to carry out the investigation.
- Finite element model is generated using 'I-DEAS' software. This model is validated against experimental and analytical cases available in the literature. To assure model accuracy experimental investigation for selected cases is carried out and compared with FEM. The model shows very good agreement with the analytical as well as the experimental one.
- It is proved that the load carrying capacity of sandwich panel can be improved by loading the panel beyond the core yield limit. This load is going to be transmitted to the face sheet.
- Increasing the stiffness of the core material to a certain extent leads to face sheet yielding before the core material. It is proved that increasing core stiffness increases the load carrying capacity of the sandwich panel.
- Loading area plays good roll in the load carrying capacity of sandwich panel. Distributing loads over large area of panel surface leads to higher load carrying capacity.

## 7. Recommendations

The following are recommendations for further extension of the FEM analysis:

- Investigate the bonding between the face sheets and the core after yielding.
- Modeling face sheets other than metal face sheets such as fiber composite materials
- Extending the FEM to include the bonding strength between the face sheets and the core so the relative dominance of core shear failure, face sheet yielding, or face sheet delamination could be determined.

## Author details

Salih Akour*
*Sustainable and Renewable Energy Engineering Program, College of Engineering,*
*University of Sharjah, Sharjah, United Arab Emirates*
*The University of Jordan, Faculty of Engineering and Technology,*
*Mechanical Engineering Department, Amman, Jordan*

Hussein Maaitah
*Jordan Aeronautical-Systems Company (JAC), Amman Civil Airport, Marka, Amman, Jordan*

## 8. References

"C 393-94 (1995). Standard Test Method for Flexural Properties of Sandwich Constructions, *Annual Book of ASTM Standards,* Vol. 15, No. 3.

"D 6419-99 (1999). Standard Test Method for Two-Dimensional Flexural Properties of Simply Supported Composite Sandwich Plates Subjected to a Distributed Load", *Annual Book of ASTM Standards,* Vol. 15, No. 3.

Airex R63 – (2008), *Typical Mechanical Properties,* Baltek Corporation, www.baltek.com.

Allen, H.G., (1969). *Analysis and Design of Structural Sandwich Panels,* Pergamon Press, ISBN 008012870X 9780080128702, Oxford.

Boyer, H.E. and Gall, T.L. (Eds.). (1991). *Metals Handbook,* American Society for Metals.

Caprino, G., Langelan, A. (2000). Study of a Three-Point Bending Specimen for Shear Characterization of Sandwich Cores, *Journal of Composite Materials,* Vol. 34, No. 9, pp. 791-814.

Chapra, Steve; Canale, Raymond (2005). *Numerical Methods for Engineers,* fifth edition pp.358, McGraw-Hill, ISBN-10: 0071244298.

Frostig, Y., Baruch, M., Vilnai, O., Sheinman, I. (1993). Higher-Order Theory for Sandwich Beam Behavior with Transversely Flexible Core, *Journal of Engineering Mechanics,* Vol. 119, No. 5, pp. 955-972.

* Corresponding Author

Gdoutos, E.E., Daniel, I.M., Wang, K.-A., Abot, J.L. (June 2001). Non-linear Behavior of Composite Sandwich Beams in Three-point Bending, *Experimental Mechanics,* Vol. 41, No. 2, pp. 182–189.

Kuang-An, W. (2001). *Failure Analysis of Sandwich Beams,* Doctor of Philosophy Dissertation, Northwestern University.

Mercado LL, Sikarskie DL, Miskioglu I. (September 2000). Higher order theory for sandwich beams with yielded core. *Proceedings of ICSS-5 Conference,* Zurich. pp. 141–53.

Mercado, L.L., Sikarskie, D.L. (1999). On Response of a Sandwich Panel with a Bilinear Core, *Mechanics of Composite Materials and Structures,* Vol. 6, No. 1, pp.57–67.

Miers, S.A. (2001). *Analysis and Design of Edge Inserts in Sandwich Beams,* Master of Science Thesis, Michigan Technological University.

Plantema, F.J. (1966). *Sandwich Construction,* John Wiley and Sons, ISBN-10: 0471691062, New York.

Rao, T. (2002). *Study of Core Compression Using Digital Image Correlation (DIC),* Master of Science Thesis, Michigan Technological University,.

Thomsen OT. (1993). Analysis of local bending effects in sandwich plates with orthotropic face layers subjected to localized loads. *Journal of Composite Structure,* vol. 25, no. 1-4, pp. 511-520.

Thomsen, O.T. (1995). Theoretical and Experimental Investigation of Local Bending Effects in Sandwich Plates, *Composite Structures,* Vol. 30, No. 1. pp. 85-101.

Triantafillou, T.C., Gibson. L.J. (1987). Minimum Weight of Foam Core Sandwich panels for a Given Strength, *Material Science and Engineering,* Vol. 95, November 1987, pp. 55–62.

Zenkert, D. (1995). *An Introduction to Sandwich Construction,* (EMAS), The Chameleon Press Ltd., ISBN 0947817778, London.

Zenkert, D. (1997). *The Handbook of Sandwich Construction,* Engineering Materials Advisory Services, ISBN 0947817964, United Kingdom.

Chapter 17

# The Finite Element Analysis of Weak Spots in Interconnects and Packages

Kirsten Weide-Zaage

Additional information is available at the end of the chapter

## 1. Introduction

Concerning the mechanical stress and the electrical-mechanical behavior ULSI multilevel metallization systems are more and more sensitive against influences of geometrical and material changes. The mechanical and electrical reliability of these metallization systems is influenced by this. The reliability of such metallization systems is investigated by thermal and thermal-electrical accelerated stress tests under high temperature load. This leads to degradation due to electro-, thermo- and stress migration. This is one major concern in reliability investigations. Generally measurements are time consuming, expensive and the time-to-market cycle is in the focus of interest too. The prediction of local weak spots in interconnects, vias and solder bumps by finite element simulations are a helpful procedure. Beside this the modern 3-d integration leads to more complex material compositions in the systems concerning the coefficient of thermal expansion (CTE) and other material properties. Higher applied currents on the interconnects and bumps result in Joule heating, high temperature gradients and mechanical stress gradients in the bump and metallization systems. The temperature gradients are much higher compared to systems with wide interconnect lines and bumps with large diameters like in conventional packages. Due to this in ball grid array (BGA) bumps as well as μ-bumps and small through silicon via (TSV) connections current induced migration effects as electromigration (EM) and as result of the high temperature gradients thermomigration (TM) can occur. In interconnects consisting of copper caused by stress gradients due to the different material properties under high temperature load also stress migration (SM) occurs.

In this chapter the degradation phenomena in dual damascene copper metallization structures as well as degradation in bumps, μ-bumps and TSV are presented. The degradation is current, temperature and mechanical stress induced under a high applied current and temperature load. The finite element analyses and the mass flux divergence calculation of these phenomena will show the suitability of the method in comparison with experimental results. For a prediction of the weakest spot the suitability of the finite element mesh as well as the modeling concerning the structure shape has to be investigated. Especially edges in the model can influence the quality of the results. The geometrical data of the different metallization or package structures can be taken from the layout as well as optical, scanning microscope or other analytical techniques. Especially for the prediction of the electromigration induced weakest spot in the system, the location of the maximum current density is a major indicator for the fault location. Due to this the maximum current density at localizations of structure inhomogeneity must be checked up. Out of this the maximum current density is calculated by conformal mapping to predict an optimized modeling concerning the shape of the edges and the use of a radius instead of edges. Geometrical variations like the thickness of the first and second metallization and a comparison of the different migration mechanisms will be presented. Concerning the mechanical stress of the DD-Cu metallization the process induced stress will be considered with different processing temperatures of the copper metallization. The influence of different dielectrics on the mechanical stress is also determined. Compared to DD-metallizations and the traces, the bumps are only exposed by electro- and thermomigration. The thermal-electrical-mechanical behavior of μ-bumps and TSV will be shown for a Wafer-on-Wafer (WoW) structure.

## 2. Calculation of the migration mechanism

To figure out the possibilities of determining the effects of migration in solder bumps, interconnects, via and conductive pathways shall be shown here. Migration, particularly in copper or aluminum metallization or migration of solder bumps of a flip-chip package, are presented in [Banas, Hou, Liu 2007, Liu2008, Tan, Wang]. Simulation algorithms are based on analogies and allow only partly- or no material or sizing variations.

Basically, migration processes are described in the metallization on the consideration of diffusion processes. These diffusion processes can lead to a change in dimension or change of geometries of the metallization [Wever]. On one side material loss or the hole formation at these locations leads to tensile stress, while it comes in places of material accumulation to a compressive stress. This leads to a reflux, also called the back stress in the material. There is a critical length existing, which compensates the electromigration related flux and the reflux. The migration of atoms leads to a change of the expansion in the metal, which then in turn changes the chemical potential. Since the chemical potential is very often influenced by a stress term, this leads to stress gradients and a reflux. In microelectronic applications the current flux pathways are subject to large thermal strain before any current load applies. This thermal strain, along with the strain caused by the electromigration can lead to a nonlinear reflux. Also, in the interconnects local self-heating take place, providing a contribution to the migration of the thermal gradients due to the local temperature increases. The components of the electrical, thermal and stress migration flux are added by superposition.

The diffusion coefficient with the activation enthalpy of $\Delta H$ is defined as follows:

$$D = D_0 \, e^{-\Delta H / k_B T} \tag{1}$$

The change of place takes place at cubic (fcc) metals such as Cu, Au and Ag across the holes. The activation enthalpy is additively composed of the enthalpy of formation and the activation energy for the hole formation. Both are nearly the same size. With equation (1) and $D_0$ as temperature independent diffusion constant as well as the melting temperature $T_m$ [Heumann, Philibert] applies:

$$5 \cdot 10^{-6} < D_0 < 5 \cdot 10^{-4} \, [m^2 / s], \quad dH \cong 1.5 \cdot 10^{-3} \, T_m \, [eV] \tag{2}$$

Due to this a diffusion coefficient of $D(T_m) \approx 10^{-8}$ cm²/s [Heumann] is resulting.

| Parameter | Al | Cu | Ag |
|---|---|---|---|
| $T_m$ | 660 | 1083 | 960 |
| $\Delta V/\Omega$ | 0.71-1.3 | 0.9 | 0.7-0.9 |
| E [eV] | 1.234 | 2.15 | 1.921 |

**Table 1.** Melting Temperature, Active Volume and Activation Energy [Philibert].

In the case of multiple components, for example solder bumps, all main materials must be involved in the material transport. In the example of a material of two components A and B, and the chemical or interdiffusion coefficients, there are the partial diffusion coefficients $D_A$ and $D_B$ [Wilkenson]. Using the Darken equation, the diffusion results as follows:

$$\upsilon = (D_A - D_B) \frac{\delta N_A}{\delta x} \quad with \quad \tilde{D} = D_B x_A + D_A x_B \tag{3}$$

From the migration velocity $\upsilon$, the chemical diffusion coefficient $\tilde{D}$ and the mole fraction, $x_A$ and $x_B$ of the considered components the partial diffusion coefficients $D_A$ and $D_B$ can be determined.

The direction of electrical transport in alloys such as leaded solder materials based on Sn, goes, depending on the proportion of Pb to the anode or the cathode. In terms of thermotransport moves the Pb in Sn to the cold part of the sample.

There are different diffusion paths in the material. These are the grain boundaries, surfaces to adjacent materials and the volume or bulk. In the aluminum as a metallization material primarily grain boundary diffusion occurs. In contrary in copper as metallization material predominantly interface or surface migration is occurring.

In the calculation of grain boundaries, the surface diffusion as well as the intermediate phase diffusion must be considered in thin layers. Atomistic transport along a grain boundary or phase boundary has a low activation energy and is therefore by orders of magnitude faster than in the crystal itself. The analytical models for the grain boundary diffusion $\delta \, D_{GB}$ are only valid for the self diffusion in pure metals [Kaur]. The part of electromigration at grain boundaries depends also on the effective width of the grain δ (d) concerning the mass transportation in relation to the average grain size.

A composition of the interconnect from almost poly crystalline parts and parts where bamboo structured grains occur, leads to the predominant grain diffusion on one hand and on the other to volume diffusion and thus to high gradients in the mass flux. If the mass flux is blocked by a large grain, then the material accumulates relating to direction of the electron or material flux, in front of the blocking grain, while it comes behind the blocking grain to a material loss. For grain boundaries in the area of 280nm the diffusion coefficient of the grain boundary diffusion is in the range of the bulk material $D_{GB} \cong D_{Bulk}$ [Kaur].

Only above a defined threshold current density $\vec{j}_{th}$, resulting from the Blech effect and denoted as 'short-length effect', it comes to effective place change and thus to the mass transportation or material flux [Blech]. The mass flux is dependent on the atomic particle density N, the Boltzmann constant $k_B$, the local temperature T, the current density $\vec{j}$, the specific resistance ϱ, the diffusion process relating activation energy $E_A$, the diffusion coefficient $D_0$, and the effective charge eZ.

$$\overrightarrow{J_{EM}} = \frac{N}{k_B T} e Z^* \left( \vec{j} - \vec{j}_{th} \right) \rho \, D_0 \exp\left( -\frac{E_A}{k_B T} \right) \tag{4}$$

The thermotransport denoted as Soret-effect is a phenomenon of overlay of diffusion and heat conduction. They closely resemble the electrotransport with the difference that the considered system is not isothermal [Wever]. There is a flux of soluted atoms and the heat flux. These fluxes are described about the chemical potential and the thermal gradients. Without the occurrence of concentration gradients the equation for thermal flux is:

$$\overrightarrow{J_{TM}} = -\frac{NQ^*}{k_B T^2} D_0 \exp\left( -\frac{E_A}{k_B T} \right) grad T \tag{5}$$

Q* represents here the transported energy at a constant temperature and is commonly referred to as reduced heat of transport or transfer. The ratio $Q^* / k_B T^2$ is referred as Soret coefficient. The heat of transport is the heat flux per unit of material without temperature gradients. Is the value of $Q^* > 0$ a heat flux is generated to keep the soluted atoms isothermal, which takes place towards the dissolved flux. Is $Q^* < 0$ the flux of dissolved particles and the heat flux are counter set. It follows that in an isothermal system, a density gradient produces a thermal flux and vice versa a temperature gradient leads to a material flux [Shewman]. The heat of transport is approximately equal to the activation energy for the material flux [Jaffe].

The differences of the coefficient of thermal expansion (CTE) between the metallization material and surrounding materials produce, depending on the ambient temperature a mechanical stress. This in turn leads to a material flux within the metallization. Under strain the enthalpy for example of tensile-stress formation in a grain boundary is reduced, which increases the concentration of holes [Heumann].

$$\overrightarrow{J_{SM}} = -\frac{N\Omega}{k_B T} D_0 \exp\left(-\frac{E_A}{k_B T}\right) grad\,\sigma_H \tag{6}$$

The $\Omega$ represents the atomic volume and $\sigma_H$ the hydrostatic stress. Out of the components of the electrical, thermal, and stress migration, the total mass flux in the metallization structure emerges through superposition from the equations (4, 5 and 6).

The formation of material accumulation or void formation requires a divergence in the flux of the total mass flux. Only divergences lead to a change in the density of the material. The calculation of the mass flux and mass flux divergence in metallization, traces and bumps in the past were done using analogies between the electrical and thermal and the thermal mechanical behavior or a variation of the materials or dimension were not possible [Banas, Hou, Liu 2007, Liu2008, Tan, Wang]. This lead to an overestimation of the temperature gradients and current density like proposed in [Ogurtani]. Without any applied current and temperature gradients the calculation of the stress migration is not possible.

With the new program code the divgrad of T and $\sigma$ are calculated directly based on the simulation results. The calculation of the mass flux is done for each element under consideration of the neighbor elements. Out of this the stress gradients are calculated. The divgrad terms are calculated on base of the super elements under consideration of an Ansatzfunction. The calculated values for the different migration mechanisms are reloaded into ANSYS for graphic display. Out of this the stress migration can be calculated under SM stress test conditions without any applied current or temperature gradients. The simplified equations neglecting concentration gradients for the different migration mechanism are:

$$div\,\overrightarrow{J_{EM}} = \left(\frac{E_A}{k_B T^2} + \frac{\alpha_\rho}{1+\alpha_\rho(T-T_0)} - \frac{1}{T}\right) \cdot \overrightarrow{J_{EM}} \cdot grad\,T \tag{7}$$

$$div\,\overrightarrow{J_{TM}} = \left(\frac{E_A}{k_B T^2} - \frac{2}{T}\right) \cdot \overrightarrow{J_{TM}} \cdot grad\,T - \frac{QND}{k_B T^2} \cdot \Delta T \tag{8}$$

$$div\,\overrightarrow{J_{SM}} = \left(\frac{E_A}{k_B T^2} - \frac{1}{T}\right) \cdot \overrightarrow{J_{SM}} \cdot grad\,T - \frac{\Omega ND}{k_B T} \cdot \Delta\sigma_H \tag{9}$$

In equation (7-9) N is the atomic concentration, the Boltzmann constant $k_B$, the local temperature T, the resistivity $\rho$, the atomic volume $\Omega$, Q* the heat of transport, the activation energy $E_A$ (taken from grain boundary and interface migration as strongest influence) and $\sigma_H$ is the hydrostatic stress. The simulation and calculation sequence is shown in figure 1.

## 3. Modeling and simulation improvements

The numbers of elements, which are determining the mesh and due to this the density of the nodes have a strong influence on the accuracy of the simulation. Out of this the mesh of the investigated metallization or metallic material as main interesting point in the simulation plays a major rule. In the metallization itself the potential as well as the temperature is calculated. The metallization is surrounded by dielectric material and the traces by FR-4 or PCB. The bumps can be surrounded by underfillers made of different plastic materials.

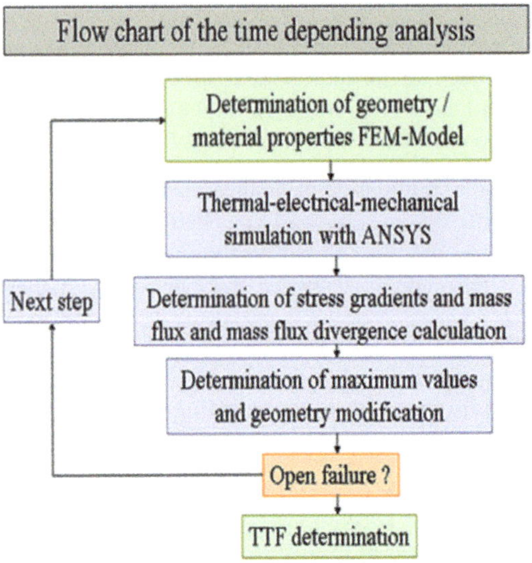


**Figure 1.** Simulation and calculation sequence for one cycle (static) and more cycle (dynamic).

Concerning the mechanical stress calculation the interfaces play also an important rule. Due to the different materials and their material properties the stress in the structures can be high and can therefore have a strong influence on the reliability of the structures. The calculation of the stress gradients and stress migration mass flux divergence is strongly affected by the accuracy of the mesh.

## 3.1. Advantages of FEM compared with other applicable methods

Simulation methods like finite difference, finite element and Monte Carlo method are common in the electrical engineering. Monte Carlo method is described for the calculation of time to failure (TTF) distributions induced by electromigration [Huang]. Also in the scope of semiconductor simulation it is widely used [Jacoboni]. The complexity of the metallization and bump structures in combination with the question of multiphysic investigations makes this approach inconvenient in use. The application of the finite difference method is described in investigations of the intermetallic phase growth for instance [Chao]. A model for electromigration calculated by concentration gradients is described in [Joo] and the void evolution and motion was investigated by [Averbuch]. The mass transport along interface caused by electromigration using the level-set method is described in [Li]. All these approaches consider only a part of the thermal-electrical and mechanical behavior.

Calculations in the scope of the coupled thermal-electrical-mechanical behavior of metallization structures and bumps can be sufficiently carried out by commercial programs like ANSYS®, ABAQUS or COMSOL using the finite element analyses. In the following investigation ANSYS® is used as simulation tool.

### 3.2. Mesh density and singularities

In the thermal electrical calculation the degree of freedom is only the temperature. The mesh of the surrounding material depends in the investigated case of the metallization and due to this of the metallization mesh. A coarse mesh of the metallization can lead to an insufficient determination of the temperature in the isolation materials. Out of this the mesh of the metallization should be investigated with regard of an optimization.

The mesh can be easily refined by varying the count of the nodes. If a convergence concerning the change of the potential voltage is found the mesh is sufficient. A predication concerning the convergence of the current density at inhomogeneities like vias or edges in the metallization cannot be done. At these positions a strong dependence of the maximum current density due to current crowding in relation to the model of this inhomogeneity occurs. The layout from the EDA tools give at such positions rectangular edges or edges with a defined angle. Taking into account such an angle in the model may lead to insufficient calculation of the local current density at this position. Due to this the relation of the current crowding and angle of the edge will be investigated here.

Considering the two-dimensional case, choosing a right angle at the edge of a metallization a singularity will occur. In this case the current density can be calculated analytical with an infinity value [Betz]. The result of the numerical analysis of the current density at this position converges to a defined value due to the fact that the neighbor elements are taken into account for the calculation. If the right angle is replaced by a radius, the calculated current density depends on the selected shape of the radius, the element size and the element shape itself. For a determination of the real current density in the structure and the optimized radius, the edge of a simplified metallization consisting of aluminum, is investigated for the 2-dimensional and 3-dimensional case in sub-models. The boundaries in the calculations were taken of the coarse model.

In the 2-dimensional as well as in the 3-dimensional case the current crowding $j_{max}/j_{in}$ increases with increasing element count cubically and the relation of the maximum current density to the applied current density $j_{max}/j_{in}$ increases with decreasing element size exponentially. Out of this the numerical solution converges in the case of infinite small elements to the theoretical expected value. In the case of a curvature of the edges, the current crowding $j_{max}/j_{in}$ has for increasing element counts as well as decreasing element size for every taken radius a constant value. Due to this the element count used in the simulations depends on the element size as well as the used radius in the model. A decreasing radius down to the range of an rectangular angle with constant element count leads to an increase of $j_{max}/j_{in}$.

Not in every case optical or scanning electron microscopy pictures are available for a determination of the structure shapes after the production process. Due to this for a determination of the optimized radius two dimensional simulations can be compared with results from the calculation of the maximum current density out of conformal mapping. The maximum current density and details about the homogeneity of the current density and resistance behavior at the investigated places can be achieved by the conformal mapping.

The calculation of the maximum current density is done by the following approach. In general for the analytical calculation of the current density distribution at the edge of an interconnection, a bent metallization with different width g and h and a radius r can be converted by two times conformal mapping into a flat conducting band with width $\pi$ in the Z′ plane. In this flat conducting band a homogenous current density distribution can be assumed (figure 2). With the back transformation of this homogenous current density distribution into the Z plane the current density distribution along the path A-B-C can be achieved and due to this the maximum current density. With the parameters S=g/h as ratio of the metallization width and P=r/h as ratio of the radius r to the smaller metallization width h and the applied current density $j_{in}$ in the vertical part of the metallization shall apply under the condition the P << 1 [Hagedorn]:

$$j_{max} = 1{,}04^{\left[\frac{S^2+1}{P\,S^2}\right]^{\frac{1}{3}}} j_{in} \tag{10}$$

With equation 10 the current crowding $j_{max}/j_{in}$ can be determined depending on the metallization width and the radius.

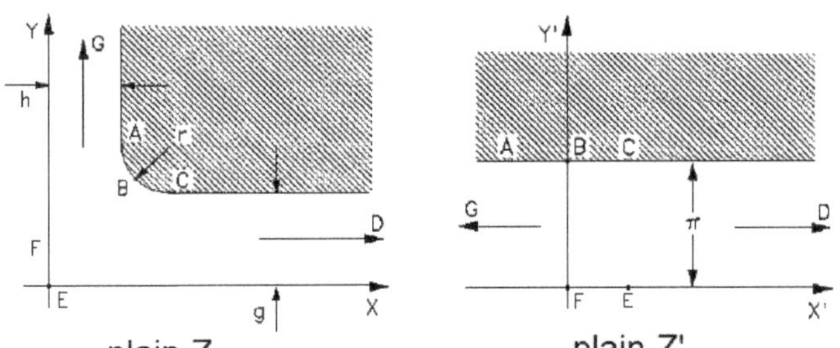


**Figure 2.** Original figure of a two dimensional edge of an metallization in the Z-plane (left). Metallization after two times of conformal mapping in the Z′ plane (right).

In the 2-dimensional case a good correlation between the analytical solution and the maximum current density determined by the simulations can be found. In the 3-dimensional case the analytical solution is valid under the assumption of a homogenous current density distribution in the via, which means that the current density distribution can be mapped on the 2-dimensional case.

The maximum current density depending on the radius for a fixed element count is compared to the simulated values. The results are shown in figure 3. For a radius of 0.03μm a good correspondence between simulation and analytical solution is found. For a radius above 30nm the simulated values are caused by the increased element size above the analytical values. A radius below 30nm has an insufficient mesh and due to this the simulated values are beneath the analytical values. To get compliance between simulation and analytical solution the mesh has to be refined. This leads to an increase of the calculation time and is therefore not useful.

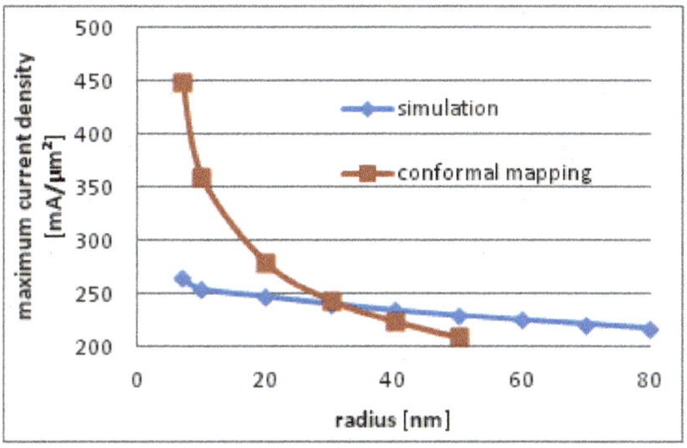


**Figure 3.** Max. current density depending on the radius for the interconnect/via edge. Analytical solution done by conformal mapping and simulation.

### 3.3. Mechanical stress and stress gradients

In the finite element method simulation of the mechanical stress, the calculated unknown parameters are the displacement of the nodes. The different forces are calculated by the derivation of the displacements. The form function is continuously but not smoothed between the elements. Due to this the values at the nodes have to be averaged. Due to this and caused by the simplified form function and the limited element expansion the FEM is an approximation method. This can lead to failures in the calculation of the dilatations. When the results of an inaccurate mesh are derived the inaccuracy concerning a second derivation will be intensified. The correlation of the accuracy and mesh refinement is shown in figure 4 [Eichelseder]. From this knowledge it is clear that the distribution of the dilatation of the components S1 to S3 is more accurate than the strain distribution resp. the Von Mises Stress (VMS). This fact also implies that a simplified mesh or a coarse mesh can deliver accurate results concerning the dilatation. The derivate strain shows accurate values only after a refinement. After an additional refinement the gradient of the strain and out of this the hydrostatic stress (HS) can be calculated with a good accuracy. The calculation of the mass flux divergence due to stress migration includes also a derivation of the stress gradients, which will affect the accuracy additional. The accuracy of the finite element mesh has to be investigated before the mass flux divergence calculation is done. And the results have to be handled with care if the mesh is not adequate tested.

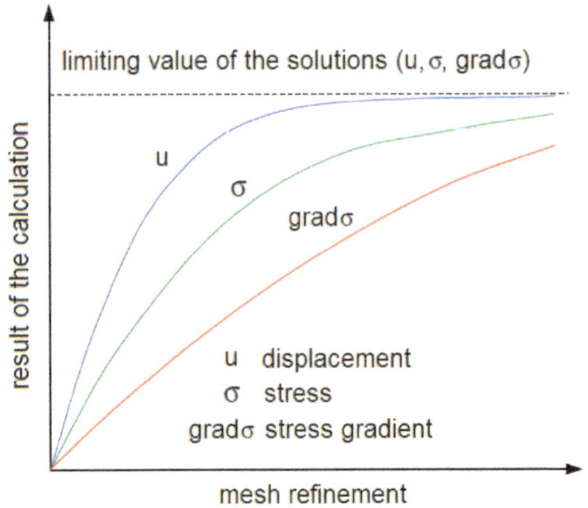


**Figure 4.** Correlation of the accuracy and mesh refinement [Eichelseder].

# 4. Metallization

## 4.1. Metallization variation in the 90nm node

The investigated DD copper structure based on the dimension of the 90nm scaled down to the 65nm process technology. The model consists of SiCN as cap layer, Ta/TaN as barrier layer and different dielectrics like Silk™, Black Diamond II™ and SiCOH. In the mechanical calculations the process temperature using the birth and die algorithm in ANSYS are included, caused by the fact that the use of a reference temperature for the stress free state is not sufficient [Weide-Zaage 2008].

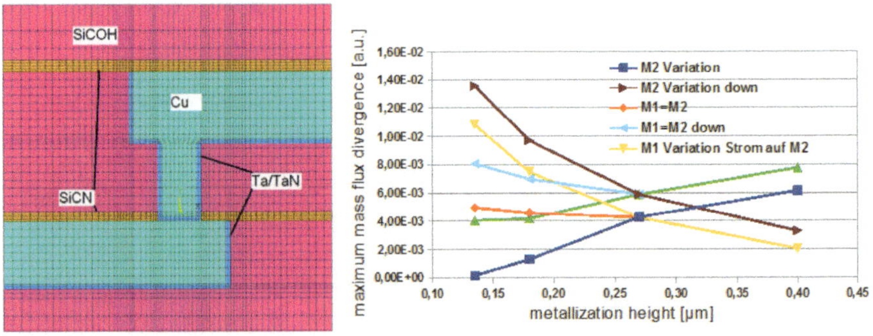


**Figure 5.** Mesh of the investigated structure (left) and mass flux divergence vs metallization height for a variation of M1 and M2 (right).

The mesh of the investigated structure is shown in figure 5 (left). For simplification one via with two metal links right and left hand was generated. As one example for this proposal the metallization height here was varied. In the investigated cases for electromigration the current flux direction was calculated downstream from the second metallization M2 to the first metallization M1 and in the opposite direction upstream. The mass flux divergence of the variation is shown in figure 5 (right). It was found that the variation of the first metallization M1 and the variation of the second metallization M2 show a big influence on the mass flux divergence and out of this on the reliability of the structure. The weakest location found here is in and beneath the via.

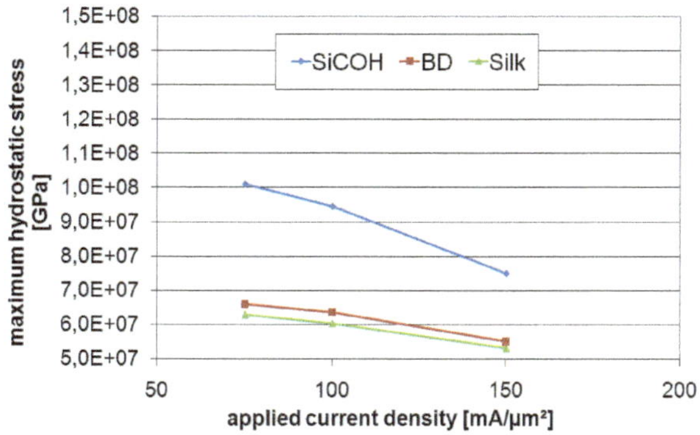


**Figure 6.** Maximum hydrostatic Stress for BDII™, Silk und SiCOH.

For technology nodes in the range of 500nm and below $SiO_2$ or combinations of USG (undoped-silicate-glass), PSG (phosphosilicated-glass) or FSG (fluorine-doped-silicate-glass) as a customary dielectric (IMD) is used. For products produced in nowadays low-κ dielectrics, with a low dielectric constant such as Silk ™, SiCOH, Black Diamond, or MSQ ™ II are used.

Figure 6 shows the maximum hydrostatic stress as a function of the applied current density for the different dielectric materials. The greatest mechanical stress occurs at a SiCOH dielectric and the lowest at Silk ™.

## 4.2. Separated migration mechanism with different via shape

A second model is based on dimensions of the 65nm technology node with a wide line and different via bottom geometries. The model is shown in figure 7 based on SEM pictures from the literature [Delsol, Lee]. Tantalum and TaN were chosen as barrier material, SiCN as capping material and SiCOH was chosen as dielectric material. The structures were investigated under test conditions with an applied current density of 1.5MA/cm² taken from [Lin2006a, Lin2006b]. The thermo-mechanical investigations were carried out considering the process-induced stress. The process temperatures are given in table 2.

The stress state in the copper metallization is strongly related to the process temperature of the copper. A high process temperature leads to a higher nearly stress free state of the whole structure compared to structures with lower temperatures [Matsuyama]. The nearly stress free state determined by investigating of the process induced stress is approx. 200°C. The calculation of the stress migration in the structures is described in [Weide-Zaage 2010].

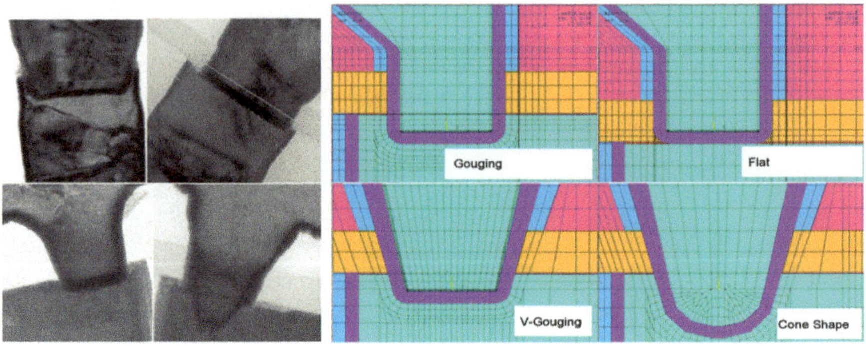


**Figure 7.** SEM pictures from [Lin2006, Lin2006(2)] (left) and Finite Element Mesh of the via region of the four different bottom geometries (right).

| | Young Modul (GPa) | Poisson | CTE (300K) | Process Temperature (°C) |
|---|---|---|---|---|
| Cu | 125 | 0.34 | $16.7\ 10^{-6}$ | 200 |
| TaN | 185 | 0.33 | $6.6\ 10^{-6}$ | 40 |
| SiCN | 100 | 0.17 | $3\ 10^{-6}$ | 300 |
| SiCOH | 15 | 0.3 | $11.6\ 10^{-6}$ | 350 |
| Si | 98 | 0.45 | $2.64\ 10^{-6}$ | 25 |

**Table 2.** Mechanical properties and process temperatures used in the simulations

### *4.2.1. Electromigration test temperature 325°C*

The electromigration behavior of the four models was investigated with current (flux in both directions) and an activation energy of 0.9eV. The different via bottom geometries were verified by comparison with investigations from literature. The calculated maximum mass flux divergences as well as the reciprocal values which are related to the MTF are given in table 3. Comparing a Gouging and a Flat via and a Cone Shape and a V-Gouging via the Gouging via is more reliable than the Flat via and the Cone shape via more reliable than the V-Gouging via. The simulated models show that the Cone Shape via has the best results concerning the electromigration behavior.

| Via-Shape | **div** $_{EM}$ | **(div** $_{EM})^{-1}$ |
|---|---|---|
| Gouging Via | $1.143\ 10^{-3}$ | 875 |
| Flat Via | $0.917\ 10^{-3}$ | 1091 |
| V-Gouging Via | $0.939\ 10^{-3}$ | 1065 |
| Cone Via | $0.672\ 10^{-3}$ | 1488 |

**Table 3.** Mass flux divergence and reciprocal mass flux divergence.

The failure locations due to the void formation after the electromigration test in the literature are found downstream at the bottom of the via and upstream in and over the via at the interface metallization cap layer. A schematically overview of the different migration effects like electro-, thermo- and stress migration mass flux electron flux upstream and downstream and supposed void formation (yellow) is drawn in the via structure shown in figure 6. The electromigration mass flux is depending on the current flux direction and temperature gradients, the thermomigration mass flux depend on the temperature gradients and the stress migration on the stress gradients, which occur at tensile regions also under electromigration test conditions. An arbitrary assumed grain distribution is grey colored in this graphic. Thermomigration mass flux (blue) in the investigated structures proceeds independent of the current flux direction from first as well as the second metallization into the via. The electromigration mass flux (green) is at the upper part of the via downstream in the same direction and at the via bottom in opposite direction of the temperature gradients and thermomigration mass flux. In the upstream case it is vice versa. Due to the process induced stress distribution in the metallization, under electromigration test conditions, the metallization is mostly compressive with some tensile regions. The stress gradients in the tensile regions may lead to a current direction independent migration. High stress migration (orange) is found at the bottom and beneath the via above and below the barrier. The stress gradients show above the barrier upwards and below the barrier downwards. At the interface of the cap layer and the metallization they show down. Only the occurrence of a migration pathway leads to voiding. For interface migration the stress gradients have to have a component into the direction of the interface. Also the existence of a grain boundary pathway supports the voiding. Out of this foot voiding as well as voids in the via itself, both indicated in yellow can be explained by this (figure 8).

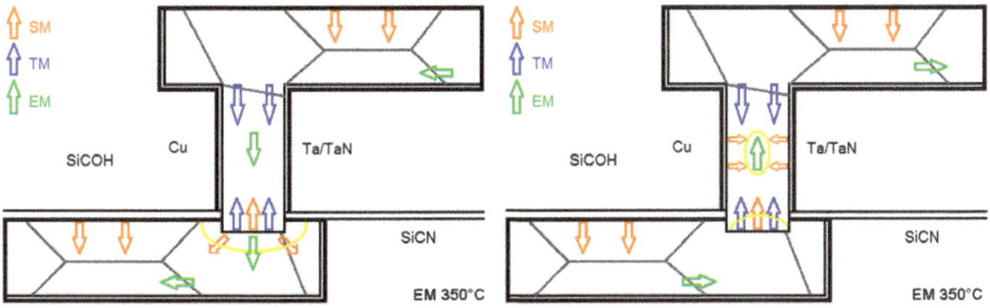


**Figure 8.** Schematically electro-, thermo- and stress migration mass flux electron flux upstream and downstream and supposed void formation (yellow) in the via structure EM>SM>TM.

## *4.2.2. Thermomigration stress temperature 325°C*

The thermomigration (Soret-Effect) was investigated under electromigration test conditions of 325°C. The calculation of the thermomigration shows high, values for the mass flux divergence. With temperature gradients of 50K/μm the gradients are high but not high enough to induce thermomigration. Gradients of 100K/μm are supposed to be an acceleration factor for the thermomigration. In figure 9 the mass flux divergence distribution due to thermomigration is shown. High values in the Gouging model are found at the via bottom in the via and at the corner of the metallization, barrier- and cap-layer. For the Cone Shape via high values are found only at the corner of the corner the metallization, barrier- and cap-layer. The activation energy for the thermomigration calculation is supposed to be too small and should be >1.2eV. Measurements of thermomigration activation energies of copper in copper are not known until now to verify this.

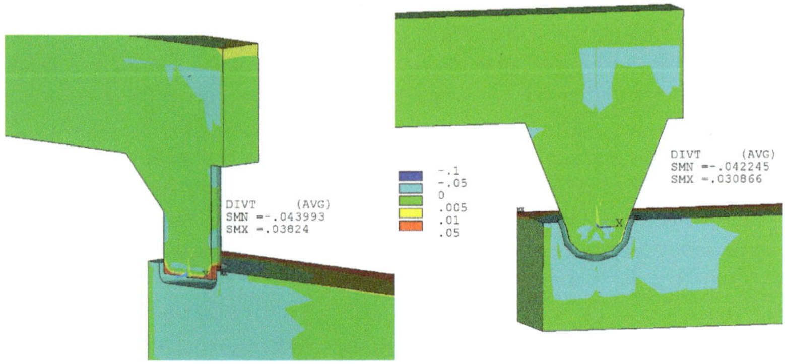


**Figure 9.** Thermomigration mass flux divergence distribution for the Gouging and the Cone Shape model.

## 5. μ-Bump, CuSn-pillar and TSV

### 5.1. μ-Bump in comparison with BGA-PoP

A variation of the applied current in a Package-on-Package (PoP) bumps [Meinshausen 2010 (a)] and μ-bump [Meinshausen 2011] was carried out and the mass flux divergence distribution was determined. The bumps in the FE model of the PoP device with one bump chain consist of SAC305 (SnAgCu). For the under-bump metallization (UBM) and the surface finishes 6μm thick Ni layers were used. As mold compound (MC) around the upper contact surfaces "Stycast 1090" was chosen. The simulations were carried out with anisotropic and temperature depending material parameters. In the FR-4 substrates five layers of Cu traces and Cu vias between the top and the bottom package are placed to connect the top and the bottom bumps. The height of the upper, the middle and the lower copper layers is 20μm, 18μm and 36μm. The width of all layers is 250μm. The model of a μ-bump between two ICs is shown in figure 10 (left). The dimensions of the μ-bumps are similar to the test structures used in [Labie]. The diameter of the μ- bump is 25μm and the height is 10μm. Over and under the μ-bump a 100μm silicon layer resp. a 50μm thick silicon layer is representing the ICs of a CoC (Chip-on-Chip) structure. The ICs are covered with a 1μm thick $Si_3N_4$ passivation layer. The copper traces at the upper and the lower contact surface have a height of 0.5μm and a width of 32μm. The pitch is 40μm.

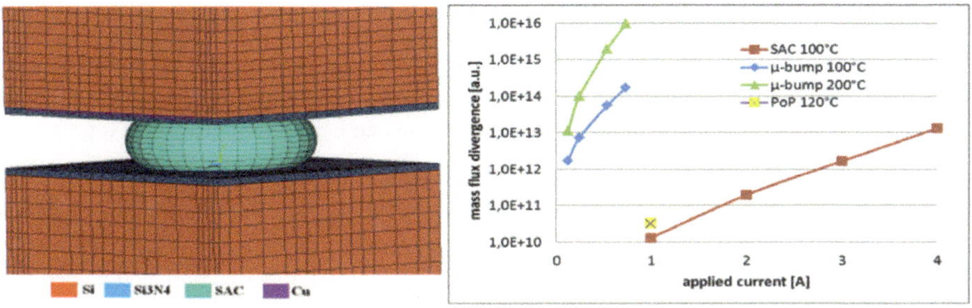


**Figure 10.** Model of the μ-bump (left). The different materials are indicated by colors. The mass flux divergence vs. the applied current for SAC bump and μ-bump (right).

In figure 10 (right) the mass flux divergence depending on the applied current for the SAC PoP bump and the μ-bump is shown. Under the same applied current the mass flux divergence of μ-bumps is compared to PoP bumps about four orders of magnitude higher. Due to this fact the reliability of the μ-bump has a high electromigration risk. The weakest point of the μ-bump was found at the top of the bump due to current crowding combined with high temperature gradients.

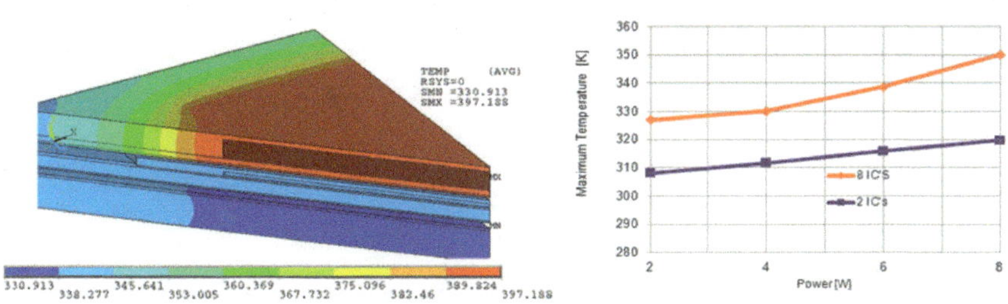


**Figure 11.** Temperature distribution in a BGA PoP Package with ICs (left) and maximum temperature in the bumps vs applied IC's power loss (right).

[Meinshausen 2010(b), Meinshausen 2012] investigated in previous simulations the heat flux density of IC stacks with a constant surface load. The heat loss of the ICs is one reason for the existence of an inhomogeneous temperature distribution in packages. An applied power of 2-8W for all ICs in the IC stack was investigated as boundary condition for the simulations. The temperature distribution in the package for a power loss of 8W is shown in figure 11, left side. The maximum temperature occurs in the IC stack. The bump temperature is much lower (figure11, right). This leads to a strong inhomogeneous temperature distribution. On the other hand the temperature gradients in the bumps are not as high compared to the EM simulation. Due to this in this investigated case thermomigration will not occur.

## 5.2. TSV with μ-bump

Through silicon vias (TSV) have a wide range of applications in the modern packaging. Three different kinds of TSV can be identified which are common under the topic of 3D IC packaging, 3D IC integration and 3D Si integration. Due to this the processing as well as the size of the TSV is completely different. The application can be divided into three topics.

a. Package on Package stacking with TSV as connections between them
b. Chip to Chip, Chip to Wafer or Wafer to Wafer stacking with $SiO_2$ to $Sio_2$ and Cu to Cu bonding.
c. Wafer to Wafer stacking; Memory chip stacking TSV are used as connection between the chips; active or passive interposer.

In the case of the 3D IC integration the TSVs are necessary. Due to the fact that the data width is limited, the use of TSV with small sizes in the range of 5-10μm with a pitch in the range of 20-40μm a much wider I/O width is possible. The TSV shortens the way of the connections compared to the common wire bonding. Using the wire bonding the chip size has to be increased and due to this the costs increase. TSV as solution will solve this problem in future. For future 3D integration Cu-TSV as well as tungsten W-TSV will be used [Murugesan] caused by the fact that W is a capable material for sub μ-vias but not usable for power and ground applications due to its high specific resistance. Therefore Cu with a Ta barrier will be used.

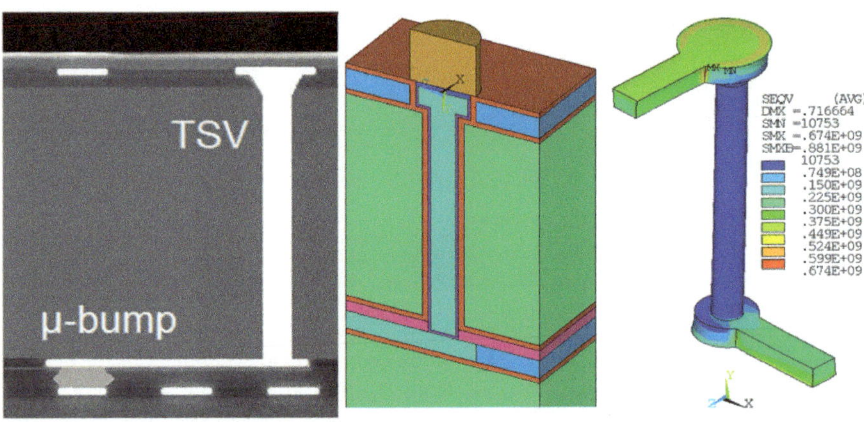


**Figure 12.** Schematic picture of a TSV and μ-bump (left) and FE-Model (right).

As an example in figure 12 (right) the mesh of a TSV with a SAC pillar is shown. The right side of figure 12 shows schematically a TSV with a μ-bump [Leduc]. The geometrical data for the simulations are taken from [Kitada, Lo]. The diameter of the TSV was set to 9μm and the height was set to 80μm. The copper barrier in the TSV consists of SiN and TiN. The thickness of the interconnections is 7μm. The passivation consists of SiN.

On top of the TSV high VMS was found caused by the fact that the copper in the TSV is under pressure and the material is pressed out of the TSV. Concerning the EM high mass fluxes were found at the bottom of the TSV to the metal trace. At this position also high thermal mass flux occurs. High hydrostatic stress occurs at the bottom of the TSV to the metal trace. The stress increases about 10% for an increase of the applied current from 0.1 to 0.3A.

### 5.3. CuSn-pillar

For a simplified assembly, a smaller pitch between the bumps and a low interconnect inductance CuSn-pillars can be used. Copper pillars are copper bumps with a thin layer of Sn on the top. These layer thicknesses can vary [Huffman, Syed]. The CuSn-pillars can be formed under pressure and a temperature load. This leads to the formation of $Cu_3Sn$ and $Cu_6Sn_5$ phases. Due to this CuSn pillars with different Sn thickness and location in the bump were investigated. The bumps had a diameter of 24µm and a complete high of 45µm. Above the bumps a Cu trace and below the bumps a TSV (figure 12) were placed in the model. The applied current was set to 175 mA, the substrate temperature was varied from -50 to 150°C and the stress free temperature was set to 135°C. In figure 13 a part of the mesh and the temperature gradient distribution in the different bumps is shown.

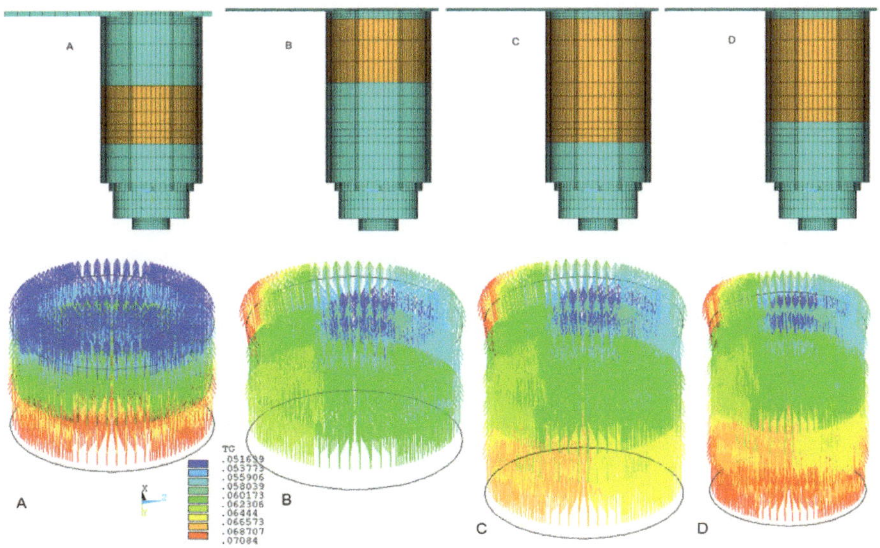


**Figure 13.** Mesh of the CuSn Pillar with Cu in magenta and Sn in yellow (above) and temperature gradient distribution for the different models (below).

The homogenous temperature distribution the bumps can be achieved by a placement of the Sn in the middle (model A). In the case of model B-C high temperature gradients occur above the Sn near the copper trace. At this position also current crowding occurs. Both can lead to a weak link at this position. Depending on the current flow direction the flux will be increased or decreased. In figure 14 the hydrostatic stress depending of the substrate temperature is shown for model A the highest stress occurs for temperatures about -50°C near the reference temperature of the stress free state the stress is the lowest. The processing temperatures should be included in the model.

| | HS [MPa] | Tmax [K] | Div $J_{max}$ [a.u.] |
|---|---|---|---|
| B | 174 | 396,86 | 0,486e12 |
| C | 175 | 396,54 | 0,459e12 |
| D | 172 | 397,02 | 0,512e12 |

**Table 4.** Hydrostatic Stress, Maximum Temperature and Maximum Mass flux Divergence.

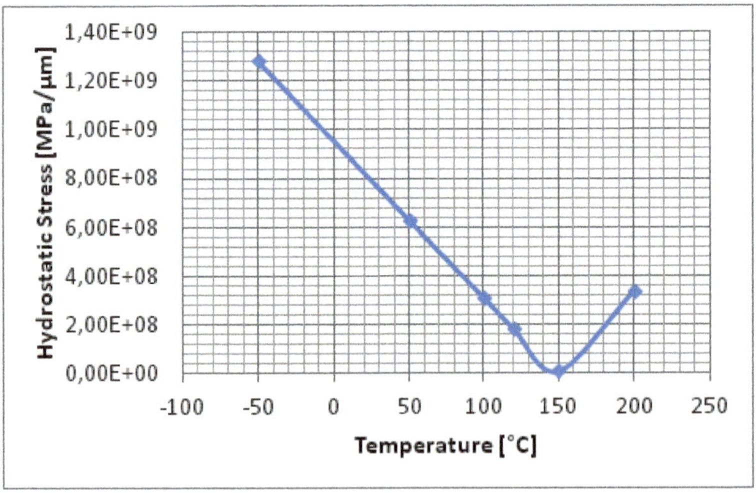


**Figure 14.** Hydrostatic stress depending of the substrate temperature (model A).

## 6. Conclusion

The finite element analysis of metallization structures or packages and bumps provide an insight into local temperatures, current density and stress gradient distributions with the possibility of mass flux divergence calculation. A prediction of weak links in the structures helps to increase the reliability during the design phase. Also the costs will be decreased and redesigns avoided.

In future phase separation in the CuSn bumps by IMC growth as well as the influence of special placed grain boundaries and concentration gradients should be included in the model. The process induced stress should be also included in the package modeling.

## Author details

Kirsten Weide-Zaage
*Gottfried Wilhelm Leibniz Universität Hannover, Information Technology Laboratory, Germany*

## 7. References

Averbuch, A.; Israeli, M.; Ravve, I.: "Electromigration of intergranular voids in metal films for microelectronic interconnects", Journal of Computational Physics 186, 2, 2003, pp. 281–304.

Banas, L.; Nürnberg, R.: "Finite Element Approximation of a Three Dimensional Phase Field Model for Void Electromigration", Springer, J. Sci. Computation, DOI 10.1007/s10915-008-9203-y.

Betz, A.: "Konforme Abbildungen", Berlin, Springer Verlag, 1964, pp.22-33.

Blech, I.A.: „Electromigration in thin aluminum films on titanium nitride", J. Appl. Phys. 47, 1976, p. 1203.

Delsol, J. R.; Jacquemin, J.-P.; et. al.: "Improved electrical and reliability performance of 65 nm interconnects with new barrier integration schemes", Microelectronic Engineering 83 (2006) pp. 2377–2380.

Chao, B; Chae, S.H.; et.al: "Electromigration enhanced intermetallic growth and void formation in Pb-free solder joints", J. App. Phys. 100, 084909 (2006)

Eichelseder, W.; Unger, B.; Dannbauer, H.: „Neue Aspekte der Submodelltechnik in der finite Elemente Methode zur Beschleunigung des Entwicklungsprozesses", FEM-Konferenz Baden Baden, 1998.

Hagedorn, F.B.; Hall, P.M.: „Right-Angle Bends in Thin Film Conductors", Journal Appl. Phys., Vol. 34, No.1, 1963, pp. 128-131.

Heumann, T.: „ Diffusion in Metallen", Springer Verlag, WFT 10, ISBN 3-540-55379-7, 1992.

Hou, Y; Tan, C.M.: "Comparison of stress-induced voiding phenomena in copper line-via structures with different dielectric materials", Semiconductor Sci. Technology, 2009, pp.1-8.

Huang, J.S.; Oates, T.S.: "Monte-Carlo simulation of electromigration failure distributions of submicron contacts and vias: a new extrapolation methodology for reliability estimate", Proceedings of the IEEE Interconnect Technology Conference 2000, pp. 208-210.

Huffman, A.; Lueck, M.; et.al: "Effects of Assembly Process Parameters on the Structure and Thermal Stability of Sn-Capped Cu Bump Bonds", ECTC Conf. 2007, pp. 1589-1596.

Jacoboni, C; Lugli, P.: "The Monte Carlo Method for Semiconductor Device Simulation", Computional Microelectronics", Edited by S. Selberherr, Vienna, Springer 1989, ISBN: 978-3-211-82110-7.

Jaffe, D.; Shewman, P.G.: „Termal Diffusion of Substitutional Impurities in Copper, Gold and Silver", Acta Metallurgica, Vol.12, 1964, pp.515-527.

Joo, Y.C.; Yang, T.Y.; et.al.: "Driving forces of mass transport in phase change materials and their effect on device failures", EPCOS 2011.

Kaur, I.; Mishin,Y.; Gust, W.;: „Fundamentals of Grain and Interphase Boundary and Diffusion", John Wiley & Sons Ltd., 1995.

Kitada, H.; Maeda, N.; et.al.: Development of Low Temperature Dielectrics down to 150°C for Multiple TSVs Structure with Wafer-on Wafer (WOW) Technology, IEEE, CMPT Tokyo 2010.

Labie, R.; Limaye, P.; et.al: „Reliability testing of Cu-Sn intermetallic micro-bump interconnections for 3D-device stacking", IEEE/Electronics-System-Integration-Conference (ESTC), Berlin, September 2010.

Leduc, P.; Assous, M.; et.al: "First integration of Cu TSV using die-to-wafer direct bonding and planarization", IEEE Int. Conf. on 3D System Integration, 3DIC, 2009, pp. 1-5.

Lee, K.-D.; Park, Y.-J.; Kim, T.: „Via Processing Effects on Electromigration in 65nm Technology, IEEE 06CH37728 44th Annual IRPS, 2006, pp.:103-106.

Li, Z; Zhao, H.; Gao, H.: "A Numerical Study of Electro-migration Voiding by Evolving Level Set Functions on a fixed Cartesian Grid", Journal of Computational Physics 152, 2, 1999, pp. 281–304.

Lin, M.H.(a); Chang, K.P.; et.al: 'Effects of Width Scaling, Length Scaling, and Layout Variation on Electromigration in Dual Damascene Copper Interconnects', IEEE RPSP, 2006, pp. 671-672.

Lin, M.H.(b); Lin, M.T.; et.al: „Copper Interconnect Electromigration Behavior in Various Structures and Precise Bimodal Fitting", Jap. J. Appl. Phys. Vol. 45, No. 2A, 2006, pp. 700–709.

Liu, Y.; Irving, S.; et.al: "3-D Modelling of Elec-tromigration Combined with Thermal-Mechanical Effect for IC Device Package", 8th Int. Conf. EuroSimE 2007, pp.1-13.

Liu, Y.; Liang, L.; et al.: "3D Modeling of electromigration combined with thermal-mechanical effect for IC device and package", Microelectronics Reliability, Vol.48 (2008), pp. 811-824.

Lo, W.-C. ; Chen, Y.-H. ; et.al: TSV and 3D Wafer Bonding Technologies For Advanced Stacking System and Application at ITRI, Symp. on VLSI Tech. Digest of Technical Papers, 2009, pp. 70-71.

Matsuyama, H.; et al.: "Investigation of stress-induced voiding inside and under vias in copper interconnects with "wing" pattern", Proc. Conf. International Reliability Physics Symposium, IEEE/IRPS 2008, pp. 683-684.

Meinshausen, L.(a); Weide-Zaage, K.; et.al: „;Virtual prototyping of PoP interconnections regarding electrically activated mechanisms" IEEE Thermal, Mechanical and Multi-Physics Simulation and Experiments in Microelectronics and Micro-Systems, EuroSimE 2010, p. 1-4.

Meinshausen, L. (b); Weide-Zaage, K.: "Exploration of Migration and Stress Effects in PoPs Considering Inhomogeneous Temperature Distribution", International Wafer-Level Packaging Conference (IWLPC), October 2010, pp. 137-145.

Meinshausen, L.; Weide-Zaage, K.; et.al: „Electro- and Thermomigration in Microbump Interconnects for 3D Integration", IEEE Electronic Components and Technology Conf., June 2011, pp. 1444-1451.

Meinshausen, L.; Weide-Zaage, K.; et.al:„Thermal Management for stackable package with stacked ICs", EuroSimE 2012, p. 1-4.

Murugesan, M; Kino, H.; et.al: High Density §D LSI Technology using W/CU Hybrid TSVs, IEEE/IEDM, 2011

Ogurtani, T.O.; Akyildiz, O.: "Morphological evolution of voids by surface drift diffusion driven by capillary, electromigration, and thermal-stress gradients induced by steady-state heat flow in passivated metallic thin films and flip chip solder joints. I. Theory, Journal of applied Physics 104, 023521, 2008

Philibert, J.: „Atom Movements Diffusion and Mass Transport in Solids", Monographies de Physique, 1991.

Shewman, P. G.: „Diffusion in Solids", Mac Graw-Hill Series in Materials Science and Engineering, 1963/1989.

Syed, A.; Dhandapani, K. et.al: "Cu Pillar and μ-bump Electromigration Reliability and Comparison with High Pb, SnPb, and SnAg bumps, IEEE Electronic Components and Technology Conf. 2011, pp. 332-339.

Tan, C.M.; Roy, A.; et.al: „Current Crowding Effect on Copper Dual Damascene Via Bottom Failure for ULSI Applications" IEEE Transactions Dev. Materials Reliability 2005; Vol. 5,198.

Wang, S.N.; Liang, L; et.al.;" Solder Joint Reliability under Electromigration and Thermal-Mechanical Load ", Elec. Comp. & Tec. Conf, 2007 pp1074-1083.

Wever, H.: „Stofftransport im metallischen Festkörper", Angewandte Chemie, Nr.7, 1963, pp.309-352.

Weide-Zaage, K.; Zhao, J.; et.al: "Determination of Migration Effects in Cu-Via Structures with Respect to Process Induced Stress", Microelectronic Reliability 48, 2008, pp. 1393–1397.

Weide-Zaage, K.; Ciptokusumo, J.; Aubel, O.: „Influence of the Activation Energy of the Different Migration Effects on Failure locations in Metallizations", AIP Conf. Proc. -- November 24, 2010 -- Volume 1300, pp. 85-90.

Wilkenson, D.S.: "Mass Transport in Solids and Fluids", Cambridge University Press, ISBN 0521624096, 2000.

# Permissions

The contributors of this book come from diverse backgrounds, making this book a truly international effort. This book will bring forth new frontiers with its revolutionizing research information and detailed analysis of the nascent developments around the world.

We would like to thank Dr. Farzad Ebrahimi, for lending his expertise to make the book truly unique. He has played a crucial role in the development of this book. Without his invaluable contribution this book wouldn't have been possible. He has made vital efforts to compile up to date information on the varied aspects of this subject to make this book a valuable addition to the collection of many professionals and students.

This book was conceptualized with the vision of imparting up-to-date information and advanced data in this field. To ensure the same, a matchless editorial board was set up. Every individual on the board went through rigorous rounds of assessment to prove their worth. After which they invested a large part of their time researching and compiling the most relevant data for our readers. Conferences and sessions were held from time to time between the editorial board and the contributing authors to present the data in the most comprehensible form. The editorial team has worked tirelessly to provide valuable and valid information to help people across the globe.

Every chapter published in this book has been scrutinized by our experts. Their significance has been extensively debated. The topics covered herein carry significant findings which will fuel the growth of the discipline. They may even be implemented as practical applications or may be referred to as a beginning point for another development. Chapters in this book were first published by InTech; hereby published with permission under the Creative Commons Attribution License or equivalent.

The editorial board has been involved in producing this book since its inception. They have spent rigorous hours researching and exploring the diverse topics which have resulted in the successful publishing of this book. They have passed on their knowledge of decades through this book. To expedite this challenging task, the publisher supported the team at every step. A small team of assistant editors was also appointed to further simplify the editing procedure and attain best results for the readers.

Our editorial team has been hand-picked from every corner of the world. Their multi-ethnicity adds dynamic inputs to the discussions which result in innovative outcomes. These outcomes are then further discussed with the researchers and contributors who give their valuable feedback and opinion regarding the same. The feedback is then collaborated with the researches and they are edited in a comprehensive manner to aid the understanding of the subject.

Apart from the editorial board, the designing team has also invested a significant amount of their time in understanding the subject and creating the most relevant covers. They scrutinized every image to scout for the most suitable representation of the subject and create an appropriate cover for the book.

The publishing team has been involved in this book since its early stages. They were actively engaged in every process, be it collecting the data, connecting with the contributors or procuring relevant information. The team has been an ardent support to the editorial, designing and production team. Their endless efforts to recruit the best for this project, has resulted in the accomplishment of this book. They are a veteran in the field of academics and their pool of knowledge is as vast as their experience in printing. Their expertise and guidance has proved useful at every step. Their uncompromising quality standards have made this book an exceptional effort. Their encouragement from time to time has been an inspiration for everyone.

The publisher and the editorial board hope that this book will prove to be a valuable piece of knowledge for researchers, students, practitioners and scholars across the globe.

# List of Contributors

**Josipa Borcic**
Department of Oral and Maxillofacial Surgery, Medical Faculty, School of Dental Medicine,
University of Rijeka, Rijeka, Croatia

**Alen Braut**
Department of Restorative Dentistry and Endodontics, Medical Faculty, School of Dental Medicine,
University of Rijeka, Rijeka, Croatia

**Chen Liangjian**
The Third Xiangya Hospital of Central South University, ChangSha, China State Key Laboratory of Powder Metallurgy Central South University, ChangSha, China

**Magnús Kjartan Gíslason and David H. Nash**
Department of Mechanical and Aerospace Engineering, University of Strathclyde, Glasgow, UK

**B. Alper Gultekin and Serdar Yalcin**
Istanbul University Faculty of Dentistry, Department of Oral Implantology, Istanbul, Turkey

**Pinar Gultekin**
Istanbul University Faculty of Dentistry, Department of Prosthodontics, Istanbul, Turkey

**Govindaraj Thangavel**
Department of Electrical and Electronics Engineering, Muthayammal Engineering College, India

**Elena Otilia Virjoghe, Diana Enescu, Mihail-Florin Stan and Marcel Ionel**
Valahia University of Targoviste, Romania

**António Martins**
Polytechnic Institute of Guarda, Portugal

**Sílvio Mariano and Maria do Rosário Calado**
University of Beira Interior, Portugal

**Hao Ren**
State Key Lab of Optical Technologies for Microfabrication, Institute of Optics and Electronics,
Chinese Academy of Sciences, Chengdu, China
School of Electrical, Computer, and Energy Engineering, Arizona State University, Tempe, AZ, USA

**Jun Yao**
state Key Lab of Optical Technologies for Microfabrication, Institute of Optics and Electronics,Chinese Academy of Sciences, Chengdu, China

**Jian Sun and Jürgen Kosel**
Physical Sciences and Engineering Division, King Abdullah University of Science and Technology,
Thuwal, Kingdom of Saudi Arabia

**Kazumi Kurihara**
Department of Electrical and Electronic Engineering, Ibaraki University, Hitachi, Japan

**Sh Mohd Firdaus**
Faculty of Mechanical Engineering, University Technology MARA (UiTM), Penang, Malaysia

**Husna Omar**
Faculty of Applied Science, University Technology MARA (UiTM), Penang, Malaysia

**Ishak Abd Azid**
Faculty of Mechanical Engineering, University Science of Malaysia (USM), Penang, Malaysia

**Marco Riva**
I.N.A.F. - Osservatorio Astronomico di Brera, Italy

**Simona Roatesi**
Military Technical Academy, Bucharest, Romania

**José Guilherme Santos da Silva, Ana Cristina Castro Fontenla Sieira, Gilvan Lunz Debona,**
**Pedro Colmar Gonçalves da Silva Vellasco and Luciano Rodrigues Ornelas de Lima**
State University of Rio de Janeiro (UERJ), Rio de Janeiro/RJ, Brazil

**S. Psarras and S.T. Pinho**
Dept. Aeronautics, Imperial College London, SW7 2AZ, United Kingdom

**B.G. Falzon**
Dept. Mechanical and Aerospace Engineering, Monash University, Victoria, Australia

**Salih Akour**
Sustainable and Renewable Energy Engineering Program, College of Engineering, University of Sharjah, Sharjah, United Arab Emirates
The University of Jordan, Faculty of Engineering and Technology, Mechanical Engineering Department, Amman, Jordan

**Hussein Maaitah**
Jordan Aeronautical-Systems Company (JAC), Amman Civil Airport, Marka, Amman, Jordan

**Kirsten Weide-Zaage**
Gottfried Wilhelm Leibniz Universität Hannover, Information Technology Laboratory, Germany